Python 机器学习算法与应用

邓立国 著

清华大学出版社
北京

内 容 简 介

本书理论与实践相结合，详细阐述机器学习数据特征与分类算法，基于 Python 3 精心编排大量的机器学习场景与开源平台应用，高效利用 Python 3 代码翔实地阐释机器学习核心算法及其工具的场景应用。

本书分为 6 章，主要内容包括机器学习概述、数据特征、分类算法、项目，以及在机器学习平台 Kaggle 与 PaddlePaddle 上实现分类、预测及推荐等实战操作。

本书适合机器学习的研究人员、计算机或数学等相关从业者参考学习，也可以作为计算机或数学等专业本科高年级或研究生专业用书。

本书封面贴有清华大学出版社防伪标签，无标签者不得销售
版权所有，侵权必究。侵权举报电话：010-62782989　13701121933

图书在版编目（CIP）数据

Python 机器学习算法与应用/邓立国著.— 北京：清华大学出版社，2020.2
ISBN 978-7-302-54899-7

Ⅰ．①P… Ⅱ．①邓… Ⅲ．①软件工具－程序设计②机器学习　Ⅳ．①TP311.561②TP181

中国版本图书馆 CIP 数据核字（2020）第 024729 号

责任编辑：夏毓彦
封面设计：王　翔
责任校对：闫秀华
责任印制：丛怀宇

出版发行：清华大学出版社
网　　址：http://www.tup.com.cn，http://www.wqbook.com
地　　址：北京清华大学学研大厦 A 座　　　邮　　编：100084
社 总 机：010-62770175　　　　　　　　　邮　　购：010-62786544
投稿与读者服务：010-62776969，c-service@tup.tsinghua.edu.cn
质量反馈：010-62772015，zhiliang@tup.tsinghua.edu.cn

印 装 者：北京鑫海金澳胶印有限公司
经　　销：全国新华书店
开　　本：190mm×260mm　　印　张：20.25　　字　数：551 千字
版　　次：2020 年 5 月第 1 版　　　　　　　印　次：2020 年 5 月第 1 次印刷
定　　价：69.00 元

产品编号：085917-01

前　言

机器学习是人工智能领域核心的研究方向，其应用遍及人工智能的各个领域。机器学习已经有了十分广泛的应用，例如数据挖掘、计算机视觉、自然语言处理、生物特征识别、搜索引擎、医学诊断、检测信用卡欺诈、证券市场分析、DNA 序列测序、语音和手写识别、战略游戏和机器人运用等。机器学习是人工智能和神经计算的核心研究课题之一，解决计算机程序如何随着经验积累自动提高性能。

作者在工作中接触和应用机器学习的相关算法过程中，发现目前没有比较完备的基于 Python 3 语言的机器学习专业图书，所以写了这本以 Python 3 为基础实践语言的机器学习工具书，仅供从事机器学习人员参阅。

本书内容

本书的目的是展现基于 Python 3 机器学习中核心的算法与实践，重点介绍与机器学习相关的知识理论与 Python 实例。

本书分为 6 章，系统地讲解机器学习的典型算法：第 1 章简要介绍有关机器学习的基础知识，第 2 章讲解机器学习的数据特征，第 3 章介绍机器学习的分类算法，第 4 章主要介绍机器学习开源项目场景应用，第 5、6 章在机器学习平台 Kaggle 与 PaddlePaddle 上进行算法应用。本书的例子都是在 Python 3 集成开发环境 Anaconda 3 中经过实际调试通过的典型案例，大部分实验数据来源于 GitHub，并且很多例子源程序都给出了网址地址，读者可以参考实现。

本书读者

本书可以作为计算机科学与工程、计算统计学和社会科学等专业的大学生或研究生的专业参考书，也可作为软件研究人员或从业人员的参考资料。由于机器学习专业素材的多学科性，读者可以根据对应的知识背景参考对应的专业书籍。

源码下载

本书配套源码下载地址请扫描下方二维码获得。如果下载有问题,请联系booksaga@163.com,邮件主题为"Python 机器学习算法与应用"。

致谢

本书完成之际,要感谢家人的支持与关爱。同时也要感谢同事,与他们的交流、探讨使得本书得以修正和完善。

由于作者水平有限,书中纰漏之处在所难免,恳请读者不吝赐教。本书中参考的网络资源均在参考文献中给出出处。

邓立国

2020 年 3 月

目 录

第1章 机器学习概述 ... 1
1.1 机器学习定义 ... 1
1.2 机器学习的发展 ... 2
1.3 机器学习的分类 ... 3
1.4 机器学习的研究领域 ... 6
1.5 本章小结 ... 8

第2章 机器学习数据特征 ... 9
2.1 数据分布性 ... 9
2.1.1 数据分布集中趋势的测定 ... 9
2.1.2 数据分布离散程度的测定 ... 14
2.1.3 数据分布偏态与峰度的测定 ... 17
2.2 数据相关性 ... 19
2.2.1 相关关系 ... 19
2.2.2 相关分析 ... 22
2.3 数据聚类性 ... 24
2.4 数据主成分分析 ... 27
2.4.1 主成分分析的原理及模型 ... 27
2.4.2 主成分分析的几何解释 ... 29
2.4.3 主成分的导出 ... 30
2.4.4 证明主成分的方差是依次递减 ... 31
2.4.5 主成分分析的计算 ... 32
2.5 数据动态性 ... 34
2.6 数据可视化 ... 37
2.7 本章小结 ... 39

第 3 章　机器学习分类算法 40

3.1　数据清洗和特征选择 40
3.1.1　数据清洗 40
3.1.2　特征选择 42
3.1.3　回归分析 45
3.2　决策树、随机森林 47
3.3　SVM 51
3.3.1　最优分类面和广义最优分类面 52
3.3.2　SVM 的非线性映射 55
3.3.3　核函数 56
3.4　聚类算法 56
3.5　EM 算法 61
3.6　贝叶斯算法 63
3.7　隐马尔可夫模型 63
3.8　LDA 主题模型 66
3.9　人工神经网络 69
3.10　KNN 算法 73
3.11　本章小结 76

第 4 章　Python 机器学习项目 77

4.1　SKlearn 78
4.1.1　SKlearn 包含的机器学习方式 78
4.1.2　SKlearn 的强大数据库 79
4.1.3　鸢尾花数据集举例 80
4.1.4　Boston 房价数据集的示例 83
4.2　TensorFlow 85
4.2.1　TensorFlow 简介 86
4.2.2　TensorFlow 的下载与安装 88
4.2.3　TensorFlow 的基本使用 91
4.3　Theano 96
4.4　Caffe 115
4.4.1　Caffe 框架与运行环境 115

	4.4.2	网络模型	119
4.5	Gensim		125
	4.5.1	Gensim 特性与核心概念	125
	4.5.2	训练语料的预处理	125
	4.5.3	主题向量的变换	126
	4.5.4	文档相似度的计算	127
4.6	Pylearn2		134
4.7	Shogun		135
4.8	Chainer		136
4.9	NuPIC		143
4.10	Neon		160
4.11	Nilearn		165
4.12	Orange3		168
4.13	PyMC 与 PyMC3		171
4.14	PyBrain		175
4.15	Fuel		181
4.16	PyMVPA		184
4.17	Annoy		186
4.18	Deap		190
4.19	Pattern		191
4.20	Requests		195
4.21	Seaborn		199
4.22	本章小结		206

第 5 章 Kaggle 平台机器学习实战 207

5.1	Kaggle 信用卡欺诈检测		207
	5.1.1	Kaggle 信用卡欺诈检测准备	207
	5.1.2	Kaggle 信用卡欺诈检测实例	210
5.2	Kaggle 机器学习案例		228
	5.2.1	Kaggle 机器学习概况	229
	5.2.2	自行车租赁数据分析与可视化案例	230
5.3	本章小结		241

第 6 章　PaddlePaddle 平台机器学习实战 242

6.1　PaddlePaddle 平台安装 242
6.2　PaddlePaddle 平台手写体数字识别 243
6.3　PaddlePaddle 平台图像分类 261
6.4　PaddlePaddle 平台词向量 277
6.5　PaddlePaddle 平台个性化推荐 289
6.6　PaddlePaddle 平台情感分析 302
6.7　本章小结 311

参考文献 312

第 1 章
机器学习概述

　　机器学习是一门从数据中发掘知识的多领域交叉学科,研究目的是如何模拟或实现人类的学习行为,根据经验和数据进行算法选择、模型构建、预测新数据,改善知识结构提升性能,是让计算机怎样模拟人类的学习行为,以获取新的知识或技能。

1.1　机器学习定义

　　机器学习(Machine Learning,ML)是一门人工智能的科学,是技术和算法的结合,致力于从数据中获取学习模式。该领域的主要研究对象是人工智能,特别是如何在经验学习中改善具体算法的性能。

　　机器学习是实现人工智能的一种途径,和数据挖掘有一定的相似性,也是一门多领域交叉学科,涉及概率论、统计学、逼近论、凸分析、计算复杂性理论等,图 1.1 给出了机器学习知识图谱。机器学习更偏重于算法的设计,让计算机能够自动地从数据中"学习"规律,并利用规律对未知数据进行预测。

　　机器学习是人工智能和神经计算的核心研究课题之一。机器学习问题解决计算机程序如何随着经验积累自动提高性能。

　　学习是人类具有的一种重要智能行为,但究竟什么是学习,长期以来却众说纷纭。社会学家、逻辑学家和心理学家都各有不同的看法。

　　比如,Langley(1996)定义的机器学习是"一门人工智能的科学,该领域的主要研究对象是人工智能,特别是如何在经验学习中改善具体算法的性能"(Machine learning is a science of the artificial. The field's main objects of study are artifacts, specifically algorithms that improve their performance with experience)。

　　Tom Mitchell 的机器学习(1997)对信息论中的一些概念有详细的解释,其中定义机器学习时提到,"机器学习是对能通过经验自动改进的计算机算法的研究"(Machine Learning is the study of computer algorithms that improve automatically through experience)。

　　Alpaydin(2004)提出"机器学习是用示例数据或过去的经验对计算机进行编程以优化性能标准"(Machine learning is programming computers to optimize a performance criterion using example data or past experience)。

尽管如此，为了便于进行讨论和估计学科的进展，有必要对机器学习给出定义，即使这种定义是不完全和不充分的。顾名思义，机器学习是研究如何使用机器来模拟人类学习活动的一门学科。稍微严格的提法是：机器学习是一门研究机器获取新知识和新技能，并识别现有知识的学问。这里所说的"机器"，指的就是计算机，包括电子计算机、中子计算机、光子计算机或神经计算机等。

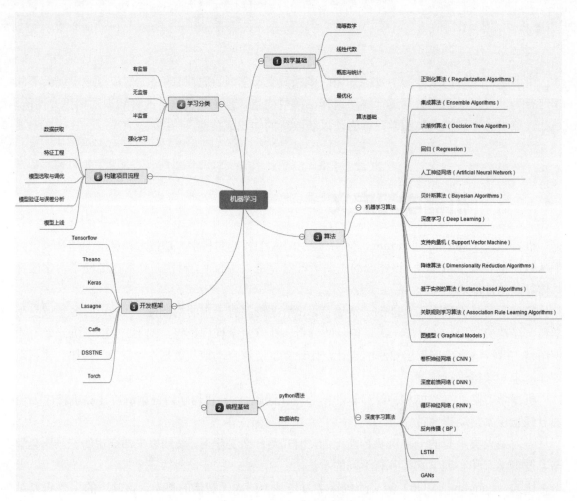

图 1.1　机器学习知识图谱

1.2　机器学习的发展

机器学习是人工智能研究较为年轻的分支，发展过程大体上可分为 4 个时期：

- 第一阶段是在 20 世纪 50 年代中叶到 60 年代中叶，属于热烈时期。
- 第二阶段是在 20 世纪 60 年代中叶至 70 年代中叶，属于冷静时期。

- 第三阶段是从20世纪70年代中叶至80年代中叶，属于复兴时期。
- 第四阶段是机器学习的最新阶段，始于1986年。

机器学习进入新阶段重要表现在下列诸方面：

- 机器学习已成为新的边缘学科并在高校形成一门课程。它综合应用心理学、生物学和神经生理学以及数学、自动化和计算机科学形成机器学习理论基础。
- 结合各种学习方法，取长补短。与符号学习的耦合可以更好地解决连续性信号处理中知识与技能的获取与求精问题。
- 机器学习与人工智能各种基础问题的统一性观点正在形成。例如，学习与问题求解结合进行、知识表达便于学习的观点产生了通用智能系统 SOAR 的组块学习。类比学习与问题求解结合的基于案例方法已成为经验学习的重要方向。
- 各种学习方法的应用范围不断扩大，一部分已形成商品。归纳学习的知识获取工具已在诊断分类型专家系统中广泛使用。连接学习在声图文识别中占优势。分析学习已用于设计综合型专家系统。遗传算法与强化学习在工程控制中有较好的应用前景。与符号系统耦合的神经网络连接学习将在企业的智能管理与智能机器人运动规划中发挥作用。
- 与机器学习有关的学术活动空前活跃。国际上除每年一次的机器学习研讨会外，还有计算机学习理论会议以及遗传算法会议。

1.3 机器学习的分类

1. 基于所获取知识的表示形式分类

学习系统获取的知识可能有行为规则、物理对象的描述、问题求解策略、各种分类及其他用于任务实现的知识类型。

根据表示的精细程度，可将知识表示形式分为两大类：泛化程度高的粗粒度符号表示、泛化程度低的精粒度亚符号（Sub-symbolic）表示。决策树、形式文法、产生式规则、形式逻辑表达式、框架和模式等属于符号表示类；代数表达式参数、图和网络、神经网络等属于亚符号表示类。

对于学习中获取的知识，主要有以下一些表示形式：

（1）代数表达式参数

学习的目标是调节一个固定函数形式的代数表达式参数或系数来达到一个理想的性能。

（2）决策树

用决策树来划分物体的类属，树中每一内部节点对应一个物体属性，而每一边对应于这些属性的可选值，树的叶节点则对应于物体的每个基本分类。

(3) 形式文法

在识别一个特定语言的学习中，通过对该语言的一系列表达式进行归纳，形成该语言的形式文法。

(4) 产生式规则

产生式规则表示为"条件-动作对"，已被极为广泛地使用。学习系统中的学习行为主要是生成、泛化、特化（Specialization）或合成产生式规则。

(5) 形式逻辑表达式

形式逻辑表达式的基本成分是命题、谓词、变量、约束变量范围的语句，以及嵌入的逻辑表达式。

(6) 图和网络

有的系统采用图匹配和图转换方案来有效地比较和索引知识。

(7) 框架和模式（Schema）

每个框架包含一组槽，用于描述事物（概念和个体）的各个方面。

(8) 计算机程序和其他的过程编码

获取这种形式的知识，目的在于取得一种能实现特定过程的能力，而不是为了推断该过程的内部结构。

(9) 神经网络

这主要用在联接学习中。学习所获取的知识，最后归纳为一个神经网络。

(10) 多种表示形式的组合

有时一个学习系统中获取的知识需要综合应用上述几种知识表示形式。

2. 按应用领域分类

最主要的应用领域有专家系统、认知模拟、规划和问题求解、数据挖掘、网络信息服务、图像识别、故障诊断、自然语言理解、机器人和博弈等。

从机器学习的执行部分所反映的任务类型上看，大部分的应用研究领域基本上集中于以下两个范畴：分类和问题求解。

（1）分类任务要求系统依据已知的分类知识对输入的未知模式（该模式的描述）做分析，以确定输入模式的类属。相应的学习目标就是学习用于分类的准则（如分类规则）。

（2）问题求解任务要求对于给定的目标状态，寻找一个将当前状态转换为目标状态的动作序列；机器学习在这一领域的研究工作大部分集中于通过学习来获取能提高问题求解效率的知识（如搜索控制知识、启发式知识等）。

3. 综合分类

综合考虑各种学习方法出现的历史渊源、知识表示、推理策略、结果评估的相似性、研究人员交流的相对集中性以及应用领域等诸因素。将机器学习方法区分为以下6类：

(1) 经验归纳学习（Empirical Inductive Learning）

经验归纳学习采用一些数据密集的经验方法（如版本空间法、ID3 法、定律发现法）对例子进行归纳学习。其例子和学习结果一般都采用属性、谓词、关系等符号表示。它相当于基于学习策略分类中的归纳学习，但扣除联接学习、遗传算法、增强学习的部分。

(2) 分析学习（Analytic Learning）

分析学习方法是从一个或少数几个实例出发，运用领域知识进行分析。其主要特征为：

- 推理策略主要是演绎，而非归纳。
- 使用过去的问题求解经验（实例）指导新的问题求解，或产生能更有效地运用领域知识的搜索控制规则。

分析学习的目标是改善系统的性能，而不是新的概念描述。分析学习包括应用解释学习、演绎学习、多级结构组块以及宏操作学习等技术。

(3) 类比学习

它相当于基于学习策略分类中的类比学习。在这一类型的学习中比较引人注目的研究是通过与过去经历的具体事例做类比来学习，称为基于范例的学习（Case-based Learning），或简称范例学习。

(4) 遗传算法（Genetic Algorithm）

遗传算法模拟生物繁殖的突变、交换和达尔文的自然选择（在每一生态环境中适者生存）。它把问题可能的解编码为一个向量（称为个体，向量的每一个元素称为基因），并利用目标函数（相应于自然选择标准）对群体（个体的集合）中的每一个个体进行评价，根据评价值（适应度）对个体进行选择、交换、变异等遗传操作，从而得到新的群体。遗传算法适用于非常复杂和困难的环境，比如带有大量噪声和无关数据、事物不断更新、问题目标不能明显和精确地定义，以及通过很长的执行过程才能确定当前行为的价值等。同神经网络一样，遗传算法的研究已经发展为人工智能的一个独立分支，其代表人物为霍勒德（J.H.Holland）。

(5) 联接学习

典型的联接模型实现为人工神经网络，其由称为神经元的一些简单计算单元以及单元间的加权联接组成。

(6) 增强学习（Reinforcement Learning）

增强学习的特点是通过与环境的试探性（Trial and Error）交互来确定和优化动作的选择，以实现所谓的序列决策任务。在这种任务中，学习机制通过选择并执行动作，导致系统状态的变化，并有可能得到某种强化信号（立即回报），从而实现与环境的交互。强化信号就是对系统行为的一种标量化的奖惩。系统学习的目标是寻找一个合适的动作选择策略，即在任一给定的状态下选择哪种动作的方法，使产生的动作序列可获得某种最优的结果。

在综合分类中，经验归纳学习、遗传算法、联接学习和增强学习均属于归纳学习，其中经验归纳学习采用符号表示方式，而遗传算法、联接学习和增强学习则采用亚符号表示方式；分析学习属于演绎学习。

4. 学习形式分类

（1）监督学习（Supervised Learning）

监督学习在机械学习过程中提供对错指示。通过算法让机器自我减少误差。这一类学习主要应用于分类和回归（Classify & Regression，也称为分类和预测）。监督学习从给定的训练数据集中学习出一个函数，当新的数据到来时，可以根据这个函数预测结果。监督学习的训练集要求是包括输入和输出，也可以说是特征和目标。训练集中的目标是由人标注的。常见的监督学习算法包括回归分析和统计分类。

（2）无监督学习（Unsupervised Learning）

无监督学习又称归纳性学习（Clustering），利用 K 方式（K-Means）建立中心（Centriole），通过循环和递减运算（Iteration & Descent）来减小误差，以达到分类的目的。

1.4 机器学习的研究领域

机器学习是继专家系统之后人工智能应用的又一重要研究领域，也是人工智能和神经计算的核心研究课题之一。现有的计算机系统和人工智能系统没有什么学习能力，至多也只有非常有限的学习能力，因而不能满足科技和生产提出的新要求。对机器学习的讨论和机器学习研究的进展，必将促使人工智能和整个科学技术的进一步发展。

其实，机器学习跟模式识别、统计学习、数据挖掘、计算机视觉、语音识别、自然语言处理等领域有着很深的联系。从范围上来说，机器学习跟模式识别、统计学习、数据挖掘是类似的，同时，机器学习与其他领域的处理技术结合，形成了计算机视觉、语音识别、自然语言处理等交叉学科。因此，一般说数据挖掘时可以等同于说机器学习。同时，我们平常所说的机器学习应用应该是通用的，不仅仅局限在结构化数据，还有图像、音频等应用。在本节对机器学习这些相关领域的介绍将有助于我们理清机器学习的应用场景与研究范围，更好地理解后面的算法与应用层次。

1. 模式识别

模式识别与机器学习的主要区别在于，前者是从工业界发展起来的概念，后者则主要源自计算机学科。在著名的 *Pattern Recognition And Machine Learning*（《模式识别与机器学习》）这本书中，Christopher M. Bishop 在开头是这样说的："模式识别源自工业界，而机器学习来自于计算机学科"。不过，它们中的活动可以被视为同一个领域的两个方面。

2. 数据挖掘

数据挖掘=机器学习+数据库。

数据挖掘（Data Mining 又称为资料探勘、数据采矿）是数据库知识发现

（Knowledge-Discovery in Databases，KDD）中的一个步骤。数据挖掘一般是指从大量数据中自动搜索隐藏于其中的有着特殊关系性（属于 Association Rule Learning，即关联规则学习）的信息的过程。数据挖掘通常与计算机科学有关，并通过统计、在线分析处理、情报检索、机器学习、专家系统（依靠过去的经验法则）和模式识别等诸多方法来实现上述目标。

3. 统计学习

统计学习近似等于机器学习。统计学习是与机器学习高度重叠的一门学科。因为机器学习中的大多数方法来自统计学，甚至可以认为，统计学的发展促进机器学习的繁荣昌盛。例如，著名的支持向量机算法就源自统计学科。但是在某种程度上两者是有区别的：统计学习者重点关注的是统计模型的发展与优化，偏数学；机器学习者更关注的是能够解决问题，偏实践，因此会重点研究学习算法在计算机上执行的效率与准确性的提升。

4. 计算机视觉

计算机视觉=图像处理+机器学习。

图像处理技术用于将图像处理为适合进入机器学习模型中的输入，机器学习则负责从图像中识别出相关的模式。计算机视觉相关的应用非常多，比如百度识图、手写字符识别、车牌识别等。

5. 语音识别

语音识别=语音处理+机器学习。

语音识别就是音频处理技术与机器学习的结合。语音识别技术一般不会单独使用，而会结合自然语言处理的相关技术。目前的相关应用有苹果的语音助手 Siri 等。

6. 自然语言处理

自然语言处理=文本处理+机器学习。

自然语言处理技术是让机器理解人类语言这个领域的一项技术。在自然语言处理技术中，大量使用了编译原理相关的技术，例如词法分析、语法分析等。除此之外，在理解层面，使用了语义理解、机器学习等技术。

7. 回归算法

在大部分机器学习课程中，回归算法都是介绍的第一个算法。原因有两个：一是回归算法比较简单，介绍它可以让人平滑地从统计学迁移到机器学习中；二是回归算法是后面若干强大算法的基石。回归算法有两个重要的子类，即线性回归和逻辑回归。

8. 神经网络

神经网络（也称为人工神经网络，ANN）算法是 20 世纪 80 年代机器学习界非常流行的算法，不过在 90 年代中途衰落。现在，携着"深度学习"之势，神经网络重装归来，重新成为最强大的机器学习算法之一。

在神经网络中，每个处理单元是一个逻辑回归模型。逻辑回归模型接收上层的输入，把模型的预测结果作为输出传输到下一个层次。通过这样的过程，神经网络可以完成非常复杂的非线性分类。

9. 支持向量机（SVM）

支持向量机算法诞生于统计学习界，是在机器学习界大放光彩的经典算法。支持向量机算法从某种意义上来说是逻辑回归算法的强化：通过给予逻辑回归算法更严格的优化条件，支持向量机算法可以获得比逻辑回归更好的分类界线。

10. 聚类算法

简单来说，聚类算法就是计算种群中的距离，根据距离的远近将数据划分为多个族群。聚类算法中最典型的代表是 K-Means 算法。训练数据都是不含分类标注的，算法的目的是通过训练推测出这些数据的分类标注。这类算法有一个统称，即无监督算法。

11. 降维算法

降维算法也是一种无监督学习算法，其主要特征是将数据从高维降低到低维层次。在这里，维度表示的是数据的特征量大小。降维算法的主要作用是压缩数据与提升机器学习其他算法的效率。通过降维算法，可以将具有几千个特征的数据压缩至若干个特征。另外，降维算法的另一个好处是数据的可视化。降维算法的主要代表是 PCA 算法（主成分分析算法）。

12. 推荐算法

推荐算法是目前业界非常火的一种算法，在电商界，如亚马逊、天猫、京东等得到了广泛的运用。推荐算法的主要特征是可以自动向用户推荐他们感兴趣的东西，从而增加购买率，提升效益。

13. 其他算法

除了以上算法之外，机器学习界还有其他算法，如高斯判别、朴素贝叶斯、决策树等。机器学习界的算法众多。

1.5 本章小结

机器学习理论由紧密联系而又自成体系的 3 个模块所构成，分别是模型、学习和推断。其中，模型为具体的问题域提供建模工具；学习是理论核心，为设定学习目标和学习效果提供理论保证；推断关注模型的使用性能和准确性。

本章比较系统地介绍了机器学习的基本概念、发展、学习分类以及机器学习的热门研究领域。

第 2 章
机器学习数据特征

机器学习专门研究计算机怎样模拟或实现人类的学习行为,以获取新的知识或技能,重新组织已有的知识结构,使之不断改善自身的性能。数据和特征决定了机器学习的上限,模型和算法是逼近这个上限的工具手段,特征工程的目的是最大限度地从原始数据中提取特征以供算法和模型使用。

2.1 数据分布性

统计数据的分布特征可以从两个方面进行描述:一是数据分布的集中趋势,二是数据分布的离散程度。集中趋势和离散程度是数据分布特征对立统一的两个方面。本节通过介绍平均指标和变异指标这两种统计指标的概念及计算来讨论反映数据集中趋势和分散程度两个方面的特征。

2.1.1 数据分布集中趋势的测定

集中趋势是指一组数据向某中心值靠拢的倾向,集中趋势的测度实际上就是对数据一般水平代表值或中心值的测度。不同类型的数据用不同的集中趋势测度值,低层次数据的集中趋势测度值适用于高层次的测量数据,反过来,高层次数据的集中趋势测度值并不适用于低层次的测量数据,选用哪一个测度值来反映数据的集中趋势,要根据所掌握的数据类型来确定。

通常用平均指标作为集中趋势测度指标,本节重点介绍众数、中位数两个位置平均数和算术平均数、调和平均数及几何平均数 3 个数值型平均数。

1. 众数

众数是指一组数据中出现次数最多的变量值,用 M_0 表示。从变量分布的角度看,众数是具有明显集中趋势点的数值,一组数据分布的最高峰点所对应的变量值即为众数。当然,如果数据的分布没有明显的集中趋势或最高峰点,众数也可以不存在;如果有多个高峰点,也就有多个众数。

(1)定类数据和定序数据众数的测定

定类数据与定序数据计算众数时,只需找出出现次数最多的组所对应的变量值即为众数。

（2）未分组数据或单变量值分组数据众数的确定

未分组数据或单变量值分组数据计算众数时，只需找出出现次数最多的变量值即为众数。

（3）组距分组数据众数的确定

组距分组数据，众数的数值与其相邻两组的频数分布有一定的关系，这种关系可做如下理解：

设众数组的频数为 f_m，众数前一组的频数为 f_{-1}，众数后一组的频数为 f_{+1}。当众数相邻两组的频数相等时，即 $f_{-1}=f_{+1}$，众数组的组中值即为众数；当众数组的前一组的频数多于众数组后一组的频数时，即 $f_{-1}>f_{+1}$，众数会向前一组靠，众数小于其组中值；当众数组后一组的频数多于众数组前一组的频数时，即 $f_{-1}<f_{+1}$，则众数会向后一组靠，众数大于其组中值。基于这种思路，借助于几何图形而导出的分组数据众数的计算公式如下：

$$\begin{aligned} M_0 &\doteq L + \frac{f_m - f_{-1}}{(f_m - f_{-1}) + (f_m - f_{+1})} \times i \\ M_0 &\doteq U - \frac{f_m - f_{+1}}{(f_m - f_{-1}) + (f_m - f_{+1})} \times i \end{aligned} \tag{2.1}$$

其中：L 表示众数所在组的下限；U 表示众数所在组的上限；i 表示众数所在组的组距；f_m 为众数组的频数；f_{-1} 为众数组前一组的频数；f_{+1} 为众数组后一组的频数。

上述下限和上限公式是假定数据分布具有明显的集中趋势，且众数组的频数在该组内是均匀分布的。若这些假定不成立，则众数的代表性就会很差。从众数的计算公式可以看出，众数是根据众数组及相邻组的频率分布信息来确定数据中心点位置的。因此，众数是一个位置代表值，不受数据中极端值的影响。

2. 中位数

中位数是将总体各单位标志值按大小顺序排列后处于中间位置的那个数值。各变量值与中位数的离差绝对值之和最小，即：

$$\sum_{i=1}^{n} |X_i - M_e| = \min \tag{2.2}$$

（1）定序数据中位数的确定

定序数据中位数确定的关键是确定中间位置，中间位置所对应的变量值即为中位数。

① 未分组原始数据中间位置的确定：

$$\begin{cases} \text{中位数位置} = \frac{N+1}{2} & N\text{为奇数} \\ \text{中位数位置} = \frac{N}{2} & N\text{为偶数} \end{cases} \tag{2.3}$$

② 分组数据中间位置的确定：

$$\text{中位数位置} = \frac{\sum f}{2} \tag{2.4}$$

（2）数值型数据中位数的确定

$$\text{数值型数据}\begin{cases}\text{未分组数据}\\ \text{分组数据}\begin{cases}\text{单变量值分组数据}\\ \text{组距分组数据}\end{cases}\end{cases}$$

① 未分组数据

将标志值按大小排序，假设排序的结果为 $x_1 \leqslant x_2 \leqslant x_3 \leqslant \cdots \leqslant x_n$，则：

$$M_e = \begin{cases} X_{\left(\frac{N+1}{2}\right)} & N\text{为奇数} \\ \frac{1}{2}\left(X_{\frac{N}{2}} + X_{\frac{N}{2}+1}\right) & N\text{为偶数} \end{cases} \tag{2.5}$$

② 单变量值分组数据

$$M_e = \begin{cases} X_{\left(\frac{\sum f+1}{2}\right)} & \sum f \text{为奇数} \\ X_{\left(\frac{\sum f}{2}\right)} & \sum f \text{为偶数} \end{cases} \tag{2.6}$$

③ 组距分组数据

根据位置公式确定中位数所在的组，假定在中位数组内的各单位是均匀分布的，就可利用下面的公式计算中位数的近似值：

$$M_e = L + \frac{\frac{\sum f}{2} - S'_{m-1}}{f_m} \cdot i$$

$$M_e = U - \frac{\frac{\sum f}{2} - S'_{m+1}}{f_m} \cdot i \tag{2.7}$$

其中，s_{m-1} 是到中位数组前面一组为止的向上累计频数，s'_{m+1} 是到中位数组后面一组为止的向下累计频数；f_m 为中位数组的频数；i 为中位数组的组距。

3. 算术平均数

算术平均数（Arithmetic Mean）也称为均值（Mean），是全部数据算术平均的结果。算术平均法是计算平均指标最基本、最常用的方法。算术平均数在统计学中具有重要的地位，是集中趋势的最主要测度值，通常用 $\bar{x}$ 表示。根据所掌握数据形式的不同，算术平均数有简单算术平均数和加权算术平均数。

（1）简单算术平均数（Simple Arithmetic Mean）

未经分组整理的原始数据，其算术平均数的计算就是直接将一组数据的各个数值相加除以数值个数。设总体数据为 $X_1, X_2, \ldots, X_n$，样本数据为 $x_1, x_2, \ldots, x_n$，则统计总体均值 $\bar{X}$ 和样本均值 $\bar{x}$

的计算公式为：

$$\overline{X} = \frac{X_1 + X_2 + \cdots + X_N}{N} = \frac{\sum_{i=1}^{N} X_i}{N}$$

$$\overline{x} = \frac{x_1 + x_2 + \cdots + x_n}{n} = \frac{\sum_{i=1}^{n} x_i}{n} \quad (2.8)$$

（2）加权算术平均数（Weighted Arithmetic Mean）

根据分组整理的数据计算的算术平均数要以各组变量值出现的次数或频数为权数计算加权的算术平均数。设原始数据（总体或样本数据）被分成 K 或 k 组，各组的变量值为 $X_1, X_2, \ldots, X_K$ 或 $x_1, x_2, \ldots, x_k$，各组变量值的次数或频数分别为 $F_1, F_2, \ldots, F_K$ 或 $f_1, f_2, \ldots, f_k$，则总体或样本的加权算术平均数为：

$$\overline{X} \doteq \frac{X_1 F_1 + X_2 F_2 + \cdots + X_K F_K}{F_1 + F_2 + \cdots + F_K} = \frac{\sum_{i=1}^{K} X_i F_i}{\sum_{i=1}^{K} F_i}$$

$$\overline{x} \doteq \frac{x_1 f_1 + x_2 f_2 + \cdots + x_k f_k}{f_1 + f_2 + \cdots + f_k} = \frac{\sum_{i=1}^{k} x_i f_i}{\sum_{i=1}^{k} f_i} \quad (2.9)$$

在公式（2.9）中，利用各组的组中值代表各组的实际数据，使用代表值时是假定各组数据在各组中是均匀分布的，但实际情况与这一假定会有一定的偏差，使得利用分组资料计算的平均数与实际的平均值会产生误差，它是实际平均值的近似值。

加权算术平均数的数值大小不仅受各组变量值 x_i 大小的影响，还受各组变量值出现的频数（权数 f_i）大小的影响。如果某一组的权数大，说明该组的数据较多，那么该组数据的大小对算术平均数的影响就越大；反之，则越小。实际上，我们将公式（2.9）变形为公式（2.10）的形式，就更能清楚地看出这一点。

$$\overline{x} = \frac{\sum_{i=1}^{K} x_i f_i}{\sum_{i=1}^{K} f_i} = \sum_{i=1}^{K} x_i \frac{f_i}{\sum_{i=1}^{K} f_i} \quad (2.10)$$

由公式（2.10）可以清楚地看出，加权算术平均数受各组变量值（x_i）和各组权数（频率 $f_i / \Sigma f_i$）大小的影响。频率越大，相应的变量值计入平均数的份额越大，对平均数的影响就越大；反之，频率越小，相应的变量值计入平均数的份额越小，对平均数的影响就越小。这就是权数权衡轻重作用的实质。

算术平均数在统计学中具有重要的地位，是进行统计分析和统计推断的基础。从统计思想上看，算术平均数是一组数据的重心所在，是消除了一些随机因素影响或者数据误差相互抵消后的必然性结果。

算术平均数具有下面一些重要的数学性质。这些数学性质在实际中有着广泛的应用，同时也体现了算术平均数的统计思想。

① 各变量值与其算术平均数的离差之和等于零，即：

$$\sum_{i=1}^{n}(x_i-\bar{x})=0 \quad 或 \quad \sum_{i=1}^{k}(x_i-\bar{x})f_i=0 \tag{2.11}$$

② 各变量值与其算术平均数的离差平方和最小，即：

$$\sum_{i=1}^{n}(x_i-\bar{x})^2=\min \quad 或 \quad \sum_{i=1}^{k}(x_i-\bar{x})^2 f_i=\min \tag{2.12}$$

4. 调和平均数（Harmonic Mean）

在实际工作中，经常会遇到只有各组变量值和各组标志总量而缺少总体单位数的情况，这时就要用调和平均数法计算平均指标。调和平均数是各个变量值倒数的算术平均数的倒数，习惯上用 H 表示，计算公式为：

$$H=\frac{m_1+m_2+\cdots+m_k}{\frac{m_1}{x_1}+\frac{m_2}{x_2}+\cdots+\frac{m_k}{x_k}}=\frac{\sum_{i=1}^{K}m_i}{\sum_{i=1}^{K}\frac{m_i}{x_i}} \tag{2.13}$$

调和平均数和算术平均数在本质上是一致的，唯一的区别是计算时使用了不同的数据。在实际应用时，可掌握这样的原则：当计算算术平均数的分子数据未知时，就采用加权算术平均数计算平均数；当分母数据未知时，就采用加权调和平均数计算平均数。

$$H=\frac{\sum_{i=1}^{K}m_i}{\sum_{i=1}^{K}\frac{m_i}{x_i}}=\frac{\sum_{i=1}^{K}x_i f_i}{\sum_{i=1}^{K}\frac{x_i f_i}{x_i}}=\frac{\sum_{i=1}^{K}x_i f_i}{\sum_{i=1}^{K}f_i}=\bar{x} \tag{2.14}$$

5. 几何平均数（Geometric Mean）

几何平均数是适应于特殊数据的一种平均数，在实际生活中通常用来计算平均比率和平均速度。当所掌握的变量值本身是比率的形式而且各比率的乘积等于总的比率时，就应采用几何平均法计算平均比率。

$$G_M=\sqrt[N]{X_1\times X_2\times\cdots\times X_N}=\sqrt[N]{\prod_{i=1}^{N}X_i} \tag{2.15}$$

也可以看作算术平均数的一种变形：

$$\log G_M = \frac{1}{N}(\log X_1 + \log X_2 + \cdots + \log X_N) = \frac{\sum_{i=1}^{N} \log X_i}{N} \tag{2.16}$$

6. 众数、中位数与算术平均数的关系

算术平均数与众数、中位数的关系取决于频数分布的状况，它们的关系如下：

① 当数据具有单一众数且频数分布对称时，算术平均数与众数、中位数三者完全相等，即 $M_0 = M_e = \bar{x}$。

② 当频数分布呈现右偏态时，说明数据存在最大值，必然拉动算术平均数向极大值一方靠，则三者之间的关系为 $\bar{x} > M_e > M_0$。

③ 当频数分布呈现左偏态时，说明数据存在最小值，必然拉动算术平均数向极小值一方靠，而众数和中位数是位置平均数，不受极值的影响，因此三者之间的关系为 $\bar{x} < M_e < M_0$。

当频数分布出现偏态时，极端值对算术平均数产生很大的影响，而对众数、中位数没有影响，此时用众数、中位数作为一组数据的中心值比算术平均数有较高的代表性。如果从数值上的关系来看，当频数分布的偏斜程度不是很大时，无论是左偏还是右偏，众数与中位数的距离约为算术平均数与中位数的距离的两倍，即：

$$|M_e - M_0| = 2|\bar{X} - M_e|$$
$$M_0 = \bar{X} - 3(\bar{X} - M_e) = 3M_e - 2\bar{X} \tag{2.17}$$

2.1.2 数据分布离散程度的测定

数据分布的离散程度是描述数据分布的另一个重要特征，反映各变量值远离其中心值的程度，因此也称为离中趋势。它从另一个侧面说明了集中趋势测度值的代表程度。不同类型的数据有不同的离散程度测度值。描述数据离散程度的测度值主要有异众比率、极差、四分位差、平均差、方差和标准差、离散系数等。这些指标又称为变异指标。

1. 异众比率

异众比率的作用是衡量众数对一组数据的代表性程度的指标。异众比率越大，说明非众数组的频数占总频数的比重越大，众数的代表性就越差；反之，异众比率越小，众数的代表性就越好。异众比率主要用于测度定类数据、定序数据的离散程度。

$$V_r = \frac{\sum F_i - F_m}{\sum F_i} = 1 - \frac{F_m}{\sum F_i} \tag{2.18}$$

其中，$\sum F_i$ 为变量值的总频数；F_m 为众数组的频数。

2. 极差

极差是一组数据的最大值与最小值之差，离散程度的最简单测度值。极差的测度如下：

- 未分组数据：$R = \max(X_i) - \min(X_i)$ （2.19）
- 组距分组数据：$R = $ 最高组上限 $-$ 最低组下限

3. 四分位差

中位数是从中间点将全部数据等分为两部分。与中位数类似的还有四分位数、八分位数、十分位数和百分位数等。它们分别是用 3 个点、7 个点、9 个点和 99 个点将数据四等分、八等分、十等分和一百等分后各分位点上的值。这里只介绍四分位数的计算，其他分位数与之类似。

一组数据排序后处于 25% 和 75% 位置上的值称为四分位数，也称四分位点。四分位数通过 3 个点将全部数据等分为 4 部分，其中每部分包含 25% 的数据。很显然，中间的分位数就是中位数，因此通常所说的四分位数是指处在 25% 位置上的数值（下四分位数）和处在 75% 位置上的数值（上四分位数）。与中位数的计算方法类似，根据未分组数据计算四分位数时，首先对数据进行排序，然后确定四分位数所在的位置。

（1）四分位数确定

设下四分位数为 Q_L，上四分位数为 Q_U。

① 未分组数据

$$Q_L = X_{\frac{n+1}{4}} \qquad Q_U = X_{\frac{3(n+1)}{4}} \qquad (2.20)$$

当四分位数的位置不在某一个位置上时，可根据四分位数的位置，按比例分摊四分位数两侧的差值。

② 单变量值分组数据

$$Q_L = X_{\frac{\sum f}{4}} \qquad Q_U = X_{\frac{3\sum f}{4}} \qquad (2.21)$$

③ 组距分组数据

$$Q_L = L + \frac{\frac{\sum f}{4} - S_L}{f_L} \cdot i \qquad Q_U = U + \frac{\frac{3\sum f}{4} - S_U}{f_U} \cdot i \qquad (2.22)$$

（2）四分位差

四分位数是离散程度的测度值之一。上四分位数与下四分位数之差称为四分位差，亦称为内距或四分间距（Inter-quartile Range），用 Q_d 表示。四分位差的计算公式为：

$$Q_d = Q_U - Q_L \qquad (2.23)$$

4. 平均差（Mean Deviation）

平均差是各变量值与其算术平均数离差绝对值的平均数，用 M_d 表示，是离散程度的测度值之一。平均差能全面反映一组数据的离散程度，但该方法的数学性质较差，实际中应用较少。

（1）简单平均法

对于未分组数据，采用简单平均法。其计算公式为：

$$M_D = \frac{\sum_{i=1}^{N}|X_i - \bar{X}|}{N} \tag{2.24}$$

（2）加权平均法

在数据分组的情况下，应采用加权平均式：

$$M_D \doteq \frac{\sum_{i=1}^{K}|X_i - \bar{X}|F_i}{\sum_{i=1}^{K}F_i} \tag{2.25}$$

5. 方差和标准差（Variance、Standard Deviation）

方差和标准差同平均差一样，也是根据全部数据计算的，反映每个数据与其算术平均数相比平均相差的数值，因此能够准确地反映出数据的差异程度。与平均差的不同之处是在计算时的处理方法不同，平均差是取离差的绝对值消除正负号，而方差、标准差是取离差的平方消除正负号，更便于数学上的处理。因此，方差、标准差是实际中应用最广泛的离中程度度量值。

① 设总体的方差为 σ^2，标准差为 σ，对于未分组整理的原始数据，方差和标准差的计算公式分别为：

$$\sigma^2 = \frac{\sum_{i=1}^{N}(X_i - \bar{X})^2}{N} \qquad \sigma = \sqrt{\frac{\sum_{i=1}^{N}(X_i - \bar{X})^2}{N}} \tag{2.26}$$

② 对于分组数据，方差和标准差的计算公式分别为：

$$\sigma^2 \doteq \frac{\sum_{i=1}^{K}(X_i - \bar{X})^2 F_i}{\sum_{i=1}^{K}F_i} \qquad \sigma \doteq \sqrt{\frac{\sum_{i=1}^{K}(X_i - \bar{X})^2 F_i}{\sum_{i=1}^{K}F_i}} \tag{2.27}$$

③ 样本的方差、标准差与总体的方差、标准差在计算上有所差别。总体的方差和标准差在对各个离差平方平均时是除以数据个数或总频数，而样本的方差和标准差在对各个离差平方平均时是用样本数据个数或总频数减 1（自由度）去除总离差平方和。

设样本的方差为 S^2、标准差为 S，对于未分组整理的原始数据，方差和标准差的计算公式为：

$$S_{n-1}^2 = \frac{\sum_{i=1}^{n}(x_i - \bar{x})^2}{n-1} \qquad S_{n-1} = \sqrt{\frac{\sum_{i=1}^{n}(x_i - \bar{x})^2}{n-1}} \tag{2.28}$$

对于分组数据，方差和标准差的计算公式为：

$$S_{n-1}^2 \doteq \frac{\sum_{i=1}^{k}(x_i-\bar{x})^2 f_i}{\sum_{i=1}^{k} f_i - 1} \qquad S_{n-1} \doteq \sqrt{\frac{\sum_{i=1}^{k}(x_i-\bar{x})^2 f_i}{\sum_{i=1}^{k} f_i - 1}}$$

(2.29)

当 n 很大时，样本方差 S^2 与总体的方差 σ^2 的计算结果相差很小，这时样本方差也可以用总体方差的公式来计算。

6. 相对离散程度：离散系数

前面介绍的全距、平均差、方差和标准差都是反映一组数值变异程度的绝对值，其数值的大小不仅取决于数值的变异程度，还与变量值水平的高低、计量单位的不同有关。所以，不宜直接利用上述变异指标对不同水平、不同计量单位的现象进行比较，应当先做无量纲化处理，即将上述反映数据的绝对差异程度的变异指标转化为反映相对差异程度的指标，然后进行对比。离散系数通常用 V 表示，常用的离散系数为标准差系数。测度了数据的相对离散程度，用于对不同组别数据离散程度的比较计算公式为：

$$V_\sigma = \frac{\sigma}{\bar{X}} \quad \text{或} \quad V_s = \frac{S}{\bar{x}}$$

(2.30)

2.1.3 数据分布偏态与峰度的测定

偏态和峰度就是对这些分布特征的描述：偏度是对数据分布在偏移方向和程度所做的进一步描述；峰度是对数据分布的扁平程度所做的描述。偏斜程度的描述用偏态系数，扁平程度的描述用峰度系数。

1. 动差法

动差又称矩，原是物理学上用以表示力与力臂对重心关系的术语，这个关系和统计学中变量与权数对平均数的关系在性质上很类似，所以统计学也用动差来说明频数分布的性质。

一般地说，取变量的 a 值为中点，所有变量值与 a 之差的 K 次方的平均数称为变量 X 关于 a 的 K 阶动差。用式子表示即为：

$$\frac{\sum(X-a)^K}{N}$$

(2.31)

当 $a=0$ 时，即变量以原点为中心，上式称为 K 阶原点动差，用大写英文字母 M 表示。
一阶原点动差：

$$M_1 = \frac{\sum X}{N}$$

(2.32)

二阶原点动差：

$$M_2 = \frac{\sum X^2}{N}$$

(2.33)

三阶原点动差：

$$M_3 = \frac{\sum X^3}{N} \tag{2.34}$$

当 $a = \bar{X}$ 时，即变量以算术平均数为中心，上式称为 K 阶中心动差，用小写英文字母 m 表示。

一阶中心动差：

$$m_1 = \frac{\sum(X-\bar{X})}{N} = 0 \tag{2.35}$$

二阶中心动差：

$$m_2 = \frac{\sum(X-\bar{X})^2}{N} = \sigma^2 \tag{2.36}$$

三阶中心动差：

$$m_3 = \frac{\sum(X-\bar{X})^3}{N} \tag{2.37}$$

2. 偏态及其测度

偏态是对分布偏斜方向及程度的度量。从前面的内容中我们已经知道，频数分布有对称的，有不对称的（偏态的）。在偏态的分布中，又有两种不同的形态，即左偏和右偏。我们可以利用众数、中位数和算术平均数之间的关系判断分布是左偏还是右偏，但要度量分布偏斜的程度，就需要计算偏态系数了。

采用动差法计算偏态系数是用变量的三阶中心动差 m_3 与 σ^3 进行对比，计算公式为：

$$\alpha = \frac{m_3}{\sigma^3} \tag{2.38}$$

当分布对称时，变量的三阶中心动差 m_3 由于离差三次方后正负相互抵消而取得 0 值，所以 $a=0$；当分布不对称时，正负离差不能抵消，就形成正的或负的三阶中心动差 m_3。当 m_3 为正值时，表示正偏离差值比负偏离差值要大，可以判断为正偏或右偏；反之，当 m_3 为负值时，表示负偏离差值比正偏离差值要大，可以判断为负偏或左偏。$|m_3|$ 越大，表示偏斜的程度就越大。由于三阶中心动差 m_3 含有计量单位，为消除计量单位的影响，就用 σ^3 去除 m_3，使其转化为相对数。同样的，a 的绝对值越大，表示偏斜的程度就越大。

3. 峰度及其测度

峰度是用来衡量分布的集中程度或分布曲线的尖峭程度的指标。计算公式如下：

$$\alpha_4 = \frac{m_4}{\sigma_4} = \frac{\sum(X-\bar{X})^4 F_i}{\sigma^4 \cdot \sum F_i} \tag{2.39}$$

分布曲线的尖峭程度与偶数阶中心动差的数值大小有直接的关系，m_2 是方差，于是就以

四阶中心动差 m_4 来度量分布曲线的尖峭程度。m_4 是一个绝对数,含有计量单位,为消除计量单位的影响,将 m_4 除以 σ^4,就得到无量纲的相对数。衡量分布的集中程度或分布曲线的尖峭程度往往是以正态分布的峰度作为比较标准的。在正态分布条件下,$m^4/\sigma^4=3$,将各种不同分布的尖峭程度与正态分布比较。

当峰度 $a_4>3$ 时,表示分布的形状比正态分布更瘦更高,意味着分布比正态分布更集中在平均数周围,这样的分布称为尖峰分布,如图 2.1(a);当 $a_4=3$ 时,分布为正态分布;当 $a_4<3$ 时,表示分布比正态分布更扁平,意味着分布比正态分布更分散,这样的分布称为扁平分布,如图 2.1(b)所示。

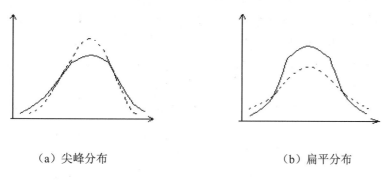

(a) 尖峰分布　　　　　　　(b) 扁平分布

图 2.1　尖峰与平峰分布示意图

2.2　数据相关性

数据相关性是指数据之间存在某种关系。大数据时代,数据相关分析因其具有可以快捷、高效地发现事物间内在关联的优势而受到广泛关注,并有效地应用于推荐系统、商业分析、公共管理、医疗诊断等领域。数据相关性可以时序分析、空间分析等方法进行分析。数据相关性分析也面对着高维数据、多变量数据、大规模数据、增长性数据及其可计算方面等挑战。

2.2.1　相关关系

数据相关关系是指 2 个或 2 个以上变量取值之间在某种意义下所存在的规律,其目的在于探寻数据集里所隐藏的相关关系网。从统计学角度看,变量之间的关系大体可分为两种类型:函数关系和相关关系。一般情况下,数据很难满足严格的函数关系,而相关关系要求宽松,所以被人们广泛接受。需要进一步说明的是,研究变量之间的相关关系主要从两个方向进行:一是相关分析,即通过引入一定的统计指标量化变量之间的相关程度;另一个是回归分析。由于回归分析不仅仅刻画相关关系,更重要的是刻画因果关系。

1. 对于不同测量尺度的变量,有不同的相关系数可用

(1)Pearson 相关系数(Pearson's r):衡量两个等距尺度或等比尺度变量的相关性,是最常见的，也是学习统计学时第一个接触的相关系数。

（2）净相关（Partial Correlation）：在模型中有多个自变量（或解释变量）时，去除其他自变量的影响，只衡量特定一个自变量与因变量之间的相关性。自变量和因变量皆为连续变量。

（3）相关比（Correlation Ratio）：衡量两个连续变量的相关性。

（4）Gamma 相关系数：衡量两个次序尺度变量的相关性。

（5）Spearman 等级相关系数：衡量两个次序尺度变量的相关性。

（6）Kendall 等级相关系数（Kendall Tau Rank Correlation Coefficient）：衡量两个人为次序尺度变量（原始数据为等距尺度）的相关性。

（7）Kendall 和谐系数：衡量两个次序尺度变量的相关性。

（8）Phi 相关系数（Phi Coefficient）：衡量两个真正名目尺度的二分变量的相关性。

（9）列联相关系数（Contingency Coefficient）：衡量两个真正名目尺度变量的相关性。

（10）四分相关（Tetrachoric Correlation）：衡量两个人为名目尺度（原始数据为等距尺度）的二分变量的相关性。

（11）Kappa 一致性系数（K Coefficient of Agreement）：衡量两个名目尺度变量的相关性。

（12）点二系列相关系数（Point-biserial Correlation）：X 变量是真正名目尺度二分变量，Y 变量是连续变量。

（13）二系列相关系数（Biserial Correlation）：X 变量是人为名目尺度二分变量，Y 变量是连续变量。

2. 数据种类

（1）高维数据的相关分析

在探索随机向量间相关性度量的研究中，随机向量的高维特征导致巨大的矩阵计算量，这也成为高维数据相关分析中的关键困难问题。面临高维特征空间的相关分析时，数据可能呈现块分布现象，如医疗数据仓库、电子商务推荐系统。探测高维特征空间中是否存在数据的块分布现象，并发现各数据块对应的特征子空间，从本质上来看，这是基于相关关系度量的特征子空间发现问题。结合子空间聚类技术，发现相关特征子空间，并以此为基础，探索新的分块矩阵计算方法，有望为高维数据相关分析与处理提供有效的求解途径。然而，面临的挑战在于：① 如果数据维度很高、数据表示非常稀疏，如何保证相关关系度量的有效性？②分块矩阵的计算可以有效提升计算效率，但是如何对分块矩阵的计算结果进行融合？

（2）多变量数据的相关分析

在现实的大数据相关分析中，往往面临多变量情况。显然，发展多变量非线性相关关系的度量方法是我们面临的一个重要挑战。

（3）大规模数据的相关分析

大数据时代，相关分析面向的是数据集的整体，因此试图高效地开展相关分析与处理仍然非常困难。为了快速计算大数据相关性，需要探索数据集整体的拆分与融合策略。显然，在这种"分而治之"的策略中,如何有效保持整体的相关性是大规模数据相关分析中必须解决的关键问题。有关学者给出了一种可行的拆分与融合策略,也指出随机拆分策略是可能的解决路径。当然,在设计拆分与融合策略时,如何确定样本子集规模、如何保持子集之间的信息传递、如何设计各子集结果的融合原理等都是具有挑战性的问题。

(4) 增长性数据的相关分析

在大数据中,数据呈现快速增长特征。更为重要的是,诸如电商精准推荐等典型增长性数据相关分析任务,迫切需要高效的在线相关分析技术。就增长性数据而言,可表现为样本规模的增长、维数规模的增长以及数据取值的动态更新。显然,对增长性数据相关分析而言,特别是对在线相关分析任务而言,每次对数据整体进行重新计算对于用户而言是难以接受的,更难以满足用户的实时性需求。无论何种类型的数据增长,往往与原始数据集存在某种关联模式,利用已有的关联模式设计具有递推关系的批增量算法是一种行之有效的计算策略。面向大数据的相关分析任务,探测增长性数据与原始数据集的关联模式,进而发展具有递推关系的高效批增量算法,可为增长性数据相关分析尤其是在线相关分析提供有效的技术手段。

3. 相关关系的种类

现象之间的相互关系很复杂,它们涉及的变动因素多少不同,作用方向不同,表现出来的形态也不同。相关关系大体有以下几种分类:

(1) 正相关与负相关

按相关关系的方向来分,可分为正相关和负相关。当两个因素(或变量)的变动方向相同时,即自变量 x 值增加(或减少),因变量 y 值也相应地增加(或减少),这样的关系就是正相关。如家庭消费支出随收入增加而增加就属于正相关。如果两个因素(或变量)变动的方向相反,即自变量 x 值增大(或减小),因变量 y 值随之减小(或增大),就称为负相关。例如,商品流通费用率随商品经营的规模增大而逐渐降低就属于负相关。

(2) 单相关与复相关

按自变量的数量来分,可分为单相关和复相关。单相关是指两个变量之间的相关关系,即所研究的问题只涉及一个自变量和一个因变量,如职工的生活水平与工资之间的关系就是单相关。复相关是指 3 个或 3 个以上变量之间的相关关系,即所研究的问题涉及若干个自变量与一个因变量,如同时研究成本、市场供求状况、消费倾向对利润的影响时,这几个因素之间的关系是复相关。

(3) 线性相关与非线性相关

按相关关系的表现形态来分,可分为线性相关与非线性相关。线性相关是指在两个变量之间,当自变量 x 值发生变动时,因变量 y 值发生大致均等的变动,在相关图的分布上,近似地表现为直线形式。比如,商品销售额与销售量即为线性相关。非线性相关是指在两个变量之间,当自变量 x 值发生变动时,因变量 y 值发生不均等的变动,在相关图的分布上,表现为抛物线、双曲线、指数曲线等非直线形式。比如,从人的生命全过程来看,年龄与医疗费支出呈非线性相关。

（4）完全相关、不完全相关与不相关

按相关程度来分，可分为完全相关、不完全相关和不相关。完全相关是指两个变量之间具有完全确定的关系，即因变量 y 值完全随自变量 x 值的变动而变动，它在相关图上表现为所有的观察点都落在同一条直线上，这时相关关系就转化为函数关系。不相关是指两个变量之间不存在相关关系，即两个变量变动彼此互不影响。自变量 x 值变动时，因变量 y 值不随之做相应变动。比如，家庭收入多少与孩子多少之间不存在相关关系。不完全相关是指介于完全相关和不相关之间的一种相关关系。比如，农作物产量与播种面积之间的关系。不完全相关关系是统计研究的主要对象。

2.2.2 相关分析

1. 相关分析的主要内容

相关分析是指对客观现象的相互依存关系进行分析、研究，这种分析方法叫相关分析法。相关分析的目的在于研究相互关系的密切程度及其变化规律，以便做出判断，进行必要的预测和控制。相关分析的主要内容包括以下几点。

（1）确定现象之间有无相关关系

这是相关与回归分析的起点，只有存在相互依存关系，才有必要进行进一步的分析。

（2）确定相关关系的密切程度和方向

确定相关关系密切程度主要是通过绘制相关图表和计算相关系数来完成。只有达到一定密切程度的相关关系，才可配合具有一定意义的回归方程。

（3）确定相关关系的数学表达式

为确定现象之间变化上的一般关系，我们必须使用函数关系的数学公式作为相关关系的数学表达式。如果现象之间表现为直线相关，我们可采用配合直线方程的方法；如果现象之间表现为曲线相关，我们可采用配合曲线方程的方法。

（4）确定因变量估计值误差程度

使用配合直线或曲线的方法可以找到现象之间一般的变化关系，也就是自变量 x 变化时因变量 y 将会发生多大的变化。根据得出的直线方程或曲线方程，我们可以给出自变量的若干数值，求得因变量的若干个估计值。估计值与实际值是有出入的，确定因变量估计值误差大小的指标是估计标准误差。估计标准误差大，表明估计不太精确；估计标准误差小，表明估计较精确。

2. 相关关系的测定

相关分析的主要方法有相关表、相关图和相关系数 3 种。现将这 3 种方法分述如下：

（1）相关表

在统计中，制作相关表或相关图，可以直观地判断现象之间大致存在的相关关系的方向、形式和密切程度。

在对现象总体中两种相关变量进行相关分析,以研究其相互依存关系时,如果将实际调查取得的一系列成对变量值的数据顺序地排列在一张表格上,那么这张表格就是相关表。相关表仍然是统计表的一种。根据数据是否分组,相关表可以分为简单相关表和分组相关表。

① 简单相关表

简单相关表是数据未经分组的相关表,是把自变量按从小到大的顺序并配合因变量一一对应平行排列起来的统计表。

② 分组相关表

在大量观察的情况下,原始数据很多,运用简单相关表表示就很难使用。这时要将原始数据进行分组,然后编制相关表,这种相关表称为分组相关表。分组相关表包括单变量分组相关表和双变量分组相关表两种。

- 单变量分组相关表。在原始数据很多时,对自变量数值进行分组,而对应的因变量不分组,只计算其平均值。根据数据具体情况,自变量可以是单项式,也可以是组距式。
- 双变量分组相关表。对两种有关变量都进行分组,交叉排列,并列出两种变量各组间的共同次数,这种统计表称为双变量分组相关表,因为表格形似棋盘,故又称棋盘式相关表。

(2)相关图

相关图又称散点图,以直角坐标系的横轴代表自变量 x,纵轴代表因变量 y,将两个变量间相对应的变量值用坐标点的形式描绘出来,用来反映两个变量之间的相关关系。

相关图可以按未经分组的原始数据来编制,也可以按分组的数据(包括按单变量分组相关表和双变量分组相关表)来编制。通过相关图将会发现,当 y 对 x 是函数关系时,所有的相关点都会分布在某一条线上。在相关关系的情况下,由于其他因素的影响,这些点并非处在一条线上,但所有相关点的分布会显示出某种趋势,因此相关图会很直观地显示现象之间相关的方向和密切程度。

(3)相关系数

相关表和相关图大体说明变量之间有无关系,但是它们的相关关系紧密程度却无法表达,因此,需运用数学解析方法构建一个恰当的数学模型来显示相关关系及其密切程度。对现象之间的相关关系的紧密程度做出确切的数量说明,就需要计算相关系数。

相关系数是在线性相关条件下说明两个现象之间关系密切程度的统计分析指标,记为 γ。

相关系数的计算公式为:

$$\gamma = \frac{\sigma_{xy}^2}{\sigma_x \sigma_y} = \frac{\frac{1}{n}\sum(x-\bar{x})\sum(y-\bar{y})}{\sqrt{\frac{1}{n}\sum(x-\bar{x})^2}\sqrt{\frac{1}{n}\sum(y-\bar{y})^2}} \tag{2.40}$$

式中 n——数据项数;

$\bar{x}$——x 变量的算术平均数;

$\bar{y}$——y 变量的算术平均数；

σ_x——x 变量的标准差；

σ_y——y 变量的标准差；

σ_{xy}——xy 变量的协方差。

在实际问题中，如果根据原始数据计算相关系数，可运用相关系数的简捷法计算，其计算公式为

$$\gamma = \frac{n\sum xy - \sum x \sum y}{\sqrt{n\sum x^2 - (\sum x)^2}\sqrt{n\sum y^2 - (\sum y)^2}} \tag{2.41}$$

（4）相关系数的分析

明晰相关系数的性质是进行相关系数分析的前提。现将相关系数的性质总结如下：

① 相关系数的数值范围是在-1 和+1 之间，即-1≤γ≤1。

② 计算结果，当γ>0 时，表示 x 与 y 为正相关；当γ<0 时，x 与 y 为负相关。

③ 相关系数 γ 的绝对值越接近于 1，表示相关关系越强；越接近于 0，表示相关关系越弱。如果|γ|=1，就表示两个现象完全线性相关。如果|γ|=0，就表示两个现象完全不相关（不是线性相关）。

④ 相关系数 γ 的绝对值在 0.3 以下是无线性相关，0.3 以上是有线性相关，0.3～0.5 是低度线性相关，0.5～0.8 是显著相关，0.8 以上是高度相关。

2.3 数据聚类性

数据聚类是指根据数据的内在性质将数据分成一些聚合类，每一聚合类中的元素尽可能具有相同的特性，不同聚合类之间的特性差别尽可能大。

聚类分析的目的是分析数据是否属于各个独立的分组，使一组中的成员彼此相似，而与其他组中的成员不同。它对一个数据对象的集合进行分析，但与分类分析不同的是，所划分的类是未知的，因此聚类分析也称为无指导或无监督（Unsupervised）学习。聚类分析的一般方法是将数据对象分组为多个类或簇（Cluster），在同一簇中的对象之间具有较高的相似度，而不同簇中的对象差异较大。由于聚类分析的上述特征，在许多应用中，对数据集进行了聚类分析后可将一个簇中的各数据对象作为一个整体对待。

数据聚类（Cluster Analysis）是对于静态数据分析的一门技术，在许多领域受到广泛应用，包括机器学习、数据挖掘、模式识别、图像分析以及生物信息。

1. 聚类应用

随着信息技术的高速发展，数据库应用的规模、范围和深度不断扩大，积累了大量的数据，而这些激增的数据后面隐藏着许多重要的信息，因此人们希望能够对其进行更高层次的分析，

以便更好地利用这些数据。目前的数据库系统可以高效、方便地实现数据的录入、查询、统计等功能，但是无法发现数据中存在的各种关系和规则，更无法根据现有的数据预测未来的发展趋势。数据聚类分析正是解决这一问题的有效途径，它是数据挖掘的重要组成部分，用于发现在数据库中未知的对象类，为数据挖掘提供有力的支持，它是近年来广为研究的问题之一。聚类分析是一个极富挑战性的研究领域，采用基于聚类分析方法的数据挖掘在实践中取得了较好的效果。聚类分析也可以作为其他一些算法的预处理步骤，聚类可以作为一个独立的工具来获知数据的分布情况，使数据形成簇，其他算法再针对生成的簇进行处理，聚类算法既可作为特征和分类算法的预处理步骤，也可将聚类结果用于进一步关联分析。迄今为止，人们提出了许多聚类算法，所有这些算法都试图解决大规模数据的聚类问题。聚类分析还成功地应用在了模式识别、图像处理、计算机视觉、模糊控制等领域，并在这些领域中取得了长足的发展。

2. 数据聚类

聚类就是将一个数据单位的集合分割成几个称为簇或类别的子集，每个类中的数据都有相似性，它的划分依据就是"物以类聚"。数据聚类分析是根据事物本身的特性，研究对被聚类的对象进行类别划分的方法。聚类分析依据的原则是使同一聚簇中的对象具有尽可能大的相似性，而不同聚簇中的对象具有尽可能大的相异性，聚类分析主要解决的问题就是如何在没有先验知识的前提下，实现满足这种要求的聚簇的聚合。聚类分析称为无监督学习（Unsupervised Study），主要体现是聚类学习的数据对象没有类别标注，需要由聚类学习算法自动计算。

3. 聚类类型

经过持续了半个多世纪的深入研究聚类算法，聚类技术已经成为最常用的数据分析技术之一。各种算法的提出、发展、演化使聚类算法家族不断壮大。下面就针对目前数据分析和数据挖掘业界主流的认知对聚类算法进行介绍。

（1）划分方法

给定具有 n 个对象的数据集，采用划分方法对数据集进行 k 个划分，每个划分（每个组）代表一个簇。其中，$k \leq n$，并且每个簇至少包含一个对象，而且每个对象一般只能属于一个组。对于给定的 k 值，一般要做一个初始划分，然后采取迭代重新定位技术，通过让对象在不同组间移动来改进划分的准确度和精度。一个好的划分原则是：同一个簇中对象之间的相似性很高（或距离很近），而不同簇的对象之间相异度很高（或距离很远）。

① K-Means 算法：又叫 K 均值算法，是目前最著名、使用最广泛的聚类算法。在给定一个数据集和需要划分的数目 k 后，该算法可以根据某个距离函数反复把数据划分到 k 个簇中，直到收敛为止。K-Means 算法用簇中对象的平均值来表示划分的每个簇，其大致的步骤是：首先将随机抽取的 k 个数据点作为初始的聚类中心（种子中心），然后计算每个数据点到每个种子中心的距离，并把每个数据点分配到距离它最近的种子中心；一旦所有的数据点都被分配完成，每个聚类的聚类中心（种子中心）就按照本聚类（本簇）的现有数据点重新计算；这个过程不断重复，直到收敛，即满足某个终止条件为止。最常见的终止条件是误差平方和 SSE（指令集的简称）局部最小。

② K-Medoids 算法：又叫 K 中心点算法，用最接近簇中心的一个对象来表示划分的每个簇。K-Medoids 算法与 K-Means 算法的划分过程相似，两者最大的区别是：K-Medoids 算法是用簇中最靠近中心点的一个真实的数据对象来代表该簇，而 K-Means 算法是用计算出来的簇中对象的平均值（这个平均值是虚拟的，并没有一个真实的数据对象具有这个平均值）来代表该簇。

（2）层次方法

在给定 n 个对象的数据集后，可用层次方法（Hierarchical Methods）对数据集进行层次分解，直到满足某种收敛条件为止。按照层次分解的形式不同，层次方法又可以分为凝聚层次聚类和分裂层次聚类。

①凝聚层次聚类：又叫自底向上方法，一开始将每个对象作为单独的一类，然后相继合并与其相近的对象或类，直到所有小的类别合并成一个类，即层次的最上面，或者达到一个收敛，即终止条件为止。

②分裂层次聚类：又叫自顶向下方法，一开始将所有对象置于一个簇中，在迭代的每一步中类会被分裂成更小的类，直到最终每个对象在一个单独的类中，或者满足一个收敛，即终止条件为止。

（3）基于密度的方法

传统的聚类算法都是基于对象之间的距离（距离作为相似性的描述指标）进行聚类划分，但是这些基于距离的方法只能发现球状类型的数据，对于非球状类型的数据来说只根据距离来描述和判断是不够的。鉴于此，人们提出了一个密度的概念——基于密度的方法（Density-Based Methods），其原理是：只要邻近区域内的密度（对象的数量）超过了某个阈值，就继续聚类。换言之，给定某个簇中的每个数据点（数据对象），在一定范围内必须包含一定数量的其他对象。该算法从数据对象的分布密度出发，把密度足够大的区域连接在一起，因此可以发现任意形状的类。该算法还可以过滤噪声数据（异常值）。基于密度的方法的典型算法包括 DBSCAN（Density-Based Spatial Clustering of Application with Noise，具有噪声的基于密度的空间聚类应用算法）以及其扩展算法 OPTICS（Ordering Points to Identify the Clustering Structure，即通过点排序识别聚类结构的密度聚类算法）。其中，DBSCAN 算法会根据一个密度阈值来控制簇的增长，将具有足够高密度的区域划分为类，并可在带有噪声的空间数据库里发现任意形状的聚类。尽管此算法优势明显，但是其最大的缺点是需要用户确定输入参数，而且对参数十分敏感。

（4）基于网格的方法

基于网格的方法（Grid-Based Methods）将把对象空间量化为有限数目的单元，这些单元再形成网格结构，让所有的聚类操作都在这个网格结构中进行。该算法的优点是处理速度快，其处理时间常常独立于数据对象的数目，只跟量化空间中每一维的单元数目有关。基于网格方法的典型算法是 STING（统计信息网格方法，Statistical Information Grid）算法。该算法是一种基于网格的多分辨率聚类技术，将空间区域划分为不同分辨率级别的矩形单元，并形成一个层次结构,且高层的低分辨率单元会被划分为多个低一层次的较高分辨率单元。这种算法从最底层的网格开始逐渐向上计算网格内数据的统计信息并储存。网格建立完成后,就用类似 DBSCAN 的方法对网格进行聚类。

4. 数据聚类需解决的问题

在聚类分析的研究中,有许多急待进一步解决的问题,比如:处理数据为大数据量、具有复杂数据类型的数据集合时,聚类分析结果的精确性问题;对高属性维数据的处理能力;数据对象分布形状不规则时的处理能力;处理噪声数据的能力,能够处理数据中包含的孤立点,未知数据、空缺或者错误的数据;对数据输入顺序的独立性,也就是对于任意的数据输入顺序产生相同的聚类结果;减少对先决知识或参数的依赖型……这些问题的存在使得我们研究高正确率、低复杂度、I/O 开销小、适合高维数据、具有高度可伸缩性的聚类方法迫在眉睫,这也是今后聚类方法研究的方向。

5. 数据聚类应用

聚类分析可以作为一个独立的工具来获得数据的分布情况,通过观察每个簇的特点,集中对特定的某些簇做进一步分析,以获得需要的信息。聚类分析应用广泛,除了在数据挖掘、模式识别、图像处理、计算机视觉、模糊控制等领域的应用,还被应用在气象分析、食品检验、生物种群划分、市场细分、业绩评估等诸多方面。例如,在商务上,聚类分析可以帮助市场分析人员从客户基本库中发现不同的客户群,并且用购买模式来刻画不同的客户群特征。聚类分析还可以应用在欺诈探测中,聚类中的孤立点就可能预示着欺诈行为的存在。聚类分析的发展过程也是聚类分析的应用过程,目前聚类分析在相关领域已经取得丰硕的成果。

2.4 数据主成分分析

在实际问题中,我们经常会遇到研究多个变量的问题,而且在多数情况下多个变量之间常常存在一定的相关性。由于变量个数较多,再加上变量之间的相关性,势必增加分析问题的复杂性。要将多个变量综合为少数几个代表性变量,既能够代表原始变量的绝大多数信息,又互不相关,并且在新的综合变量基础上可以进一步进行统计分析,需要进行主成分分析。

2.4.1 主成分分析的原理及模型

1. 主成分分析原理

主成分分析采取一种数学降维的方法,找出几个综合变量来代替原来众多的变量,使这些综合变量能尽可能地代表原来变量的信息量,而且彼此之间互不相关。这种将把多个变量化为少数几个互相无关的综合变量的统计分析方法就叫作主成分分析或主分量分析。

主成分分析所要做的就是设法将原来众多具有一定相关性的变量重新组合为一组新的相互无关的综合变量来代替原来的变量。通常,数学上的处理方法就是将原来的变量进行线性组合，作为新的综合变量，但是这种组合如果不加以限制，就可以有很多，应该如何选择呢？如果将选取的第一个线性组合（第一个综合变量）记为 F_1，自然希望它尽可能多地反映原来变量的信息，这里"信息"用方差来测量，即希望 $Var(F_1)$ 越大，表示 F_1 包含的信息越多。在所有的线性组合中所选取的 F_1 应该是方差最大的，故称 F_1 为第一主成分。如果第一主成分不足以代表原来 p 个变量的信息，再考虑选取 F_2（第二个线性组合）。为了有效地反映原来的信息，F_1 已有的信息不需要出现在 F_2 中，用数学语言表达就是要求 $Cov(F_1, F_2)=0$，称 F_2 为第二主成分，以此类推，可以构造出第三、四、…p 个主成分。

2. 主成分分析的数学模型

对于一个样本数据，观测 p 个变量 $x_1, x_2, \cdots, x_p$，n 个样品的数据矩阵为：

$$X = \begin{pmatrix} x_{11} & x_{12} & \cdots & x_{1p} \\ x_{21} & x_{22} & \cdots & x_{2p} \\ \vdots & \vdots & \vdots & \vdots \\ x_{n1} & x_{n2} & \cdots & x_{np} \end{pmatrix} = (x_1, x_2, \cdots, x_p) \quad (2.42)$$

其中：$x_j = \begin{pmatrix} x_{1j} \\ x_{2j} \\ \vdots \\ x_{nj} \end{pmatrix}, j = 1, 2, \cdots, P$。

主成分分析就是将 p 个观测变量综合成为 p 个新的变量（综合变量），即

$$\begin{cases} F_1 = a_{11}x_1 + a_{12}x_2 + \cdots + a_{1p}x_p \\ F_2 = a_{21}x_1 + a_{22}x_2 + \cdots + a_{2p}x_p \\ \cdots \\ F_p = a_{p1}x_1 + a_{p2}x_2 + \cdots + a_{pp}x_p \end{cases} \quad (2.43)$$

简写为：

$$F_j = a_{j1}x_1 + a_{j2}x_2 + \cdots + a_{jp}x_p \quad (2.44)$$
$$j = 1, 2, \cdots, p$$

要求模型满足以下条件：

① F_i 和 F_j 互不相关（$i \neq j$, $i,j=1,2,\cdots,P$）；
② F_1 的方差大于 F_2 的方差，大于 F_j 的方差；
③ $a_{k1}^2 + a_{k2}^2 + \cdots + a_{kp}^2 = 1$。其中，$k = 1, 2, \cdots, P$。

于是，称 F_1 为第一主成分，F_2 为第二主成分，以此类推，有第 p 个主成分。主成分又叫主分量。这里 a_{ij} 称为主成分系数。

上述模型可用矩阵表示为：
其中 $F = AX$

其中：

$$F = \begin{pmatrix} F_1 \\ F_2 \\ \vdots \\ F_p \end{pmatrix} \quad X = \begin{pmatrix} x_1 \\ x_2 \\ \vdots \\ x_p \end{pmatrix} \tag{2.45}$$

$$A = \begin{pmatrix} a_{11} & a_{12} & \cdots & a_{1p} \\ a_{21} & a_{22} & \cdots & a_{2p} \\ \vdots & \vdots & \vdots & \vdots \\ a_{p1} & a_{p2} & \cdots & a_{pp} \end{pmatrix} = \begin{pmatrix} a_1 \\ a_2 \\ \vdots \\ a_p \end{pmatrix} \tag{2.46}$$

A 称为主成分系数矩阵。

2.4.2 主成分分析的几何解释

假设有 n 个样品，每个样品有两个变量，即在二维空间中讨论主成分的几何意义。设 n 个样品在二维空间中的分布大致为一个椭圆，如图 2.2 所示。

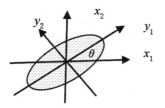

图 2.2 主成分几何解释图

将图 2.2 中的坐标系进行正交旋转一个角度 θ，使其椭圆长轴方向取坐标 y_1，在椭圆短轴方向取坐标 y_2，旋转公式为

$$\begin{cases} y_{1j} = x_{1j}\cos\theta + x_{2j}\sin\theta \\ y_{2j} = x_{1j}(-\sin\theta) + x_{2j}\cos\theta \end{cases} \quad j=1,2,\cdots,n \tag{2.47}$$

写成矩阵形式为：

$$\begin{aligned} Y &= \begin{bmatrix} y_{11} & y_{12} & \cdots & y_{1n} \\ y_{21} & y_{22} & \cdots & y_{2n} \end{bmatrix} \\ &= \begin{bmatrix} \cos\theta & \sin\theta \\ -\sin\theta & \cos\theta \end{bmatrix} \cdot \begin{bmatrix} x_{11} & x_{12} & \cdots & x_{1n} \\ x_{21} & x_{22} & \cdots & x_{2n} \end{bmatrix} = U \cdot X \end{aligned} \tag{2.48}$$

其中 U 为坐标旋转变换矩阵，是正交矩阵，即有 $U'=U^{-1}$，$UU'=I$，即满足 $\sin^2\theta + \cos^2\theta = 1$。经过旋转变换后，得到图 2.3 所示的新坐标系。

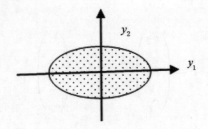

图 2.3　经过旋转变换后得到的新坐标系

新坐标系中的 y_1、y_2 有如下性质：

（1）n 个点的坐标 y_1 和 y_2 的相关几乎为零。

（2）二维平面上 n 个点的方差大部分都归结到 y_1 轴上，而 y_2 轴上的方差较小。

y_1 和 y_2 称为原始变量 x_1 和 x_2 的综合变量。由于 n 个点在 y_1 轴上的方差最大，因而将二维空间的点用 y_1 轴上的一维综合变量来代替，所损失的信息量最小，由此称 y_1 轴为第一主成分。y_2 轴与 y_1 轴正交，有较小的方差，称为第二主成分。

2.4.3　主成分的导出

根据主成分分析的数学模型的定义，要进行主成分分析，就需要根据原始数据以及模型 3 个条件的要求，求出主成分系数，以便得到主成分模型。这就是导出主成分所要解决的问题。

（1）根据主成分数学模型的条件①要求主成分之间互不相关，为此主成分之间的协差阵应该是一个对角阵。对于主成分，

$$F = AX \tag{2.49}$$

其协差阵应为，

$$Var(F) = Var(AX) = (AX) \cdot (AX)' = AXX'A'$$

$$= \Lambda = \begin{pmatrix} \lambda_1 & & & \\ & \lambda_2 & & \\ & & \ddots & \\ & & & \lambda_p \end{pmatrix} \tag{2.50}$$

（2）设原始数据的协方差阵为 V，如果原始数据进行了标准化处理，那么协方差阵等于相关矩阵，即有

$$V = R = XX' \tag{2.51}$$

（3）根据主成分数学模型条件③和正交矩阵的性质，若能够满足条件③最好要求 A 为正交矩阵，即满足

$$AA' = I \tag{2.52}$$

将原始数据的协方差代入主成分的协差阵公式，得

$$Var(F)=AXX'A'=ARA'=\Lambda$$
$$ARA'=\Lambda \quad RA'=A'\Lambda \tag{2.53}$$

展开上式，得

$$\begin{pmatrix} r_{11} & r_{12} & \cdots & r_{1p} \\ r_{21} & r_{22} & \cdots & r_{2p} \\ \vdots & \vdots & \vdots & \vdots \\ r_{p1} & r_{p2} & \cdots & r_{pp} \end{pmatrix} \cdot \begin{pmatrix} a_{11} & a_{21} & \cdots & a_{p1} \\ a_{12} & a_{22} & \cdots & a_{p2} \\ \vdots & \vdots & \vdots & \vdots \\ a_{1p} & a_{2p} & \cdots & a_{pp} \end{pmatrix} = \\ \begin{pmatrix} a_{11} & a_{21} & \cdots & a_{p1} \\ a_{12} & a_{22} & \cdots & a_{p2} \\ \vdots & \vdots & \vdots & \vdots \\ a_{1p} & a_{2p} & \cdots & a_{pp} \end{pmatrix} \cdot \begin{pmatrix} \lambda_1 & & & \\ & \lambda_2 & & \\ & & \ddots & \\ & & & \lambda_p \end{pmatrix} \tag{2.54}$$

展开等式两边，根据矩阵相等的性质，这里只根据第一列得出的方程为：

$$\begin{cases} (r_{11}-\lambda_1)a_{11}+r_{12}a_{12}+\cdots+r_{1p}a_{1p}=0 \\ r_{21}a_{11}+(r_{22}-\lambda_1)a_{12}+\cdots+r_{2p}a_{1p}=0 \\ \cdots\cdots \\ r_{p1}a_{11}+r_{p2}a_{12}+\cdots+(r_{pp}-\lambda_1)a_{1p}=0 \end{cases} \tag{2.55}$$

为了得到该齐次方程的解，要求其系数矩阵行列式为 0，即

$$\begin{vmatrix} r_{11}-\lambda_1 & r_{12} & \cdots & r_{1p} \\ r_{21} & r_{22}-\lambda_1 & \cdots & r_{2p} \\ \vdots & \vdots & \vdots & \vdots \\ r_{1p} & r_{p2} & \cdots & r_{pp}-\lambda_1 \end{vmatrix}=0 \tag{2.56}$$

$$|\boldsymbol{R}\text{-}\lambda_1\boldsymbol{I}|=0$$

显然，λ_1 是相关系数矩阵的特征值，$a_1=(a_{11},a_{12},\cdots,a_{1p})$ 是相应的特征向量。根据第二列、第三列等可以得到类似的方程，于是 λ_i 是方程 $|\boldsymbol{R}\text{-}\lambda\boldsymbol{I}|=0$ 的 p 个根，λ_i 为特征方程的特征根，a_j 是其特征向量的分量。

2.4.4 证明主成分的方差是依次递减

设相关系数矩阵 $\boldsymbol{R}$ 的 p 个特征根为 $\lambda_1 \geqslant \lambda_2 \geqslant \cdots \geqslant \lambda_p$，相应的特征向量为 a_j，得

$$A = \begin{pmatrix} a_{11} & a_{12} & \cdots & a_{1p} \\ a_{21} & a_{22} & \cdots & a_{2p} \\ \vdots & \vdots & \vdots & \vdots \\ a_{p1} & a_{p2} & \cdots & a_{pp} \end{pmatrix} = \begin{pmatrix} a_1 \\ a_2 \\ \vdots \\ a_p \end{pmatrix} \tag{2.57}$$

相对于 F_1 的方差为：

$$Var(F_1) = a_1 XX' a'_1 = a_1 R a'_1 = \lambda_1 \tag{2.58}$$

同样有 $Var(F_i) = \lambda_1$，即主成分的方差依次递减，并且协方差为：

$$\begin{aligned} Cov(a'_i X', a_j X) &= a'_i R a_j \\ &= a'_i \left(\sum_{a=1}^{P} \lambda_a a_a a'_a \right) a_j \\ &= \sum_{a=1}^{P} \lambda_a (a'_a a_a)(a'_a a_j) = 0 \quad i \neq j \end{aligned} \tag{2.59}$$

综上所述，根据证明可知，主成分分析中的主成分协方差是对角矩阵，其对角线上的元素恰好是原始数据相关矩阵的特征值，而主成分系数矩阵 A 的元素则是原始数据相关矩阵特征值相应的特征向量。矩阵 A 是一个正交矩阵。

于是，变量 $(x_1, x_2 \cdots x_p)$ 经过变换后得到新的综合变量为

$$\begin{cases} F_1 = a_{11} x_1 + a_{12} x_2 + \cdots + a_{1p} x_p \\ F_2 = a_{21} x_1 + a_{22} x_2 + \cdots + a_{2p} x_p \\ \quad \cdots \\ F_p = a_{p1} x_1 + a_{p2} x_2 + \cdots + a_{pp} x_p \end{cases} \tag{2.60}$$

新的随机变量彼此不相关，且方差依次递减。

2.4.5 主成分分析的计算

样本观测数据矩阵为：

$$X = \begin{pmatrix} x_{11} & x_{12} & \cdots & x_{1p} \\ x_{21} & x_{22} & \cdots & x_{2p} \\ \vdots & \vdots & \vdots & \vdots \\ x_{n1} & x_{n2} & \cdots & x_{np} \end{pmatrix} \tag{2.61}$$

（1）对原始数据进行标准化处理。

$$x_{ij}^* = \frac{x_{ij} - \bar{x}_j}{\sqrt{\text{var}(x_j)}} \quad (i=1,2,\cdots,n; j=1,2,\cdots,p) \tag{2.62}$$

其中，$\bar{x}_j = \frac{1}{n} \sum_{i=1}^{n} x_{ij}$，$\text{var}(x_j) = \frac{1}{n-1} \sum_{i=1}^{n} (x_{ij} - \bar{x}_j)^2 \quad (j=1,2,\cdots,p) \tag{2.63}$

（2）计算样本相关系数矩阵。

$$R = \begin{bmatrix} r_{11} & r_{12} & \cdots & r_{1p} \\ r_{21} & r_{22} & \cdots & r_{2p} \\ \vdots & \vdots & \cdots & \vdots \\ r_{p1} & r_{p2} & \cdots & r_{pp} \end{bmatrix} \qquad (2.64)$$

为方便，假定原始数据标准化后仍用 X 表示，则经标准化处理后的数据相关系数为：

$$r_{ij} = \frac{1}{n-1}\sum_{t=1}^{n} x_{ti} x_{tj} \qquad (i,j=1,2,\cdots,p) \qquad (2.65)$$

（3）用雅克比方法求相关系数矩阵 R 的特征值（$\lambda_1, \lambda_2, \lambda_P$）和相应的特征向量 $a_i = (a_{i1}, a_{i2}, \cdots a_{ip})$（其中，$i=1,2,\cdots,p$）。

（4）选择重要的主成分，并写出主成分表达式。主成分分析可以得到 p 个主成分，但是由于各个主成分的方差是递减的，包含的信息量也是递减的，因此实际分析时一般不是选取 p 个主成分，而是根据各个主成分累计贡献率的大小选取前 k 个主成分，这里贡献率就是指某个主成分的方差占全部方差的比重，实际也就是某个特征值占全部特征值合计的比重，即

$$\text{贡献率} = \frac{\lambda_i}{\sum_{i=1}^{p}\lambda_i} \qquad (2.66)$$

贡献率越大，说明该主成分所包含的原始变量的信息越强。主成分个数 k 的选取主要根据主成分的累积贡献率来决定，即一般要求累计贡献率达到 85%以上，这样才能保证综合变量能包括原始变量的绝大多数信息。

另外，在实际应用中，选择了重要的主成分后，还要注意主成分实际含义的解释。主成分分析中一个关键的问题是如何给主成分赋予新的意义，给出合理的解释。一般而言，这个解释是根据主成分表达式的系数结合定性分析来进行的。主成分是原来变量的线性组合，在这个线性组合中各变量的系数有大有小、有正有负，有的大小相当，因而不能简单地认为这个主成分是某个原变量属性的作用。线性组合中各变量系数的绝对值大者表明该主成分主要综合了绝对值大的变量。有几个变量系数大小相当时，应认为这一主成分是几个变量的总和。这几个变量综合在一起应赋予怎样的实际意义，则要结合具体实际问题和专业给出恰当的解释，进而才能达到深刻分析的目的。

（5）计算主成分得分。根据标准化的原始数据，按照各个样品分别代入主成分表达式，就可以得到各主成分下各个样品的新数据，即为主成分得分，具体形式可如下：

$$\begin{pmatrix} F_{11} & F_{12} & \cdots & F_{1k} \\ F_{21} & F_{22} & \cdots & F_{2k} \\ \vdots & \vdots & \vdots & \vdots \\ F_{n1} & F_{n2} & \cdots & F_{nk} \end{pmatrix} \qquad (2.67)$$

（6）依据主成分得分的数据，可以进行进一步的统计分析。其中，常见的应用有主成分回归、变量子集合的选择、综合评价等。

2.5 数据动态性

动态数据是指观察或记录下来的一组按时间先后顺序排列起来的数据序列。

1. 数据特征

（1）构成
- 时间。
- 反映现象在一定时间条件下数量特征的指标值。

（2）表示
- $x(t)$：时间 t 为自变量。
- 整数：离散的，等间距的。
- 非整数：连续的，实际分析时必须进行采样处理。
- 时间单位：秒，分，小时，日，周，月，年。

2. 动态数据分类——按照指标值的表现形式

（1）绝对数序列
- 时期序列。
- 可加性。
- 时点序列。
- 不可加性。

（2）相对数/平均数序列
- 把一系列同种相对数指标按时间先后顺序排列而成的时间序列叫作相对数时间序列。
- 平均数时间序列是指由一系列同类平均指标按时间先后顺序排列的时间序列。

3. 时间数据分类——按照时间的表现形式

- 连续。
- 离散。
- 时间序列中，时间必须是等间隔的。

4. 动态数据的特点

- 数据取值随时间变化。
- 在每一时刻取什么值，不可能完全准确地用历史值预报。

- 前后时刻（不一定是相邻时刻）的数值或数据点有一定的相关性。
- 整体存在某种趋势或周期性。

5. 动态数据的构成与分解

时间序列=趋势+周期+平稳随机成分+白噪声。

6. 动态数据分析模型分类

（1）研究单变量或少数几个变量的变化

- 随机过程。
- 周期分析和时间序列分析。
- 灰色系统。
- 关联分析，GM 模型。

（2）研究多变量的变化

- 系统动力学建模。

7. 时间序列模型

- 研究一个或多个被解释变量随时间变化规律的模型。
- 模型主要用于预测分析。
- 目的——精确预测未来变化。
- 数据要求——序列平稳。
- 研究角度。
- 时间域。
- 频率域。
- 模型内容。
- 周期分析。
- 时间序列预测。

时间序列模型可表示为

$$x_t = f(x_{t-1}, x_{t-2}, \cdots) + \varepsilon_t \tag{2.68}$$

其中，ε_t 表示白噪声。

8. 动态系统模型

- 研究具有时变特点的多个因素之间的相互作用，以及这些作用与系统整体发展之间的关系的模型。
- 模型主要用于模拟和情景分析。
- 重点。
- 各种因素是如何相互作用影响系统总体发展的。

9. 模型表示
- 因果反馈逻辑图。
- 未来系统要素变化趋势图。

10. 建模步骤
- 分析数据的动态特征。
- 进行数据序列分解。
- 数据预处理。
- 模型构建，模型确认。

11. 建模方法

（1）时间序列模型

- 统计学方法。
- 随机过程理论。
- 灰色系统方法。

（2）动态系统模型

- 动态系统仿真方法。

12. 时间序列模型

（1）平稳随机过程

如果一个随机过程的均值和方差在时间过程上是常数，并且在任何两个时期之间的协方差值仅依赖于这两个时期间的距离和滞后，而不依赖于计算这个协方差的实际时间，那么这个随机过程称为平稳随机过程。

- 严平稳。一种条件比较苛刻的平稳性定义。认为只有当序列所有的统计性质都不会随着时间的推移而发生变化时，该序列才能被认为平稳。
- 宽平稳。宽平稳是使用序列的特征统计量来定义的一种平稳性。认为序列的统计性质主要由它的低阶矩决定，所以只要保证序列低阶矩平稳（二阶），就能保证序列的主要性质近似稳定。

（2）平稳序列的统计性质

- 常数均值。
- 自协方差函数和自相关函数只依赖于时间的平移长度而与时间的起止点无关。

(3) 自相关函数

$$\hat{\rho}_k = \frac{\sum_{t=1}^{n-k}(x_t - \bar{x})(x_{t+k} - \bar{x})}{\sum_{t=1}^{n}(x_t - \bar{x})^2} \quad (2.69)$$

其他的动态数据模型还有线性模型法、非线性趋势等。

13. 时间序列建模

任何时间序列都可以看作是一个平稳的过程。所看到的数据集可以看作该平稳过程的一个实现。主要方法有自回归 AR(p)、移动平均 MA(q) 及自回归移动平均 ARMA(p,q) 等。

(1) 自回归模型 AR

时间序列可以表示成它的先前值和一个冲击值的函数：

$$x_t = \phi_1 x_{t-1} + \phi_{12} x_{t-2} + \cdots + \phi_p x_{t-p} + \varepsilon_t \quad (2.70)$$

(2) 滑动平均模型 MA

序列值是现在和过去的误差或冲击值的线性组合：

$$x_t = \varepsilon_t - \theta_1 \varepsilon_{t-1} - \theta_2 \varepsilon_{t-2} - \cdots - \theta_q \varepsilon_{t-q} \quad (2.71)$$

(3) 自回归滑动平均模型 ARMA

序列值是现在和过去的误差或冲击值以及先前的序列值的线性组合：

$$x_t = \varphi_1 x_{t-1} + \varphi_2 x_{t-2} + \cdots + \varphi_p x_{t-p} + \varepsilon_t - \theta_1 \varepsilon_{t-1} - \theta_2 \varepsilon_{t-2} - \cdots - \theta_q \varepsilon_{t-q} \quad (2.72)$$

2.6 数据可视化

数据可视化是关于数据视觉表现形式的科学技术研究。其中，这种数据的视觉表现形式被定义为一种以某种概要形式抽提出来的信息，包括相应信息单位的各种属性和变量。它是一个处于不断演变之中的概念，其边界在不断扩大，主要指的是技术上较为高级的技术方法，而这些技术方法允许利用图形图像处理、计算机视觉以及用户界面，通过表达、建模以及对立体、表面、属性以及动画的显示，对数据加以可视化解释。与立体建模之类的特殊技术方法相比，数据可视化所涵盖的技术方法要广泛得多。为了有效地传达思维概念，美学形式与功能需要齐头并进，通过直观地传达关键的方面与特征，从而实现对于相当稀疏而又复杂的数据集的深入洞察。

数据可视化与信息图形、信息可视化、科学可视化以及统计图形密切相关。当前，在研究、教学和开发领域，数据可视化是一个极为活跃而又关键的方面。"数据可视化"这条术语实现了成熟的科学可视化领域与较年轻的信息可视化领域的统一。

数据可视化技术包含以下几个基本概念：

①数据空间：由 n 维属性和 m 个元素组成的数据集所构成的多维信息空间。

②数据开发：利用一定的算法和工具对数据进行定量的推演和计算。

③数据分析：对多维数据进行切片、分块、旋转等动作剖析数据，从而能多角度多侧面观察数据。

④数据可视化：将大型数据集中的数据以图形图像形式表示，并利用数据分析和开发工具发现其中未知信息的处理过程。

数据可视化已经提出了许多方法，这些方法根据其可视化的原理不同可以划分为基于几何的技术、面向像素技术、基于图标的技术、基于层次的技术、基于图像的技术和分布式技术等。数据可视化的适用范围存在着不同的划分方法。一个常见的关注焦点就是信息的呈现。数据可视化的两个主要组成部分是统计图形和主题图。

先理解数据，再掌握可视化方法，才能实现高效的数据可视化。在设计时，你可能会遇到以下几种常见的数据类型：

①量性：数据是可以计量的，所有的值都是数字。

②离散型：数字类数据可能在有限范围内取值。

③持续性：数据可以测量，且在有限范围内。

④范围性：数据可以根据编组和分类而分类。

可视化的意义是帮助人更好地分析数据，也就是说它是一种高效的手段，并不是数据分析的必要条件；如果我们采用了可视化方案，就意味着机器并不能精确分析。当然，也要明确可视化不能直接带来结果，需要人为介入来分析结论。

代表性的图形化数据可视化方法有：

- 图表类型。
- 图表的创建。
- 使用图表。
- 散点图的显示。
- 条形图的绘制。
- 绘制直方图。
- 收集图显示。
- 多重散点图。
- 网络图显示。
- 评估节点图。
- 时间散点图的显示。
- 编程语言类数据可视化工具有 R、Scala、Python、Java 等。

①R 经常被称为是"为统计人员开发的一种语言"。如果需要深奥的统计模型用于计算，可在 CRAN 上找到它，CRAN 叫综合 R 档案网络（Comprehensive R Archive Network）并非无缘无故。说到用于分析和标绘，没有什么比得过 ggplot2。如果想利用比你机器提供的功能还强大的功能，可以使用 SparkR 绑定，在 R 上运行 Spark。

②Scala 是最轻松的语言，因为大家都欣赏其类型系统。Scala 在 JVM 上运行，基本上成功地结合了函数范式和面向对象范式，目前它在金融界和需要处理海量数据的公司企业中取得了巨大进展，常常采用一种大规模分布式方式来处理（比如 Twitter 和 LinkedIn）。它还是驱动 Spark 和 Kafka 的一种语言。

③Python 在学术界中一直很流行，尤其是在自然语言处理（NLP）等领域。因而，如果你有一个需要 NLP 处理的项目，就会面临数量多得让人眼花缭乱的选择，包括经典的 NTLK、使用 Gensim 的主题建模，或者超快、准确的 spaCy。同样，说到神经网络，Python 同样游刃有余，有 Theano 和 TensorFlow；随后还有面向机器学习的 Scikit-learn，以及面向数据分析的 NumPy 和 Pandas。

④Java 很适合大数据项目。Hadoop MapReduce 是用 Java 编写的，HDFS 也是用 Java 来编写的，Storm、Kafka 和 Spark 也都可以在 JVM 上运行（使用 Clojure 和 Scala）。另外，还有 Google Cloud Dataflow（Apache Beam）等新技术，它们还只支持 Java。

在大数据时代，可视化图表工具不可能"单独作战"，而我们都知道大数据的价值在于数据挖掘，一般数据可视化都是和数据分析功能组合的，数据分析又需要数据接入整合、数据处理、ETL 等数据功能，发展成为一站式的大数据分析平台。

2.7 本章小结

数据和特征决定了机器学习的上限，而模型和算法只是逼近这个上限而已。机器学习数据分析的目的其实就是直观地展现数据，例如让花费数小时甚至更久才能归纳的数据量转化成一眼就能读懂的指标；通过加减乘除、各类公式权衡计算得到的两组数据差异，在图中颜色敏感、长短大小即能形成对比。

本章从机器学习数据分布性、数据相关性、数据聚类性、数据成分、动态及数据可视化等方面介绍了机器学习的数据特征。

第 3 章
机器学习分类算法

机器学习这门技术是多种技术的结合。在这个结合体中，如何进行数据分析处理是最核心的内容。通常在机器学习中指的数据分析是，从一大堆数据中筛选出一些有意义的数据，并推断出一个潜在的可能结论。要得到这样的结论，经过的步骤通常是原始数据、数据特征、数据特征映射和结论。

- 预处理：把数据处理成一些有意义的特征，这一步的目的主要是为了降维。
- 建模：主要是建立模型（通常是曲线的拟合），为分类器搭建一个可能的边界。
- 分类器处理：根据模型把数据分类并进行数据结论的预测。

3.1 数据清洗和特征选择

3.1.1 数据清洗

数据清洗（Data Cleaning）是对数据进行重新审查和校验的过程，目的在于删除重复信息、纠正存在的错误，并提供数据一致性。数据清洗就是把"脏"的"洗掉"，即发现并纠正数据文件中可识别的错误的最后一道程序，包括检查数据一致性、处理无效值和缺失值等。因为数据仓库中的数据是面向某一主题的数据集合，这些数据从多个业务系统中抽取而来，而且包含历史数据，避免不了有的数据是错误数据、有的数据相互之间有冲突，这些错误的或有冲突的数据显然是我们不想要的，称为"脏数据"。我们要按照一定的规则把"脏数据""洗掉"，这就是数据清洗。数据清洗的任务是过滤那些不符合要求的数据，将过滤的结果交给业务主管部门，确认是过滤掉还是由业务单位修正之后再进行抽取。不符合要求的数据主要有不完整的数据、错误的数据、重复的数据 3 大类。数据清洗与问卷审核不同，录入后的数据清理一般是由计算机而不是人工完成的。

1. 一致性检查

一致性检查（Consistency Check）是根据每个变量的合理取值范围和相互关系，检查数据是否合乎要求，发现超出正常范围、逻辑上不合理或者相互矛盾的数据。

2. 无效值和缺失值的处理

由于调查、编码和录入误差，数据中可能存在一些无效值和缺失值，需要给予适当的处理。常用的处理方法有估算、整例删除、变量删除和成对删除。采用不同的处理方法可能对分析结果产生影响，尤其是当缺失值的出现并非随机且变量之间明显相关时。因此，在调查中应当尽量避免出现无效值和缺失值，保证数据的完整性。

3. 数据清洗原理

利用有关技术如数理统计、数据挖掘或预定义的清理规则将脏数据转化为满足数据质量要求的数据。

4. 主要类型

（1）残缺数据

这一类数据主要是一些应该有的信息缺失，将这一类数据过滤出来，按缺失的内容分别写入不同 Excel 文件向客户提交，要求在规定的时间内补全。补全后才写入数据仓库。

（2）错误数据

这一类错误产生的原因是业务系统不够健全，在接收输入后没有进行判断而直接写入后台数据库造成的，比如数值数据输成全角数字字符、字符串数据后面有一个回车操作、日期格式不正确、日期越界等。这一类数据也要分类，对于类似于全角字符、数据前后有不可见字符的问题，只能通过编写 SQL 语句的方式找出来，然后要求客户在业务系统修正之后抽取。日期格式不正确或者是日期越界的这一类错误会导致 ETL 运行失败，这一类错误需要去业务系统数据库用 SQL 的方式挑出来，交给业务主管部门要求限期修正，修正之后再抽取。

（3）重复数据

对于这一类数据（特别是维表中会出现这种情况），将重复数据记录的所有字段导出来，让客户确认并整理。

数据清洗是一个反复的过程，不可能在几天内完成，只有不断地发现问题、解决问题。对于是否过滤、是否修正一般要求客户确认，对于过滤掉的数据，写入 Excel 文件或者将过滤数据写入数据表，在 ETL 开发的初期可以每天向业务单位发送过滤数据的邮件，促使他们尽快修正错误，同时也可以作为将来验证数据的依据。数据清洗需要注意的是不要将有用的数据过滤掉，对于每个过滤规则要认真进行验证，并要用户确认。

5. 数据清洗方法

一般来说，数据清理是将数据库精简以除去重复记录，并使剩余部分转换成标准可接收格式的过程。数据清理标准模型是将数据输入到数据清理处理器，通过一系列步骤"清理"数据，然后以期望的格式输出清理过的数据。数据清理从数据的准确性、完整性、一致性、唯一性、适时性、有效性几个方面来处理数据的丢失值、越界值、不一致代码、重复数据等问题。

数据清理一般针对具体应用，因而难以归纳统一的方法和步骤，但是根据数据不同，可以给出相应的数据清理方法。

（1）解决不完整数据（值缺失）的方法

大多数情况下，缺失的值必须手工填入（手工清理）。当然，某些缺失值可以从本数据源或其他数据源推导出来，这就可以用平均值、最大值、最小值或更为复杂的概率估计代替缺失的值，从而达到清理的目的。

（2）错误值的检测及解决方法

用统计分析的方法识别可能的错误值或异常值，如偏差分析、识别不遵守分布或回归方程的值，也可以用简单规则库（常识性规则、业务特定规则等）来检查数据值，或使用不同属性间的约束、外部的数据来检测和清理数据。

（3）重复记录的检测及消除方法

数据库中属性值相同的记录被认为是重复记录，通过判断记录间的属性值是否相等来检测记录是否相等，相等的记录合并为一条记录（合并/清除）。合并/清除是消重的基本方法。

（4）不一致性（数据源内部及数据源之间）的检测及解决方法

从多数据源集成的数据可能有语义冲突，可定义完整性约束用于检测不一致性，也可通过分析数据发现联系，从而使数据保持一致。目前开发的数据清理工具大致可分为3类。

3.1.2　特征选择

特征选择（Feature Selection）也称特征子集选择（Feature Subset Selection，FSS），或属性选择（Attribute Selection），是指从已有的 M 个特征（Feature）中选择 N 个特征，使得系统的特定指标最优化，是从原始特征中选择一些最有效特征以降低数据集维度的过程，是提高学习算法性能的一个重要手段，也是模式识别中关键的数据预处理步骤。对于一个学习算法来说，好的学习样本是训练模型的关键。

此外，需要区分特征选择与特征提取。特征提取？（Feature Extraction）是指利用已有的特征计算出一个抽象程度更高的特征集，也指计算得到某个特征的算法。

特征选择过程一般包括产生过程、评价函数、停止准则、验证过程4个部分。

1. 基本框架

迄今为止，已有很多学者从不同角度对特征选择进行过定义。Kira 等人定义理想情况下特征选择是寻找必要的、足以识别目标的最小特征子集。John 等人从提高预测精度的角度定义特征选择是一个能够增加分类精度，或者在不降低分类精度的条件下降低特征维数的过程。Koller 等人从分布的角度定义特征选择为：在保证结果类分布尽可能与原始数据类分布相似的条件下，选择尽可能小的特征子集。Dash 等人给出的定义是选择尽量小的特征子集，并满足不显著降低分类精度和不显著改变类分布两个条件。上述各种定义的出发点不同，各有侧重点，但是目标都是寻找一个能够有效识别目标的最小特征子集。图3.1为特征选择的基本框架。

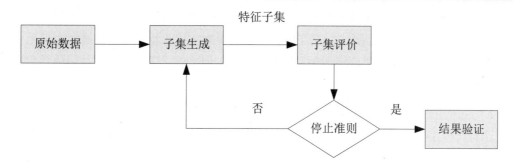

图 3.1　特征选择的基本框架

子集搜索是一个比较费时的步骤，Yu 等人基于相关和冗余分析，给出了另一种特征选择框架，避免了子集搜索，可以高效快速地寻找最优子集。图 3.2 为改进的特征选择框架。

图 3.2　改进的特征选择框架

从特征选择的基本框架可以看出，特征选择方法中有 4 个基本步骤：候选特征子集的生成（搜索策略）、评价准则、停止准则和验证方法。目前对特征选择方法的研究主要集中于搜索策略和评价准则，因而一般从搜索策略和评价准则两个角度对特征选择方法进行分类。

2. 特征选择的一般过程

（1）产生过程（Generation Procedure）

产生过程是搜索特征子集的过程，负责为评价函数提供特征子集。

（2）评价函数（Evaluation Function）

评价函数是评价一个特征子集好坏程度的准则。

（3）停止准则（Stopping Criterion）

停止准则是与评价函数相关的，一般是一个阈值，当评价函数值达到这个阈值后就可停止搜索。

（4）验证过程（Validation Procedure）

在验证数据集上验证选出来的特征子集的有效性。

3. 基于搜索策略的方法分类

基本的搜索策略按照特征子集的形成过程可分为 3 种：全局最优、随机搜索和启发式搜索。一个具体的搜索算法会采用两种或多种基本搜索策略，例如遗传算法是一种随机搜索算法，同时也是一种启发式搜索算法。下面对 3 种基本的搜索策略进行分析比较。

（1）采用全局最优搜索策略的特征选择方法

迄今为止，唯一得到最优结果的搜索方法是分支定界法。这种算法能保证在事先确定优化特征子集中特征数目的情况下找到相对于所设计的可分性判据而言的最优子集。它的搜索空间是 $O(2N)$（其中 N 为特征的维数）。存在的问题：很难确定优化特征子集的数目；满足单调性的可分性判据难以设计；处理高维多类问题时，算法的时间复杂度较高。因此，虽然全局最优搜索策略能得到最优解，但是因为诸多因素限制，无法被广泛应用。

（2）采用随机搜索策略的特征选择方法

在计算过程中把特征选择问题与模拟退火算法、禁忌搜索算法、遗传算法等，或者仅仅是一个随机重采样过程结合起来，以概率推理和采样过程作为算法的基础，基于对分类估计的有效性，在算法运行中对每个特征赋予一定的权重；然后根据用户所定义的或自适应的阈值来对特征重要性进行评价。当特征所对应的权重超出了这个阈值，它便被选中作为重要的特征来训练分类器。Relief 系列算法是一种典型的根据权重选择特征的随机搜索方法，能有效地去掉无关特征，但不能去除冗余，而且只能用于两类分类。随机方法可以细分为完全随机方法和概率随机方法两种。虽然搜索空间仍为 $O(2N)$，但是可以通过设置最大迭代次数限制搜索空间小于 $O(2N)$。例如，遗传算法采用启发式搜索策略，搜索空间远远小于 $O(2N)$。存在的问题是：具有较高的不确定性，只有当总循环次数较大时，才可能找到较好的结果。在随机搜索策略中，可能需对一些参数进行设置，参数选择的合适与否对最终结果的好坏起着很大的作用。因此，参数选择是一个关键步骤。

（3）采用启发式搜索策略的特征选择方法

这类特征选择方法主要有单独最优特征组合、序列前向选择方法（SFS）、广义序列前向选择方法（GSFS）、序列后向选择方法（SBS）、广义序列后向选择方法（GSBS）、增 l 去 r 选择方法、广义增 l 去 r 选择方法、浮动搜索方法。这类方法易于实现且快速，搜索空间是 $O(N2)$。一般认为采用浮动广义后向选择方法（FGSBS）是较为有利于实际应用的一种特征选择搜索策略，既考虑到特征之间的统计相关性，又用浮动方法保证算法运行的快速稳定性。存在的问题是：启发式搜索策略虽然效率高，但是它以牺牲全局最优为代价。

每种搜索策略都有各自的优缺点，在实际应用过程中，可以根据具体环境和准则函数来寻找一个最佳的平衡点。例如，如果特征数较少，可采用全局最优搜索策略；若不要求全局最优但要求计算速度快，则可采用启发式策略；若需要高性能的子集，而不介意计算时间，则可采用随机搜索策略。

4. 基于评价准则划分特征选择方法

特征选择方法依据是否独立于后续的学习算法，可分为过滤式（Filter）和封装式（Wrapper）两种。Filter 与后续学习算法无关，一般直接利用所有训练数据的统计性能评估特征，速度快，但评估与后续学习算法的性能偏差较大。Wrapper 利用后续学习算法的训练准确率评估特征子集，偏差小，计算量大，不适合大数据集。

（1）过滤式（Filter）评价策略的特征选择方法

Filter 特征选择方法一般使用评价准则来增强特征与类的相关性，削减特征之间的相关性。可将评价函数分成 4 类：距离度量、信息度量、依赖性度量以及一致性度量。

（2）封装式（Wrapper）评价策略的特征选择方法

Wrapper 模型将特征选择算法作为学习算法的一个组成部分，并且直接使用分类性能作为特征重要性程度的评价标准。它的依据是选择子集最终被用于构造分类模型。因此，若在构造分类模型时，直接采用那些能取得较高分类性能的特征即可，从而获得一个分类性能较高的分类模型。该方法在速度上要比 Filter 方法慢，但是它所选择的优化特征子集的规模相对要小得多，非常有利于关键特征的辨识；同时它的准确率比较高，但泛化能力比较差，时间复杂度较高。

3.1.3 回归分析

就一般意义而言，相关分析包括回归和相关两方面内容，因为回归与相关都是研究两变量相互关系的分析方法，但就具体方法而言，回归分析和相关分析是有明显差别的。相关图表、相关系数能判定两个变量之间相关的方向和密切程度，但不能指出两个变量相互关系的具体表现形式，也无法从一个变量的变化来推测另一个变量的变化情况。回归分析就是对具有相关关系的两个或两个以上变量的数量变化规律进行测定，确立一个相应的数学表达式，并进行估算和预测的一种统计方法。

回归分析和相关分析是互相补充、密切联系的。相关分析需要回归分析来表明数量关系的具体表现形式，而回归分析则应该建立在相关分析的基础上。只有依靠相关分析，对现象的数量变化规律判明具有密切相关关系后再进行回归分析，求其相关的具体表现形式，这样才具有实际意义。

回归分析建立的数学表达式称为回归方程（或回归模型）。回归方程为线性方程的称为线性回归，回归方程为非线性方程的称为非线性回归。两个变量之间的回归称为一元回归（简单回归）；3 个或 3 个以上变量之间的回归称为多元回归。本章只介绍一元线性回归，即简单线性回归分析方法。

1. 回归分析的主要内容

（1）建立相关关系的回归方程

利用回归分析方法，配合一个表明变量之间数量上相关的方程式，而且根据自变量 x 的变动来预测因变量 y 的变动。

（2）测定因变量的估计值与实际值的误差程度

通过计算估计标准误差指标，可以反映因变量估计值的准确程度，从而将误差控制在一定范围内。

2. 回归分析的特点

回归分析与相关分析比较具有以下特点：

①在相关分析中，各变量都是随机变量；在回归分析中，因变量是随机变量，自变量不是随机的，而是给定的数值。

②在相关分析中，各变量之间是对等关系，调换变量的位置，不影响计算的结果；而在回归分析中，自变量与因变量之间不是对等的关系，调换其位置，将得到不同的回归方程。因此，在进行回归分析时，必须根据研究目的先确定哪一个是自变量、哪一个是因变量。

③相关分析计算的相关系数是一个绝对值在 0 与 1 之间的抽象系数，其数值的大小反映变量之间相关关系的程度；回归分析建立的回归方程反映的是变量之间的具体变动关系，不是抽象的系数。根据回归方程，利用自变量的给定值可以估计或推算出因变量的数值。

3. 一元线性回归方程的拟合

在回归分析中，最简单、最基本的形式就是一元线性回归，也就是通常所说的配合直线方程式的问题。若通过观察或实验，得到 n 对数据$(x_1,y_1),(x_2,y_2),\cdots,(x_n,y_n)$相关图上的散布点接近分布在一条直线上，就可以认为变量 x 与 y 之间存在着线性关系，可设经验公式为

$$\hat{y} = a + bx \tag{3.1}$$

式中，a 与 b 为待定参数，也就是需要根据实际数据求解的数值，a 为直线的截距，b 为直线的斜率，也称回归系数，表示自变量 x 每变动一个单位时因变量 y 的平均变动量。a、b 值一旦确定，这条直线就被唯一确定了。用于描述这 n 组数据的直线有许多条，究竟用哪条直线来代表两个变量之间的关系需要有一个明确的原则。我们希望选择距离各散布点最近的一条直线来代表 x 与 y 之间的关系，以便更好地反映变量之间的关系。根据这一思想确定未知参数 a、b 的方法，称为最小二乘法，也就是通过使得下面的 Q 值为最小值来确定 a、b 的方法。

$$Q = \sum(y - \hat{y})^2 = \sum(y - a - bx)^2 \tag{3.2}$$

可见，用最小二乘法得到的直线与所有数据(x_i, y_i)的离差平方和为最小。

要使 Q 为最小值，就要用数学中对二元函数求极值的原理，求 Q 关于 a 和 b 的偏导数，并令其等于 0，整理得出直线回归方程中求解参数 a、b 的标准方程组为

$$\begin{cases} \sum y = na + b\sum x \\ \sum xy = a\sum x + b\sum x^2 \end{cases} \tag{3.3}$$

解方程组得：

$$b = \frac{\sum(x-\bar{x})(y-\bar{y})}{\sum(x-\bar{x})^2} = \frac{n\sum xy - \sum x \sum y}{n\sum x^2 - (\sum x)^2} \tag{3.4}$$

4. 估计标准误差

（1）估计标准误差的意义

回归方程的一个重要作用在于根据自变量的已知值推算因变量的可能值$\hat{y}$，这个可能值或称估计值、理论值、平均值，它和真正的实际值 y 可能一致，也可能不一致，因而产生了估计值的代表性问题。当$\hat{y}$值与 y 值一致时，表明推断准确；当$\hat{y}$值与 y 值不一致时，表明推断不够准确。显而易见，将一系列$\hat{y}$值与 y 值加以比较，可以发现其中存在着一系列离差，有的是正差，有的是负差，还有的为零。回归方程的代表性如何，一般是通过计算估计标准误差指标

来加以检验的。估计标准误差指标是用来说明回归方程代表性大小的统计分析指标，简称为估计标准差或估计标准误差，其计算原理与标准差基本相同。估计标准误差说明理论值（回归直线）的代表性。若估计标准误差小，就说明回归方程准确性高，代表性大；反之，估计不够准确，代表性小。

（2）估计标准误差的计算

估计标准误差是指因变量实际值与理论值离差的平均数，计算公式为

$$S_{yx} = \sqrt{\frac{\sum(y-\hat{y})^2}{n-2}} \tag{3.5}$$

式中　S_{yx}——估计标准差，其下标 yx 代表 y 依 x 而回归的方程；

　　　$\hat{y}$——根据回归方程推算出来的因变量的估计值；

　　　y——因变量的实际值；

　　　n——数据的项数。

估计标准误差的简化计算公式为

$$S_{yx} = \sqrt{\frac{\sum y^2 - a\sum y - b\sum xy}{n-2}} \tag{3.6}$$

（3）估计标准误差与相关系数的关系

二者在数量上具有如下关系：

$$\gamma = \sqrt{1 - \frac{S_{yx}^2}{\sigma_y^2}} \tag{3.7}$$

$$S_{yx} = \sigma_y\sqrt{1-\gamma^2} \tag{3.8}$$

式中　γ——相关系数；

　　　σ_y——因变量数列的标准差；

　　　S_{yx}——估计标准误差。

从上面的计算公式中可以看出 γ 和 S_{yx} 的变化方向是相反的。当 γ 越大时，S_{yx} 越小，这时相关密切程度较高，回归直线的代表性较大；当 γ 越小时，S_{yx} 越大，这时相关密切程度较低，回归直线的代表性较小。

3.2 决策树、随机森林

1. 决策树

决策树（Decision Tree）是在已知各种情况发生概率的基础上，通过构成决策树来求取净

现值的期望值大于等于零的概率，评价项目风险，判断其可行性的决策分析方法，是直观运用概率分析的一种图解法。由于这种决策分支画成图形很像一棵树的枝干，故称决策树。在机器学习中，决策树是一个预测模型，代表的是对象属性与对象值之间的一种映射关系。熵（Entropy）表示系统的凌乱程度，使用算法 ID3、C4.5 和 C5.0 生成树算法得出熵，这一度量是基于信息学理论中熵的概念。

决策树是一种树形结构，其中每个内部节点表示一个属性上的测试，每个分支代表一个测试输出，每个叶节点代表一种类别。

分类树（决策树）是一种十分常用的分类方法，是一种监督学习。所谓监督学习，就是给定一堆样本，每个样本都有一组属性和一个类别，这些类别是事先确定的，通过学习得到一个分类器，对新出现的对象给出正确的分类。

（1）组成

- 决策点，是对几种可能方案的选择，即最后选择的最佳方案。如果决策属于多级决策，则决策树的中间可以有多个决策点，以决策树根部的决策点为最终决策方案。
- 状态节点，代表备选方案的经济效果（期望值），通过各状态节点经济效果的对比，按照一定的决策标准就可以选出最佳方案。由状态节点引出的分支称为概率枝，概率枝的数目表示可能出现的自然状态数目，每个分支上要注明该状态出现的概率。
- 结果节点，将每个方案在各种自然状态下取得的损益值标注于结果节点的右端。

在机器学习中，决策树是一个预测模型，代表的是对象属性与对象值之间的一种映射关系。树中每个节点表示某个对象，而每个分叉路径则代表某个可能的属性值，而每个叶节点则对应从根节点到该叶节点所经历的路径所表示的对象的值。决策树仅有单一输出，若想有复数输出，可以建立独立的决策树，以处理不同输出。

决策树是一种简单且广泛使用的分类器。通过训练数据构建决策树，可以高效地对未知的数据进行分类。

（2）信息熵

熵代表信息的不确定性。信息的不确定性越大，熵越大。例如，"明天太阳从东方升起"这一句话代表的信息可以认为是 0。因为太阳从东方升起是一个特定的规律，可以把这个事件的信息熵约等于 0；也就是说，信息熵和事件发生的概率成反比。数学上把信息熵定义如下：

$$H(X)=H(P_1, P_2, \ldots, P_n)=-\sum P(x_i)\log P(x_i) \tag{3.9}$$

其中，n 代表 X 的 n 种不同离散取值，$P(x_i)$ 代表 X 取值为 x_i 的概率。熟悉了一个变量 X 的熵，很容易推广到多个变量的联合熵，这里给出两个变量 X 和 Y 的联合熵表达式：

$$H(X,Y)=-\sum_{i=1}^{n} p(x_i,y_i)\log p(x_i,y_i) \tag{3.10}$$

条件熵类似于条件概率，度量 X 在知道 Y 以后剩下的不确定性，表达式如下：

$$H(X|Y)=-\sum_{i=1}^{n}p(x_i,y_i)\log p(x_i|y_i)=\sum_{j=1}^{n}p(y_j)H(X|y_j) \qquad (3.11)$$

（3）互信息

互信息指的是两个随机变量之间的关联程度，即给定一个随机变量后，另一个随机变量不确定性的削弱程度，因而互信息取值最小为0，意味着给定一个随机变量对确定另一个随机变量没有关系，最大取值为随机变量的熵，意味着给定一个随机变量能完全消除另一个随机变量的不确定性。

（4）ID3算法

ID3算法是由Quinlan首先提出的。该算法以信息论为基础，以信息熵和信息增益度为衡量标准，从而实现对数据的归纳分类。以下是一些信息论的基本概念：

定义1：若存在 n 个相同概率的消息，则每个消息的概率 p 是 $1/n$，一个消息传递的信息量为

$$\text{Log}2(1/n) \qquad (3.12)$$

定义2：若有 n 个消息，其给定概率分布为 $P=(p_1,p_2,…,p_n)$，则由该分布传递的信息量称为 P 的熵。

定义3：若一个记录集合 T 根据类别属性的值被分成互相独立的类 $C_1C_2…C_k$，则识别 T 的一个元素所属哪个类所需要的信息量为 $\text{Info}(T)=I(p)$，其中 P 为 $C_1C_2…C_k$ 的概率分布，即

$$P=(|C_1|/|T|,…,|C_k|/|T|) \qquad (3.13)$$

定义4：若我们先根据非类别属性 X 的值将 T 分成集合 $T_1,T_2…T_n$，则确定 T 中一个元素类的信息量可通过确定 T_i 的加权平均值来得到，即 $\text{Info}(T_i)$ 的加权平均值为

$$\text{Info}(X,\ T)=(i=1\ \text{to}\ n\ \text{求和})((|T_i|/|T|)\text{Info}(T_i)) \qquad (3.14)$$

定义5：信息增益度是两个信息量之间的差值，其中一个信息量是需确定 T 的一个元素的信息量，另一个信息量是在已得到的属性 X 的值后需确定 T 的一个元素的信息量，信息增益度公式为

$$\text{Gain}(X,\ T)=\text{Info}(T)-\text{Info}(X,\ T) \qquad (3.15)$$

ID3算法计算每个属性的信息增益，并选取具有最高增益的属性作为给定集合的测试属性。对被选取的测试属性创建一个节点，并以该节点的属性标记，对该属性的每个值创建一个分支，据此划分样本。

（5）C4.5算法

C4.5是Ross Quinlan于1993年在ID3的基础上改进而提出的。ID3采用的信息增益度量存在一个缺点，一般会优先选择有较多属性值的特征（Feature），因为属性值多的特征会有相对较大的信息增益（信息增益反映的是给定一个条件以后不确定性减少的程度，必然是分得越

细的数据集确定性更高，也就是条件熵越小，信息增益越大）。为了避免这个不足，C4.5 中用信息增益比率（Gain Ratio）来作为选择分支的准则。信息增益比率通过引入一个被称作分裂信息（Split Information）的项来惩罚取值较多的特征。除此之外，C4.5 还弥补了 ID3 中不能处理特征属性值连续的问题。但是，连续属性值需要扫描排序，会使 C4.5 性能下降。

分类信息：

$$\text{SplitInformation}(D, A) = -\sum_{i=1}^{n} \frac{|D_i|}{|D|} \log \frac{|D_i|}{|D|} \tag{3.16}$$

信息增益率：

$$\text{GainRatio}(D, A) = \frac{g(D, A)}{\text{SplitInformation}(D, A)} \tag{3.17}$$

（6）CART 算法

CART（分类回归树）由 L.Breiman、J.Friedman、R.Olshen 和 C.Stone 于 1984 年提出。ID3 中根据属性值分割数据，之后该特征不会再起作用，这种快速切割的方式会影响算法的准确率。CART 是一棵二叉树，采用二分法，每次把数据切成两份，分别进入左子树、右子树；而且每个非叶子节点都有两个孩子，所以 CART 的叶子节点比非叶子节点多 1。相比 ID3 和 C4.5，CART 应用要多一些，既可以用于分类，也可以用于回归。CART 分类时，使用基尼指数（Gini）来选择最好的数据分割的特征，Gini 描述的是纯度，与信息熵的含义相似。CART 中每一次迭代都会降低 Gini 系数。基尼系数的计算与信息熵增益的方式非常类似，公式如下：

$$\text{Gini}(D) = 1 - \sum_{i=0}^{n} \left(\frac{D_i}{D}\right)^2 \tag{3.18}$$

$$\text{Gini}(D/A) = \sum_{I=0}^{N} \frac{D_i}{D} \text{Gini}(D_i) \tag{3.19}$$

提到决策树算法，很多想到的就是上面提到的 ID3、C4.5、CART 分类决策树。其实决策树分为分类树和回归树，前者用于分类，后者用于预测实数值。

2. 随机森林

在机器学习中，随机森林是一个包含多个决策树的分类器，并且其输出的类别是由个别树输出的类别众数而定的。Leo Breiman 和 Adele Cutler 发展出推论随机森林的算法。Random Forests（随机森林）是他们的商标。这个术语是由 1995 年贝尔实验室的 Tin Kam Ho 所提出的随机决策森林（Random Decision Forests）而来的。这个方法结合 Breiman 的"Bootstrap aggregating"（自举汇聚法）想法和 Ho 的"Random Subspace Method"（随机子空间方法）来建造决策树的集合。

（1）学习算法

根据下列算法而建造每棵树：

- 用 N 来表示训练用例（样本）的个数，M 表示特征数目。
- 输入特征数目 m，用于确定决策树上一个节点的决策结果。其中，m 应远小于 M。
- 从 N 个训练用例（样本）中以有放回抽样的方式取样 N 次，形成一个训练集（即 Bootstrap 取样），并用未抽到的用例（样本）做预测，评估其误差。
- 对于每一个节点，随机选择 m 个特征，决策树上每个节点的决定都是基于这些特征确定的。根据这 m 个特征，计算其最佳的分裂方式。
- 每棵树都会完整成长而不会剪枝，这有可能在建完一棵正常树状分类器后会被采用。

（2）相关概念

- 分裂：在决策树的训练过程中，需要一次次地将训练数据集分裂成两个子数据集，这个过程就叫作分裂。
- 特征：在分类问题中，输入到分类器中的数据叫作特征。
- 待选特征：在决策树的构建过程中，需要按照一定的次序从全部特征中选取特征。待选特征就是在目前的步骤之前还没有被选择的特征集合。
- 分裂特征：待选特征的定义，每一次选取的特征就是分裂特征。

（3）随机森林的构建

随机森林具体构建有两个方面：数据的随机性选取，待选特征的随机选取。

① 数据的随机选取

首先，从原始的数据集中采取有放回的抽样，构造子数据集。子数据集的数据量是和原始数据集相同的。不同子数据集的元素可以重复，同一个子数据集中的元素也可以重复。然后，利用子数据集来构建子决策树，将这个数据放到每个子决策树中，每个子决策树输出一个结果。最后，如果有了新的数据需要通过随机森林得到分类结果，就可以通过对子决策树的判断结果的投票得到随机森林的输出结果。

② 待选特征的随机选取

与数据集的随机选取类似，随机森林中子树的每一个分裂过程并未用到所有的待选特征，而是从所有的待选特征中随机选取一定的特征，之后再在随机选取的特征中选取最优的特征。这样能够使得随机森林中的决策树都能够彼此不同，提升系统的多样性，从而提升分类性能。

3.3 SVM

支持向量机（SVM）方法是建立在统计学习理论的 VC 维理论和结构风险最小原理基础上的，根据有限的样本信息在模型的复杂性（对特定训练样本的学习精度）和学习能力（无错误地识别任意样本的能力）之间寻求最佳折中。在机器学习中，支持向量机（SVM）是与相关的学习算法有关的监督学习模型，可以分析数据，识别模式，用于分类和回归分析。给定一组训练样本，标注为两类，一个 SVM 训练算法建立了一个模型，分配新的实例为一类或其他类，使其成为非概率二元线性分类。

支持向量机（SVM）是一个类分类器，正式的定义是一个能够将不同类样本在样本空间分隔的超平面。换句话说，给定一些标注（Label）好的训练样本（监督学习），SVM 算法输出一个最优化的分隔超平面。

给定训练样本，支持向量机建立一个超平面作为决策曲面，使得正例和反例的隔离边界最大化。用以下例子对 SVM 快速建立一个认知。

如图 3.3A（来源网络）所示，想象红色和蓝色的球为球台上的桌球，目的是找到一条曲线将蓝色和红色的球分开，于是得到一条黑色的曲线。为了使黑色的曲线离任意的蓝色球和红色球距离（也就是我们后面要提到的 margin）最大化，需要找到一条最优的曲线，如图 3.3B 所示。如图 3.3C 所示，想象一下如果这些球不是在球桌上，而是被抛向了空中，我们仍然需要将红色球和蓝色球分开，这时就需要一个曲面，而且需要这个曲面仍然满足任意红色球和蓝色球的间距最大化。需要找到的曲面就是后面要详细介绍的最优超平面。离曲面最近的红色球和蓝色球就是支持向量（Support Vector）。

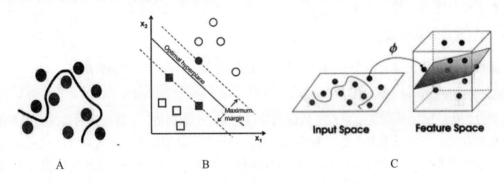

图 3.3　SVM 算法建立决策曲面

3.3.1　最优分类面和广义最优分类面

1. 线性可分和线性不可分

线性可分（Linearly Separable）在二维空间可以理解为用一条直线（一个函数）把两个类型的样本隔开，被隔离开来的两类样本即为线性可分样本。同理，在高维空间，可以理解为可以被一个曲面（高维函数）隔开的两类样本。

线性不可分，可以理解为自变量和因变量之间的关系不是线性的。

实际上，线性不可分的情况更多，即使是非线性的样本通常也是通过高斯核函数将其映射到高维空间，将高维空间非线性的问题转化为线性可分的问题。

2. 函数间隔和几何间隔

函数间隔（Functional Margin）：给定一个训练样本 $(x^{(i)}, y^{(i)})$，有

$$\tilde{\gamma}^{(i)} = y^{(i)}(\boldsymbol{w}^\mathrm{T} x + b) \tag{3.20}$$

函数间隔代表了特征是正例或者反例的确信度。

几何间隔（Geometrical Margin）：

$$\gamma^{(i)} = \frac{(w^T x + b)}{\|w\|} \tag{3.21}$$

向量点到超平面的距 $r = \frac{g(x)}{\|w\|}$。（其中，$g(x) = w^T x + b$）

3. 最优分类面和广义最优分类面

最优分类面要求分类面不但能将两类正确分开，而且使分类间隔最大。将两类正确分开是为了保证训练错误率为 0，也就是经验风险最小（为 0）。确切地讲就是要使推广性的界的置信区间最小,真正的风险才会降到最低限度。推广到高维空间,最优分类线就成为最优分类面。

设线性可分样本集 (x_i, y_i)，$i = 1, \dots, n$，$x \in R^d$，$y \in \{+1, -1\}$ 为归类符号。d 维空间中线性判别函数的一般形式为归类符号。d 维空间中线性判别函数的常规形式为 $g(x) = w \cdot x + b$（w 代表 Hilbert 空间中权向量，b 代表阈值），分类线方程为 $w \cdot x + b = 0$。将判别函数进行归一化，使两类所有样本都满足 $|g(x)| = 1$，也就是使离分类面最近的样本的 $|g(x)| = 1$，此时分类间隔等于 $2/\|w\|$，因此使间隔最大等价于使 $\|w\|$（或 $\|w\|^2$）最小。要求分类线对所有样本正确分类，就是要求它满足

$$y_i[(w \cdot x) + b] - 1 \geq 0 \quad (i = 1, 2, \cdots, n) \tag{3.22}$$

满足上述条件（公式 3.22），并且使 $\|w\|^2$ 最小的分类面就叫作最优分类面，过两类样本中离分类面最近的点且平行于最优分类面的超平面 H_1、H_2 上的训练样本点就称作支持向量（Support Vector），因为它们"支持"了最优分类面。

利用 Lagrange（拉格朗日）优化方法可以把上述最优分类面问题转化为如下这种较简单的对偶问题，即在约束条件

$$\sum_{i=1}^{n} y_i \alpha_i = 0 \quad (a_i \geq 0, \; i=1,2,\cdots,n) \tag{3.23}$$

下面对 a_i（对偶变量，即拉格朗日乘子）求解下列函数的最大值：

$$Q(\alpha) = \sum_{i=1}^{n} \alpha_i - \frac{1}{2} \sum_{i,j=1}^{n} \alpha_i \alpha_j y_i y_j (x_i x_j) \tag{3.24}$$

若 a^* 为最优解，则

$$w^* = \sum_{i=1}^{n} a^* y a_i \tag{3.25}$$

即最优分类面的权系数向量是训练样本向量的线性组合。

公式（3.24）的由来（利用 Lagrange 函数计算）：

$$L(\boldsymbol{w},b,\alpha) = \tfrac{1}{2}\|\boldsymbol{w}\|^2 - \sum_{i=1}^{l}\alpha_i(y_i \cdot ((x_i \cdot \boldsymbol{w})+b)-1)$$

$$\frac{\partial}{\partial b}L(\boldsymbol{w},b,\alpha)=0 \quad \frac{\partial}{\partial \boldsymbol{w}}L(\boldsymbol{w},b,\alpha)=0$$

$$\sum_{i=1}^{l}a_i y_i = 0 \quad \boldsymbol{w}=\sum_{i=1}^{l}\alpha_i y_i x_i \tag{3.26}$$

$$W(\alpha)=\sum_{i=1}^{l}\alpha_i - \tfrac{1}{2}\sum_{i,j=1}^{l}\alpha_i\alpha_j y_i y_j(x_i \cdot x_j)$$

$$\alpha_i \geqslant 0, \ i=1,\ldots,l, \ \sum_{i=1}^{l}\alpha_i y_i = 0$$

$$f(x)=\operatorname{sgn}(\sum_{i=1}^{l} y_i \alpha_i \cdot (x \cdot x_i)+b)$$

从前面的分析可以看出,最优分类面是在线性可分的前提下讨论的,在线性不可分的情况下,就是某些训练样本不能满足式(3.26)的条件,因此可以在条件中增加一个松弛项参数,$\varepsilon_i \geqslant 0$ 变成:

$$y_i[(\boldsymbol{w}\cdot x_i)+b]-1+\varepsilon_i \geqslant 0 \ (i=1,2,\ldots,n) \tag{3.27}$$

对于足够小的 $S>0$,只要使下面的公式值最小就可以使得错分样本数最小,对应线性可分情况下的分类间隔最大。

$$F_\sigma(\varepsilon) = \sum_{i=1}^{n}\varepsilon_i^\sigma \tag{3.28}$$

在线性不可分情况下可引入约束:

$$\|\boldsymbol{w}\|^2 \leqslant c_k \tag{3.29}$$

在约束条件(3.27)和(3.29)下对函数(3.28)求极小,就得到了线性不可分情况下的最优分类面,称作广义最优分类面。为方便计算,取 $S=1$。

为使计算进一步简化,广义最优分类面问题可以进一步演化成在约束条件(3.27)下求下列函数的极小值:

$$\phi(\boldsymbol{w},\varepsilon)=\tfrac{1}{2}(\boldsymbol{w},\boldsymbol{w})+C(\sum_{i=1}^{n}\varepsilon_i) \tag{3.30}$$

其中,C 为某个指定的常数,实际上起控制对错分样本惩罚程度的作用,实现在错分样本的比例与算法复杂度之间的折中。

求解这一优化问题的方法与求解最优分类面时的方法相同,都是转化为一个二次函数极值问题,其结果与可分情况下得到的(3.23)、(3.24)几乎完全相同,但是(3.23)条件变为:

$$0 \leqslant a_i \leqslant C, \ i=1,\ldots,n \tag{3.31}$$

3.3.2 SVM 的非线性映射

对于非线性问题，可以通过非线性交换转化为某个高维空间中的线性问题，在变换空间求最优分类超平面。这种变换可能比较复杂，因此在一般情况下不易实现。但是在上面的对偶问题中，不论是寻优目标函数（3.24）还是分类函数（3.26）都只涉及训练样本之间的内积运算 $(x \cdot x_i)$。设有非线性映射 $\Phi: R^d \to H$ 将输入空间的样本映射到高维（可能是无穷维）的特征空间 H 中，当在特征空间 H 中构造最优超平面时，训练算法仅使用空间中的点积，即 $\phi(x_i) \cdot \phi(x_j)$，而没有单独的 $\phi(x_i)$ 出现。因此，如果能够找到一个函数 K 使得

$$K(x_i \cdot x_j) = \phi(x_i) \cdot \phi(x_j) \tag{3.32}$$

这样在高维空间中实际上只需进行内积运算，而这种内积运算是可以用原空间中的函数实现的，我们甚至没有必要知道变换中的形式。根据泛函的有关理论，只要有一种核函数 $K(x_i \cdot x_j)$ 满足 Mercer 条件，它就对应某一变换空间中的内积。因此，在最优超平面中采用适当的内积函数 $K(x_i \cdot x_j)$ 就可以实现某一非线性变换后的线性分类，而计算复杂度却没有增加。此时目标函数（3.24）变为：

$$Q(\alpha) = \sum_{i=1}^{n} \alpha_i - \frac{1}{2} \sum_{i,j=1}^{n} \alpha_i \alpha_j y_i y_j K(x_i \cdot x_j) \tag{3.33}$$

而相应的分类函数也变为

$$f(x) = \text{sgn}\{\sum_{i=1}^{n} \alpha_i^* y_i K(x_i \cdot x_j) + b^*\} \tag{3.34}$$

算法的其他条件不变，这就是 SVM。

概括地说，SVM 就是通过某种事先选择的非线性映射将输入向量映射到一个高维特征空间，在这个特征空间中构造最优分类超平面。在形式上，SVM 分类函数类似于一个神经网络，输出是中间节点的线性组合，每个中间节点对应于一个支持向量，如图 3.4 所示。

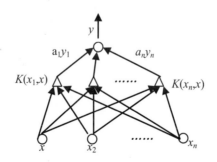

图 3.4 SVM 示意图

其中，输出（决策规则）$y = \text{sgn}\{\sum_{i=1}^{n} \alpha_i y_i K(\boldsymbol{x} \cdot \boldsymbol{x}_i) + b\}$，权值 $w_i = \alpha_i y_i$，$K(\boldsymbol{x} \cdot \boldsymbol{x}_i)$ 为基于 S 个支持向量 $x_1, x_2, ..., x_s$ 的非线性变换（内积），$\boldsymbol{x} = (x_1, x_2, ..., x_d)$ 为输入向量。

3.3.3 核函数

选择满足 Mercer 条件的不同内积核函数,就构造了不同的 SVM,这样也就形成了不同的算法。目前研究最多的核函数主要有以下 3 类。

1. 多项式核函数

$$K(x, x_i) = [(x \cdot x_i) + 1]^q \tag{3.35}$$

其中,q 是多项式的阶次,所得到的是 q 阶多项式分类器。

2. 径向基函数(RBF)

$$K(x, x_i) = \exp\left\{-\frac{|x - x_i|^2}{\sigma^2}\right\} \tag{3.36}$$

所得的 SVM 是一种径向基分类器,与传统径向基函数方法的基本区别是,这里每一个基函数的中心对应于一个支持向量,它们以及输出权值都是由算法自动确定的。径向基形式的内积函数类似人的视觉特性,在实际应用中经常用到,需要注意的是,选择不同的 S 参数值,相应的分类面会有很大差别。

3. S 形核函数

$$K(x, x_i) = \tanh[v(x \cdot x_i) + c] \tag{3.37}$$

这时的 SVM 算法中包含了一个隐藏层的多层感知器网络,网络的权值和网络的隐藏层结点数都是由算法自动确定的,而不像传统的感知器网络那样由人凭借经验确定。此外,该算法不存在困扰神经网络的局部极小点的问题。

在上述几种常用的核函数中,最为常用的是多项式核函数和径向基核函数。除了上面提到的 3 种核函数外,还有指数径向基核函数、小波核函数等其他一些核函数,应用相对较少。事实上,需要进行训练的样本集各式各样,核函数也各有优劣。B.Bacsens 和 S.Viaene 等人曾利用 LS-SVM 分类器,采用 UCI 数据库对线性核函数、多项式核函数和径向基核函数进行实验比较,从实验结果来看,对于不同的数据库,不同的核函数各有优劣,而径向基核函数在多数数据库上得到略为优良的性能。

3.4 聚类算法

聚类分析起源于分类学,在古老的分类学中,人们主要依靠经验和专业知识来实现分类,很少利用数学工具进行定量的分类。随着人类科学技术的发展,对分类的要求越来越高,以致有时仅凭经验和专业知识难以确切地进行分类,于是人们逐渐把数学工具引用到分类学中,形成了数值分类学,之后又将多元分析的技术引入到数值分类学形成了聚类分析。聚类分析内容非常丰富,有系统聚类法、有序样品聚类法、动态聚类法、模糊聚类法、图论聚类法、聚类预报法等。

聚类分析作为数据挖掘中的一个模块,可以作为一个单独的工具以发现数据库中分布的一些深层的信息,并且概括出每一类的特点,或者把注意力放在某一个特定的类上以做进一步的分析;并且,聚类分析也可以作为数据挖掘算法中其他分析算法的一个预处理步骤。

聚类分析的算法可以分为划分法(Partitioning Methods)、层次法(Hierarchical Methods)、基于密度的方法(Density-based Methods)、基于网格的方法(Grid-based Methods)、基于模型的方法(Model-based Methods)和图论聚类法。

1. 算法分类

很难对聚类方法提出一个简洁的分类,因为这些类别可能重叠,从而使得一种方法具有几类的特征。尽管如此,对于各种不同的聚类方法提供一个相对有组织的描述依然是有用的,为聚类分析计算方法主要有如下几种。

(1)划分法

给定一个有 N 个元组或者记录的数据集,构造 K($K<N$)个分组,每一个分组就代表一个聚类。这 K 个分组满足下列条件:

① 每一个分组至少包含一个数据记录。

② 每一个数据记录属于且仅属于一个分组(注意:这个要求在某些模糊聚类算法中可以放宽)。

对于给定的 K,算法首先给出一个初始的分组方法,以后通过反复迭代的方法改变分组,使得每一次改进之后的分组方案都较前一次好。所谓好的标准就是:同一分组中的记录越近越好,不同分组中的记录越远越好。

大部分划分方法是基于距离的。给定要构建的分区数 k,划分方法首先创建一个初始化划分。然后,采用一种迭代的重定位技术,通过把对象从一个组移动到另一个组来进行划分。一个好的划分的一般准备是:同一个簇中的对象尽可能相互接近或相关,而不同簇中的对象尽可能远离或不同。还有许多评判划分质量的其他准则。传统的划分方法可以扩展到子空间聚类,而不是搜索整个数据空间。当存在很多属性并且数据稀疏时,这是有用的。为了达到全局最优,基于划分的聚类可能需要穷举所有可能的划分,计算量极大。实际上,大多数应用都采用了流行的启发式方法,如 K 均值和 K 中心点算法,渐近地提高聚类质量,逼近局部最优解。这些启发式聚类方法很适合发现中小规模的数据库中的球状簇。为了发现具有复杂形状的簇和对超大型数据集进行聚类,需要进一步扩展基于划分的方法。

使用这个基本思想的算法有 K-Means 算法、K-Medoids 算法、CLARANS 算法。

(2)基于密度的方法

基于密度的方法与其他方法的一个根本区别是,它不是基于各种各样距离的,而是基于密度的。这样就能克服基于距离的算法只能发现"类圆形"的聚类的缺点。

这个方法的指导思想就是,只要一个区域中的点的密度大于某个阈值,就把它加到与之相近的聚类中去。

代表算法有 DBSCAN 算法、OPTICS 算法、DENCLUE 算法等。

（3）图论聚类方法

图论聚类方法解决的第一步是建立与问题相适应的图，图的节点对应于被分析数据的最小单元，图的边（或弧）对应于最小处理单元数据之间的相似性度量。因此，每一个最小处理单元数据之间都会有一个度量表达，这就确保了数据的局部特性比较易于处理。图论聚类法是以样本数据的局域连接特征作为聚类的主要信息源，因而其主要优点是易于处理局部数据的特性。

（4）网格算法

这种方法首先将数据空间划分成为有限个单元（Cell）的网格结构，所有的处理都是以单个的单元为对象的。这么处理的一个突出优点就是处理速度很快，通常这是与目标数据库中记录的个数无关的，只与把数据空间分为多少个单元有关。

代表算法有 STING 算法、CLIQUE 算法、WAVE-CLUSTER 算法。

（5）基于模型的方法

基于模型的方法给每一个聚类假定一个模型，然后去寻找能够很好地满足这个模型的数据集。这样一个模型可能是数据点在空间中的密度分布函数或者其他。它的一个潜在假定就是：目标数据集是由一系列的概率分布所决定的。

通常有两种方向：统计的方案和神经网络的方案。

2. 聚类的原理

① 先随机产生（选取数据集中的地方，收敛会比较快）每个类别的中心点（设定多少类别就产生多少个类的中心点）。

② 计算每个样本和中心点之间的距离，离哪个最近，就将它归为哪一类。

③ 每一类都会有很多样本，计算这些样本的平均值作为新的中心点。

④ 如果新的中心点和旧的中心点差别不大，就完成聚类，否则重新跳至第二步。

3. 四种常用聚类算法研究

（1）K-Means 聚类算法

K-Means 是划分方法中较经典的聚类算法之一。由于该算法的效率高，因此在对大规模数据进行聚类时被广泛应用。目前，许多算法均围绕着该算法进行扩展和改进。

K-Means 算法以 k 为参数，把 n 个对象分成 k 个簇，使簇内具有较高的相似度，而簇间的相似度较低。K-Means 算法的处理过程如下：首先，随机地选择 k 个对象，每个对象初始地代表了一个簇的平均值或中心；对剩余的每个对象，根据其与各簇中心的距离，将其赋给最近的簇；然后重新计算每个簇的平均值。这个过程不断重复，直到准则函数收敛。通常，采用平方误差准则，其定义如下：

$$E = \sum_{i=1}^{k} \sum_{p \in C} |p - m_i|^2 \tag{3.38}$$

这里 E 是数据库中所有对象平方误差的总和，p 是空间中的点，m_i 是簇 C_i 的平均值。该目标函数使生成的簇尽可能紧凑独立，使用的距离度量是欧几里得距离，当然也可以用其他距离度量。K-Means 聚类算法的流程如下：

输入：包含 n 个对象的数据库和簇的数目 k。

输出：k 个簇，使平方误差准则最小。

步骤：

① 任意选择 k 个对象作为初始的簇中心。
② 重复。
③ 根据簇中对象的平均值，将每个对象（重新）赋予最类似的簇。
④ 更新簇的平均值，即计算每个簇中对象的平均值。
⑤ 直到不再发生变化。

（2）层次聚类算法

根据层次分解的顺序是自底向上的还是自顶向下的，层次聚类算法分为凝聚型层次聚类算法和分裂型层次聚类算法。

凝聚型层次聚类的策略是先将每个对象作为一个簇，然后合并这些原子簇为越来越大的簇，直到所有对象都在一个簇中或者某个终结条件被满足。绝大多数层次聚类属于凝聚型层次聚类，它们只是在簇间相似度的定义上有所不同。4 种广泛采用的簇间距离度量方法如下：

最小距离：

$$d_{\min}(c_i, c_j) = \min_{p \in c_i, p' \in c_j} |p - p'| \tag{3.39}$$

最大距离：

$$d_{\max}(c_i, c_j) = \max_{p \in c_i, p' \in c_j} |p - p'| \tag{3.40}$$

平均值距离：

$$d_{\mathrm{mean}}(c_i, c_j) = |m_i - m_j| \tag{3.41}$$

平均距离：

$$d_{\arg}(c_i, c_j) = \frac{1}{n_i n_j} \sum_{p \in c_i} \sum_{p' \in c_j} |p - p'| \tag{3.42}$$

这里，$|p-p'|$ 是两个对象 p 和 p' 之间的距离，m_i 是簇 c_i 的平均值，n_i 是簇 c_i 中对象的数目。这里给出采用最小距离的凝聚层次聚类算法流程：

① 将每个对象看作一类，计算两两之间的最小距离。
② 将距离最小的两个类合并成一个新类。
③ 重新计算新类与所有类之间的距离。

④ 重复②、③，直到所有类最后合并成一类。

（3）SOM 聚类算法

SOM 神经网络是由芬兰神经网络专家 Kohonen 教授提出的，假设在输入对象中存在一些拓扑结构或顺序，可以实现从输入空间（n 维）到输出平面（2 维）的降维映射，其映射具有拓扑特征保持性质，与实际的大脑处理有很强的理论联系。

SOM 网络包含输入层和输出层。输入层对应一个高维的输入向量，输出层由一系列组织在 2 维网格上的有序节点构成，输入节点与输出节点通过权重向量连接。在学习过程中，找到与之距离最短的输出层单元，即获胜单元，对其更新。同时，将邻近区域的权值更新，使输出节点保持输入向量的拓扑特征。

算法流程：

① 网络初始化，对输出层每个节点权重赋初值。
② 在输入样本中随机选取输入向量，找到与输入向量距离最小的权重向量。
③ 定义获胜单元，在获胜单元的邻近区域调整权重使其向输入向量靠拢。
④ 提供新样本、进行训练。
⑤ 收缩邻域半径，减小学习率，重复，直到小于允许值，输出聚类结果。

（4）FCM 聚类算法

1965 年美国加州大学柏克莱分校的扎德教授第一次提出了'集合'的概念。经过十多年的发展，模糊集合理论渐渐被应用到各个实际应用方面。为克服非此即彼的分类缺点，出现了以模糊集合论为数学基础的聚类分析。用模糊数学的方法进行聚类分析，就是模糊聚类分析。

FCM 算法是一种以隶属度来确定每个数据点属于某个聚类程度的算法。该聚类算法是传统硬聚类算法的一种改进。

设数据集 $X=\{x_1,x_2,\ldots,x_n\}$，它的模糊 c 划分可用模糊矩阵 $U=[u_{ij}]$ 表示，矩阵 U 的元素 u_{ij} 表示第 j（$j=1,2,\ldots,n$）个数据点属于第 i（$i=1,2,\ldots,c$）类的隶属度，u_{ij} 满足如下条件：

$$\forall_j, \sum_{i=1}^{c} u_{ij} = 1; \quad \forall i, j u_{ij} \in [0,1]; \quad \forall i, \sum_{j=1}^{n} u_{ij} > 0 \tag{3.43}$$

目前被广泛使用的聚类准则为聚类内加权误差平方和的极小值，即

$$(\min)J_m(U,V) = \sum_{j=1}^{n} \sum_{i=1}^{c} u_{ij}^{m} d_{ij}^{2}(x_j, v_i) \tag{3.44}$$

其中，V 为聚类中心，$m \in [1, +\infty)$ 为加权指数，又称作平滑参数。

$$d_{ij}(x_j, v_i) = \|v_i - x_j\| \tag{3.45}$$

算法流程：

① 标准化数据矩阵。

② 建立模糊相似矩阵，初始化隶属矩阵。
③ 算法开始迭代，直到目标函数收敛到极小值。
④ 根据迭代结果，由最后的隶属矩阵确定数据所属的类，显示最后的聚类结果。

3.5 EM 算法

EM 算法（Expectation Maximization Algorithm）是一种迭代优化策略。由于它的计算方法中每一次迭代都分两步，其中一个为期望步（E 步），另一个为极大步（M 步），因此被称为 EM 算法。EM 算法受到缺失思想影响，最初是为了解决数据缺失情况下的参数估计问题，其算法基础和收敛有效性等问题在 Dempster、Laird 和 Rubin 三人于 1977 年所写的文章 "Maximum likelihood from incomplete data via the EM algorithm" 中给出了详细的阐述。其基本思想是：首先根据已经给出的观测数据，估计出模型参数的值；然后依据上一步估计出的参数值来估计缺失数据的值，再根据估计出的缺失数据加上之前已经观测到的数据重新对参数值进行估计，并反复迭代，直至最后收敛，迭代结束。

EM 算法作为一种数据添加算法，在近几十年得到迅速发展。在当前科学研究以及各方面实际应用中数据量越来越大的情况下，经常存在数据缺失或者不可用的问题。这时直接处理数据比较困难。数据添加办法有很多种，常用的有神经网络拟合、添补法、卡尔曼滤波法等，EM 算法之所以能迅速普及，主要源于算法简单、稳定上升的步骤能非常可靠地找到"最优的收敛值"。随着理论的发展，EM 算法已经不单单用于处理缺失数据的问题，有时缺失数据并非是真的缺少了，而是为了简化问题而采取的策略。这时 EM 算法被称为数据添加技术，所添加的数据通常被称为"潜在数据"。复杂的问题通过引入恰当的潜在数据得到有效解决。

假设 X 为观测变量，Z 表示未观测变量，θ 为模型参数。在概率模型中，我们常常根据最大似然估计方法进行参数的估计，即极大化观测数据关于参数的对数似然函数：

$$\log P(X|\theta) = \log \sum_Z (X, Z|\theta) \tag{3.46}$$

上式包含隐变量以及和的对数，而无法有效求解。EM 算法这一利器避开上式的求解，通过 Jensen 不等式（$\log \sum_j \lambda_j y_j \geq \sum_j \lambda_j \log y_j$，其中 $\lambda_j \geq 0$，且 $\sum_j \lambda_j = 1$）找到其下界（Lower Bound），通过不断求解下界的极大化来逼近求解对数似然函数极大化。为了使用 Jensen 不等式，必须把对数似然函数构造成类似的式子：

$$\log \sum_Z P(X, Z|\theta) = \log \sum_Z q(Z) \frac{P(X, Z|\theta)}{q(Z)} \tag{3.47}$$

其中，$q(Z)$ 为概率分布，满足 Jensen 不等式的条件，因此进一步得到：

$$\log \sum_Z P(X, Z|\theta) = \log \sum_Z q(Z) \frac{P(X, Z|\theta)}{q(Z)}$$

$$\geq \sum_Z q(Z)\log \frac{P(X,Z|\theta)}{q(Z)} \tag{3.48}$$

对数似然函数的下界是 $L(q,\theta)=\sum_Z q(Z)\log(P(X,Z|\theta)/q(Z))$。可以看出下界是关于 $q(Z)$ 和 θ 的函数，因此将 EM 算法分为两步：

① 固定 θ，得到 $q(Z)$ 的分布。如果进一步求隐变量 Z 的期望，则对应 EM 算法的 E-step。这里值得注意的是 $q(Z)$ 的分布该如何确定。在 Jensen 不等式中，只有当 y 为常数时，等式成立，即 $P(X,Z|\theta)/q(Z)=C$。通过这个条件，能轻易得到 $q(Z)=P(Z|X,\theta)$。

② 固定 $q(Z)$，优化 θ，对应 EM 算法的 M-step。这两个步骤不断重复，直至收敛到局部最优解。

一般来讲，EM 算法主要用于含有隐变量的概率模型学习，针对不完全数据 X 的最大对数似然函数找到局部最优解。下面从另一个角度解释 EM 算法：在现实中，很难得到完全数据 $\{X,Z\}$，但是能根据已有知识得到隐变量 Z 的后验概率分布 $P(Z|X,\theta)$。虽然我们不能建立完全数据的对数似然函数，但是可以考虑隐变量在其后验分布下的期望值，这样有了隐变量的期望值，就可以构建完全数据的对数似然函数。进一步，参数的优化目标为

$$\begin{aligned}&\max_\theta E_{p(Z|X,\theta^{old})}[\log p(X,Z|\theta)]\\&=\max_\theta \sum_Z p(Z|X,\theta^{old})\log p(X,Z|\theta)\end{aligned} \tag{3.49}$$

先利用当前参数 θ^{old} 得到隐变量的后验分布，再根据完全数据的对数似然函数在后验分布下的期望下更新 θ。

如果对对数似然函数的 Jensen 不等式进行进一步分析，就会发现如下等式成立：

$$\log p(X|\theta)=L(q,\theta)+KL(q\|p) \tag{3.50}$$

其中

$$L(q,\theta)=\sum_Z q(Z)\log \frac{p(X,Z|\theta)}{q(Z)} \tag{3.51}$$

$$KL(q\|p)=-\sum_Z q(Z)\log \frac{p(Z|X,\theta)}{q(Z)} \tag{3.52}$$

$L(q,\theta)$ 为下界，包含 X 与 Z 的联合概率分布；而 $KL(q\|p)$ 包含 Z 的条件分布。Kullback-Leibler divergence 具有非负性，$KL(q\|p)\geq 0$，且只有当 $q(Z)=p(Z|X,\theta)$ 时，等号才成立。此时下界 $L(q,\theta)$ 等于不完全数据的对数似然函数。

$$\log p(\theta|X)=\log p(X|\theta)+\log p(\theta)-\log p(X) \tag{3.53}$$

根据前面的 Jensen 不等式，能够得到 $\log p(X|\theta)$ 的下界 $L(q,\theta)$，则

$$\log p(\theta|X)\geq L(q,\theta)+\log p(\theta)-\log p(X) \tag{3.54}$$

其中，log$p(X)$为常数项。上式的优化又可交替地优化 q 和 θ。与标准的最大似然对比，只增加了参数的先验项。

3.6 贝叶斯算法

贝叶斯定理是关于随机事件 A 和 B 的条件概率（或边缘概率）的一则定理。其中，$P(A|B)$ 是在 B 发生的情况下 A 发生的概率（即可能性）。

贝叶斯定理也称贝叶斯推理，早在 18 世纪，英国学者贝叶斯（1702—1763）曾提出计算条件概率的公式用来解决如下一类问题：假设 H[1],H[2],…,H[n]互斥且构成一个完全事件，已知它们的概率 $P(H[i])$，$i=1,2,…,n$，现观察到某事件 A 与 H[1],H[2],…,H[n]相伴随机出现，且已知条件概率 $P(A/H[i])$，求 $P(H[i]/A)$。

人们根据不确定性信息做出推理和决策需要对各种结论的概率做出估计，这类推理称为概率推理。概率推理既是概率学和逻辑学的研究对象，也是心理学的研究对象，但研究的角度是不同的。概率学和逻辑学研究的是客观概率推算的公式或规则；心理学研究人们主观概率估计的认知加工过程的规律。贝叶斯推理的问题是条件概率推理问题，这一领域的探讨对揭示人们对概率信息的认知加工过程与规律、指导人们进行有效的学习和判断决策具有十分重要的理论意义和实践意义。

贝叶斯公式：

$$P(B_i|A) = \frac{P(B_i)P(A|B_i)}{\sum_{j=1}^{n} P(B_j)P(A|B_j)} \tag{3.55}$$

贝叶斯定理用于投资决策分析是在已知相关项目 B 的资料而缺乏论证项目 A 的直接资料时，通过对 B 项目的有关状态及发生概率分析推导项目 A 的状态及发生概率。如果用数学语言描绘，即当已知事件 B_i 的概率 $P(B_i)$ 和事件 B_i 已发生条件下事件 A 的概率 $P(A|B_i)$，则可运用贝叶斯定理计算出在事件 A 发生条件下事件 B_i 的概率 $P(B_i|A)$。按贝叶斯定理进行投资决策的基本步骤是：

- 列出在已知项目 B 条件下项目 A 的发生概率，即将 $P(A|B)$ 转换为 $P(B|A)$。
- 绘制树形图。
- 求各状态结点的期望收益值，并将结果填入树形图。
- 根据对树形图的分析，进行投资项目决策。

3.7 隐马尔可夫模型

隐马尔可夫模型是马尔可夫链的一种。它的状态不能直接观察到，但能通过观测向量序列观察到。每个观测向量都是通过某些概率密度分布表现为各种状态，每一个观测向量是由一个具有相应概率密度分布的状态序列产生的。所以，隐马尔可夫模型是一个双重随机过程——具有一定状态数的隐马尔可夫链和显示随机函数集。自 20 世纪 80 年代以来，HMM 被应用于语音识别，取得重大成功。到了 20 世纪 90 年代，HMM 还被引入计算机文字识别和移动通信核心技术"多用户的检测"。HMM 在生物信息科学、故障诊断等领域也开始得到应用。

马尔可夫模型有两个假设：

- 系统在时刻 t 的状态只与时刻 $t-1$ 处的状态相关（也称为无后效性）。
- 状态转移概率与时间无关（也称为齐次性或时齐性）。

第一条具体可以用如下公式表示：

$$P(q_t=S_j|q_t-1=S_i,q_t-2=S_k,\ldots)= P(q_t=S_j|q_t-1=S_i) \tag{3.56}$$

其中，t 为大于 1 的任意数值，S_k 为任意状态。

第二个假设可以用如下公式表示：

$$P(q_t=S_j|q_t-1=S_i)= P(q_k=S_j|q_k-1=S_i) \tag{3.57}$$

其中，k 为任意时刻。

隐马尔可夫过程如图 3.5。

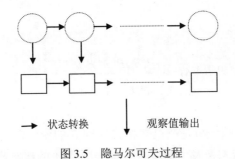

图 3.5　隐马尔可夫过程

对于 HMM 模型，首先假设 Q 是所有可能的隐藏状态的集合，V 是所有可能的观测状态的集合，即：

$$Q=\{q_1,q_2,\ldots,q_n\},\quad V=\{v_1,v_2,\ldots,v_m\} \tag{3.58}$$

其中，n 是可能的隐藏状态数，m 是所有可能的观察状态数。对于一个长度为 t 的序列，I 是对应的状态序列，O 是对应的观察序列，即：

$$I=\{i_1,i_2,\ldots,i_t\},\quad O=\{o_1, o_2,\ldots, o_t\} \tag{3.59}$$

其中，任意一个隐藏状态 $i_t\in Q$，任意一个观察状态 $o_t\in V$。

（1）HMM 模型两个很重要的假设

① 齐次马尔可夫链假设：任意时刻的隐藏状态只依赖于前一个隐藏状态，当然这样假设有点极端，因为很多时候某一个隐藏状态不仅仅只依赖于前一个隐藏状态，可能是前两个或者

前三个。这样假设的好处是模型简单，便于求解。如果在时刻 t 的隐藏状态是 $i_t=q_i$，在时刻 $t+1$ 的隐藏状态是 $i_t+1=q_j$，那么从时刻 t 到时刻 $t+1$ 的 HMM 状态转移概率 a_{ij} 可以表示为：

$$a_{ij}=P(i_t+1=q_j|i_t=q_i) \tag{3.60}$$

这样 a_{ij} 可以组成马尔可夫链的状态转移矩阵 A：

$$A_{ij}=\left[a_{ij}\right]_{n\times m} \tag{3.61}$$

② 观测独立性假设：任意时刻的观察状态仅仅依赖于当前时刻的隐藏状态，也是一个为了简化模型的假设。如果在时刻 t 的隐藏状态是 $i_t=q_j$，而对应的观察状态为 $o_t=v_k$，那么该时刻观察状态 v_k 在隐藏状态 q_j 下生成的概率为 $b_j(k)$，满足：

$$b_j(k)=P(o_t=v_k|i_t=q_j) \tag{3.62}$$

这样 $b_j(k)$ 可以组成观测状态生成的概率矩阵 B：

$$B=\left[b_j(k)\right]_{n\times m} \tag{3.63}$$

除此之外，我们需要一组在时刻 $t=1$ 的隐藏状态概率分布 $\mathit{\Pi}$：

$$\mathit{\Pi}=\left[\pi(i)\right]_n \tag{3.64}$$

其中，$\pi(i)=P(i_1=q_i)$。

一个 HMM 模型，可以由隐藏状态初始概率分布 $\mathit{\Pi}$、状态转移概率矩阵 A 和观测状态概率矩阵 B 决定。$\mathit{\Pi}$ 和 A 决定状态序列，B 决定观测序列。因此，HMM 模型可以由一个三元组 λ 表示如下：

$$\lambda=(A,B,\mathit{\Pi}) \tag{3.65}$$

（2）HMM 模型的 3 个基本问题

① 评估观察序列概率。给定模型 $\lambda=(A,B,\mathit{\Pi})$ 和观测序列 $O=\{o_1,o_2,...,o_t\}$，计算在模型 λ 下观测序列 O 出现的概率 $P(O|\lambda)$。这个问题的求解需要用到前向后向算法，是 HMM 模型 3 个问题中最简单的。

② 模型参数学习问题。给定观测序列 $O=\{o_1,o_2,...,o_t\}$，估计模型 $\lambda=(A,B,\mathit{\Pi})$ 的参数，使该模型下观测序列的条件概率 $P(O|\lambda)$ 最大。这个问题的求解需要用到基于 EM 算法的鲍姆-韦尔奇算法。

③ 预测问题，也称为解码问题。给定模型 $\lambda=(A,B,\mathit{\Pi})$ 和观测序列 $O=\{o_1,o_2,...,o_t\}$，求给定观测序列条件下最可能出现的对应的状态序列。这个问题的求解需要用到基于动态规划的维特比算法，是 HMM 模型 3 个问题中复杂度居中的算法。

3.8 LDA 主题模型

LDA 由 Blei、David M.、Ng、Andrew Y.、Jordan 于 2003 年提出，用来推测文档的主题分布。它可以将文档集中每篇文档的主题以概率分布的形式给出，从而通过分析一些文档抽取出它们的主题分布，之后根据主题分布进行主题聚类或文本分类。在机器学习领域，LDA 是两个常用模型 Linear Discriminant Analysis（线性判别分析）和 Latent Dirichlet Allocation（隐含狄利克雷分布）的简称。

LDA 是一种文档主题生成模型，也称为一个 3 层贝叶斯概率模型，包含词、主题和文档 3 层结构。所谓生成模型，是指一篇文章的每个词都是通过"以一定概率选择了某个主题，并从这个主题中以一定概率选择某个词语"这样的过程得到的。文档到主题服从多项式分布，主题到词服从多项式分布。

LDA 是一种无监督机器学习技术，可以识别大规模文档集（Document Collection）或语料库（Corpus）中潜藏的主题信息。它采用了词袋（Bag of Words）的方法，将每一篇文档视为一个词频向量，从而将文本信息转化为易于建模的数字信息。词袋方法没有考虑词与词之间的顺序，简化了问题的复杂性，同时也为模型的改进提供了契机。每一篇文档代表了一些主题所构成的一个概率分布，每一个主题又代表了很多单词所构成的一个概率分布。

1. 先验知识

LDA 模型涉及很多数学知识，也许就是 LDA 晦涩难懂的主要原因。LDA 涉及的先验知识有二项分布、Gamma 函数、Beta 分布、多项分布、Dirichlet 分布、马尔可夫链、MCMC、Gibs Sampling、EM 算法等。限于篇幅，本文仅会有的放矢地介绍部分概念，不会每个概念都仔细介绍，亦不会涉及每个概念的数学公式推导。

（1）词袋模型

LDA 采用词袋模型。所谓词袋模型，是指针对一篇文档，仅考虑一个词汇是否出现，而不考虑其出现的顺序。与词袋模型相反的一个模型是 n-gram。n-gram 考虑了词汇出现的先后顺序。

（2）二项分布

二项分布是 N 重伯努利分布，即为 $X \sim B(n, p)$。概率密度公式为：

$$P(K=k) = \binom{n}{k} p^k (1-p)^{n-k} \tag{3.66}$$

（3）多项分布

多项分布是二项分布扩展到多维的情况。多项分布是指单次试验中随机变量的取值不再是 0~1 的，而是有多种离散值可能（1,2,3,...,k）。概率密度函数为：

$$P(x_1, x_2, \ldots, x_k; n, p_1, p_2, \ldots, p_k) = \frac{n!}{x_1! \cdots x_k!} p_1^{x1} \cdots p_k^{xk} \tag{3.67}$$

（4）Gamma 分布

Gamma 函数的定义：

$$\Gamma(x) = \int_0^\infty t^{x-1} e^{-t} dt \tag{3.68}$$

（5）Beta 分布

对于参数 $\alpha > 0$、$\beta > 0$，取值范围为[0, 1]的随机变量 x 的概率密度函数为：

$$f(x;\alpha,\beta) = \frac{1}{B(\alpha,\beta)} x^{\alpha-1}(1-x)^{\beta-1} \tag{3.69}$$

其中，$\dfrac{1}{B(\alpha,\beta)} = \dfrac{\Gamma(\alpha,\beta)}{\Gamma(\alpha)\Gamma(\beta)}$。

（6）共轭先验分布

在贝叶斯概率理论中，如果后验概率 $P(\theta|x)$ 和先验概率 $P(\theta)$ 满足同样的分布律，那么先验分布和后验分布被叫作共轭分布，同时先验分布叫作似然函数的共轭先验分布。

$$P(\theta|x) = \frac{P(\theta,x)}{P(x)} \tag{3.70}$$

Beta 分布是二项式分布的共轭先验分布，而狄利克雷（Dirichlet）分布是多项式分布的共轭分布。共轭的意思是，以 Beta 分布和二项式分布为例来理解共轭的意思：数据符合二项分布的时候，参数的先验分布和后验分布都能保持 Beta 分布的形式。这种形式不变的好处是，能够在先验分布中赋予参数很明确的物理意义，这个物理意义可以延续到后续分布中进行解释，同时从先验变换到后验过程中从数据中补充的知识也容易有物理解释。

（7）Dirichlet 分布

Dirichlet 的概率密度函数为：

$$f(x_1, x_2, \ldots, x_k; \alpha_1, \alpha_2, \ldots, \alpha_k) = \frac{1}{B(\alpha)} \prod_{i=1}^k x_i^{\alpha_i - 1} \tag{3.71}$$

其中，$B(\alpha) = \dfrac{\prod_{i=1}^k \Gamma(\alpha^i)}{\Gamma(\sum_{i=1}^k \alpha^i)}, \sum_{i=1}^k x^i = 1$。

（8）Beta / Dirichlet 分布的一个性质

若 $p\ Beta(t|\alpha,\beta)$，则

$$\begin{aligned} E(p) &= \int_0^1 t * Beta(t|\alpha,\beta) dt \\ &= \int_0^1 t * \frac{\Gamma(\alpha+\beta)}{\Gamma(\alpha)\Gamma(\beta)} t^{(\alpha-1)}(1-t)^{\beta-1} dt \\ &= \frac{\Gamma(\alpha+\beta)}{\Gamma(\alpha)\Gamma(\beta)} \int_0^1 t^\alpha (1-t)^{\beta-1} dt \end{aligned} \tag{3.72}$$

上式右边的积分对应到概率分布 $Beta(t|\alpha+1,\beta)$，对于这个分布，有

$$\int_0^1 \frac{\Gamma(\alpha+\beta+1)}{\Gamma(\alpha)\Gamma(\beta)} t^\alpha (1-t)^{\beta-1} \mathrm{d}t = 1 \tag{3.73}$$

把式（3.75）带入 $E(p)$ 的计算式（3.74），得到

$$\begin{aligned} E(p) &= \frac{\Gamma(\alpha+\beta)}{\Gamma(\alpha)\Gamma(\beta)} \cdot \frac{\Gamma(\alpha+1)\Gamma(\beta)}{\Gamma(\alpha+\beta+1)} \\ &= \frac{\Gamma(\alpha+\beta)}{\Gamma(\alpha+\beta+1)} \cdot \frac{\Gamma(\alpha+1)}{\Gamma(\alpha)} \\ &= \frac{\alpha}{\alpha+\beta} \end{aligned} \tag{3.74}$$

这说明，对于 $Beta$ 分布的随机变量，其均值可以用 $\alpha/(\alpha+\beta)$ 来估计。Dirichlet 分布也有类似的结论，即如果 $\vec{p} \sim Dir(\vec{t}\,|\,\vec{\alpha})$，同样可以证明：

$$E(p) = \left(\frac{\alpha^1}{\sum_{i=1}^K \alpha_i}, \frac{\alpha^1}{\sum_{i=2}^K \alpha_i}, \cdots, \frac{\alpha^K}{\sum_{i=1}^K \alpha_i} \right) \tag{3.75}$$

2. LDA 生成过程

对于语料库中的每篇文档，LDA 定义了如下生成过程（Generative Process）：

- 对每一篇文档，从主题分布中抽取一个主题。
- 从上述被抽到的主题所对应的单词分布中抽取一个单词。
- 重复上述过程，直至遍历文档中的每一个单词。

语料库中的每一篇文档与 T（通过反复试验等方法事先给定）个主题的一个多项分布（Multinomial Distribution）相对应，将该多项分布记为 θ。每个主题又与词汇表（Vocabulary）中 V 个单词的一个多项分布相对应，将这个多项分布记为 φ。

3. LDA 整体流程

先定义一些字母的含义：文档集合 D，主题（topic）集合 T。

D 中每个文档 d 看作一个单词序列 $<w_1,w_2,...,w_n>$，w_i 表示第 i 个单词，设 d 有 n 个单词。（LDA 里面称之为 Word of Bag，即词袋，实际上每个单词的出现位置对 LDA 算法无影响。）

- D 中涉及的所有不同单词组成一个大集合 VOCABULARY（简称 VOC），LDA 以文档集合 D 作为输入，希望训练出两个结果向量（设聚成 k 个 topic，VOC 中共包含 m 个词）。
- 对每个 D 中的文档 d，对应到不同 topic 的概率 $\theta d<pt_1,...,pt_k>$。其中，pt_i 表示 d 对应 T 中第 i 个 topic 的概率。计算方法是直观的，$pt_i=nt_i/n$。其中，nt_i 表示 d 对应第 i 个 topic 的词的数目，n 是 d 中所有词的总数。
- 对每个 T 中的 topic，生成不同单词的概率 $\varphi t<pw_1,...,pw_m>$。其中，pw_i 表示 t 生成 VOC

中第 i 个单词的概率。计算方法同样很直观，$pw_i=Nw_i/N$。其中，Nw_i 表示对应到 topic 的 VOC 中第 i 个单词的数目，N 表示所有对应到 topic 的单词总数。

LDA 的核心公式如下：

$$p(w|d)=p(w|t)*p(t|d) \qquad (3.76)$$

直观地看这个公式，就是以 topic 作为中间层，可以通过当前的 $θd$ 和 $φt$ 给出文档 d 中出现单词 w 的概率。其中，$p(t|d)$ 利用 $θd$ 计算得到，$p(w|t)$ 利用 $φt$ 计算得到。

实际上，利用当前的 $θd$ 和 $φt$ 可以为一个文档中的一个单词计算它对应任意一个 topic 时的 $p(w|d)$，然后根据这些结果来更新这个词应该对应的 topic。然后，如果这个更新改变了单词所对应的 topic，就会反过来影响 $θd$ 和 $φt$。

LDA 算法开始时先随机给 $θd$ 和 $φt$ 赋值（对所有的 d 和 t），然后不断重复上述过程，得到的最终收敛结果就是 LDA 的输出。

3.9 人工神经网络

人工神经网络（Artificial Neural Network，ANN）是 20 世纪 80 年代以来人工智能领域兴起的研究热点。它从信息处理角度对人脑神经元网络进行抽象，建立某种简单模型，按不同的连接方式组成不同的网络。在工程与学术界也常直接简称为神经网络或类神经网络。神经网络是一种运算模型，由大量的节点（或称神经元）之间相互连接构成。每个节点代表一种特定的输出函数，称为激励函数（Activation Function）。每两个节点间的连接都代表一个对于通过该连接信号的加权值，称之为权重，相当于人工神经网络的记忆。网络的输出依网络的连接方式、权重值和激励函数的不同而不同。网络自身通常都是对自然界某种算法或者函数的逼近，也可能是对一种逻辑策略的表达。

最近十多年来，人工神经网络的研究工作不断深入，取得了很大进展，在模式识别、智能机器人、自动控制、预测估计、生物、医学、经济等领域已成功地解决了许多现代计算机难以解决的实际问题，表现出了良好的智能特性。

人工神经网络模型主要考虑网络连接的拓扑结构、神经元的特征、学习规则等。目前，已有近 40 种神经网络模型，其中有反传网络、感知器、自组织映射、Hopfield 网络、波耳兹曼机、适应谐振理论等。根据连接的拓扑结构，神经网络模型可以分为前向网络和反馈网络。

- 前向网络

网络中各个神经元接受前一级的输入，并输出到下一级，网络中没有反馈，可以用一个有向无环路图表示。这种网络实现信号从输入空间到输出空间的变换，信息处理能力来自于简单非线性函数的多次复合。网络结构简单，易于实现。反传网络是一种典型的前向网络。

- 反馈网络

网络内神经元间有反馈，可以用一个无向的完备图表示。这种神经网络的信息处理是状态的变换，可以用动力学系统理论处理。系统的稳定性与联想记忆功能有密切关系。Hopfield 网络、波耳兹曼机均属于这种类型。

1. 人工神经网络原理

由大量简单元件广泛互连而成的复杂网络系统。所谓简单元件，即人工神经元。每一个神经元有许多输入键和输出键，各神经元之间以连接键（又称突触）相连，连接键决定神经元之间的连接强度（突触强度）和性质（兴奋或抑制），即决定神经元间相互作用的强弱和正负，共有 3 种类型：兴奋型连接、抑制型连接、无连接。这样，N 个神经元（一般 N 很大）构成一个相互影响的复杂网络系统，通过调整网络参数，可使人工神经网络具有所需要的特定功能，即学习、训练或自组织过程。一个简单的人工神经网络结构图如图 3.6 所示。

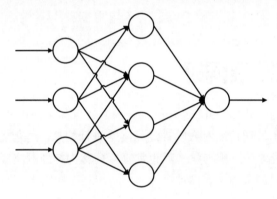

图 3.6　人工神经网络结构图

在图 3.6 中，左侧为输入层（输入层的神经元个数由输入的维度决定），右侧为输出层（输出层的神经元个数由输出的维度决定），输入层与输出层之间即为隐藏层。

输入层节点上的神经元接收外部环境的输入模式，并由它传递给相连隐藏层上的各个神经元。隐藏层是神经元网络的内部处理层，这些神经元在网络内部构成中间层，不直接与外部输入、输出打交道。人工神经网络所具有的模式变换能力主要体现在隐藏层的神经元上。输出层用于产生神经网络的输出模式。

2. BP 神经网络

目前，在这一基本原理上已发展了几十种神经网络，例如 Hopfield 模型、Feldmann 等的连接型网络模型、Hinton 等的玻尔兹曼机模型、Rumelhart 等的多层感知机模型和 Kohonen 的自组织网络模型等。在众多神经网络模型中，应用最广泛的是多层感知机神经网络。这里我们重点讲述一下 BP 神经网络。多层感知机神经网络的研究始于 20 世纪 50 年代，但一直进展不大。直到 1985 年，Rumelhart 等人提出误差反向传递学习算法（BP 算法），实现了 Minsky 的多层网络设想。它可以分为输入层、隐藏层（也叫中间层）和输出层。其中，中间层可以是一层，也可以是多层，视实际情况而定。

输入模式由输入层进入网络，经中间各隐藏层的顺序变换，最后由输出层产生一个输出模式，如图 3.7 所示。

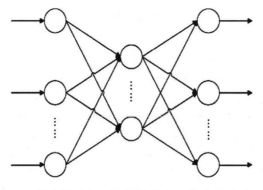

图 3.7　BP 神经网络连接方式

BP 神经网络由隐藏层神经元的非线性处理衍生它的能力，这个任务的关键在于将神经元的加权输入非线性转换成一个输出的非线性激励函数。在 BP 神经网络中，输入层和输出层的节点个数都是确定的，而隐藏层节点个数不确定，那么应该设置为多少才合适呢？隐藏层节点个数的多少对神经网络的性能是有影响的，有一个经验公式可以确定隐藏层节点数目：

$$h = \sqrt{m+n} + a \tag{3.77}$$

其中，m 为隐藏层节点数目，n 为输入层节点数目，1 为输出层节点数目，a 为之间的调节常数。

图 3.8 给出了一个接收 n 个输入 $x_1, x_2, \ldots, x_n$ 的神经元。

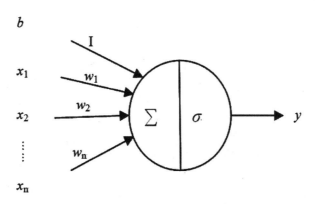

图 3.8　接收 n 个输入 $x_1, x_2, \ldots, x_n$ 的神经元

神经元的输出由式（3.78）给出：

$$y = \sigma(\sum_{j=1}^{n} w_j x + b) \tag{3.78}$$

这里输入的加权和（括号内部分）由一个非线性函数传递，b 表示与偏差输入相关的权值，w_j 表示与第 j 个输入相关的权值。

对于输入信号，要先向前传播到隐藏层节点，经激励函数后，再把隐藏层节点的输出信号传播到输出节点，最后给出输出结果。节点的激励函数通常选取 S 型函数：

$$f(x) = \frac{1}{1+e^{-x/Q}} \tag{3.79}$$

其中，Q 为调整激励函数形式的 Sigmoid 参数。该算法的学习过程由正向传播和反向传播组成。在正向传播过程中，输入信息从输入层经隐藏层逐层处理，并传向输出层。每一层神经元的状态只影响下一层神经元的状态。如果输出层得不到期望的输出，则转入反向传播，将误差信号沿原来的连接通道返回，通过修改各层神经元的权值，使误差信号最小。

设含有 n 个节点的任意网络，各节点的特性为 Sigmoid 型。为简便起见，指定网络只有一个输出 y，任一节点 i 的输出为 O_i，并设有 N 个样本 $(x_k, y_k)(k=1,2,3,\cdots,N)$，对某一输入 x_k，网络输出为 y_k 节点 i 的输出为 O_{ik}，节点 j 的输入为

$$net_{ik} = \sum_i W_{ij} O_{ik} \tag{3.80}$$

并将误差函数定义为

$$E = \frac{1}{2}\sum_{k=1}^{N}(y_k - \widehat{y}_k)^2 \tag{3.81}$$

其中，$\widehat{y}_k$ 为网络实际输出。定义 $E_k=(y_k-\widehat{y}_k)^2$，$\delta_{jk} = \dfrac{\partial E_k}{\partial net_{jk}}$，$O_{jk}=f(net_{jk})$，于是

$$\begin{aligned}\frac{\partial E_k}{\partial W_{ij}} &= \frac{\partial E_k}{\partial \text{net}_{jk}}\frac{\partial \text{net}_{jk}}{\partial W_{ij}} = \frac{\partial E_k}{\partial \text{net}_{jk}}O_{ik}\\ &= \delta_{jk}O_{ik}\end{aligned} \tag{3.82}$$

当 j 为输出节点时：$\quad O_{jk}=\widehat{y}_k$

$$\delta_{jk} = \frac{\partial E_k}{\partial \widehat{y}_k}\frac{\partial \widehat{y}_k}{\partial \text{net}_{jk}} = -(y_k - \widehat{y}_k)f'(\text{net}_{jk}) \tag{3.83}$$

若 j 不是输出节点，则有

$$\delta_{jk} = \frac{\partial E_k}{\partial \text{net}_{jk}} = \frac{\partial E_k}{\partial O_{jk}} \frac{\partial O_{jk}}{\partial \text{net}_{jk}} = \frac{\partial E_k}{\partial O_{jk}} f'(\text{net}_{jk})$$

$$\begin{aligned}\frac{\partial E_k}{\partial O_{jk}} &= \sum_m \frac{\partial E_k}{\partial \text{net}_{mk}} \frac{\partial \text{net}_{mk}}{\partial O_{jk}} \\ &= \sum_m \frac{\partial E_k}{\partial \text{net}_{mk}} \frac{\partial}{\partial O_{jk}} \sum_i W_{mi} O_{ik} \\ &= \sum_m \frac{\partial E_k}{\partial \text{net}_{mk}} \sum_i W_{mj} = \sum_m \delta_{mk} W_{mj}\end{aligned} \quad (3.84)$$

因此,

$$\begin{cases} \delta_{jk} = f'(\text{net}_{jk}) \sum_m \delta_{mk} W_{mj} \\ \dfrac{\partial E_k}{\partial W_{ij}} = \delta_{mk} O_{ik} \end{cases} \quad (3.85)$$

从上述 BP 算法可以看出,BP 模型把一组样本的 I/O 问题变为一个非线性优化问题,它使用的是优化中最普通的梯度下降法。若把神经网络看成输入到输出的映射,则这个映射是一个高度非线性映射。

设计一个神经网络专家系统,重点在于模型的构成和学习算法的选择。一般来说,结构是根据所研究领域及要解决的问题确定的。通过对所研究问题的大量历史资料的分析及目前的神经网络理论发展水平,建立合适的模型,并针对所选的模型采用相应的学习算法,在网络学习过程中不断地调整网络参数,直到输出结果满足要求。

3.10 KNN 算法

KNN 算法又称 K 近邻分类(K-nearest Neighbor Classification)算法。它是根据不同特征值之间的距离来进行分类的一种简单的机器学习方法,是一种简单但是"懒惰"的算法。它的训练数据都是有标记的数据,即训练的数据都有自己的类别。KNN 算法的主要应用领域是对未知事物进行分类,即判断未知事物属于哪一类,判断思想是,基于欧几里得定理,判断未知事物的特征和哪一类已知事物的特征最接近。它也可以用于回归,通过找出一个样本的 k 个最近邻居,将这些邻居的属性平均值赋给样本,就可以得到该样本的属性。

以下面一个 KNN 算法的例子(见图 3.9)来说明 KNN 算法的特点。

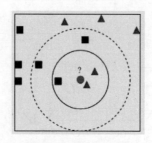

图 3.9 KNN 算法举例

在图 3.9 中，要判断圆形属于哪个类，是三角形还是四方形：$K=3$ 时，三角形所占比例为 2/3，圆形将被赋予三角形类；$K=5$ 时，四方形比例为 3/5，因此圆形被赋予四方形类。

在利用 KNN 算法判断类别时 K 的取值很重要。KNN 算法主要依据邻近的 k 个样本来进行类别的判断，然后依据 k 个样本中出现次数最多的类别作为未知样本的类别。这也就是我们常说的"物以类聚"。

1. KNN 算法的思想和步骤

（1）KNN 算法的思想

KNN 算法用于分类的核心思想是：存在一个样本数据集合，也称训练样本集，并且样本集中每个数据都存在标注，即我们知道样本集中每个数据与其所属分类的关系。输入没有标记的新数据后，将新数据的每个特征与样本集中数据对应的特征进行比较，然后算法提取样本集中特征最相似数据（最近邻）的分类标注。一般来说，我们只选择样本数据集中前 k 个最相似的数据，这就是 k 近邻算法中 k 的出处（通常 $k<20$）。最后，我们选择 k 个最相似数据中出现次数最多的分类，作为新数据的分类。

KNN 算法用于回归的核心思想是：找到近邻的 k 个样本，然后取平均值作为未知样本的值，对其进行预测。

（2）KNN 算法的步骤

- 算距离：给定未知对象，计算它与训练集中每个对象的距离。
- 找近邻：圈定距离最近的 k 个训练对象，作为未知对象的近邻。
- 做分类：在 k 个近邻中出现次数最多的类别就是测试对象的预测类别。

2. KNN 算法的核心知识

（1）距离或相似度的衡量

在 KNN 算法中常使用欧氏距离、曼哈顿距离和夹角余弦来计算距离，从而来衡量各个对象之间的非相似度。在实际中，使用哪一种衡量方法需要具体情况具体分析。对于关系型数据，常使用欧氏距离；对于文本分类来说，使用夹角余弦（Cosine）来计算相似度就比欧式（Euclidean）距离更合适。

欧几里得度量（Euclidean Metric，也称欧氏距离）是一个通常采用的距离定义，指在 m 维空间中两个点之间的真实距离，或者向量的自然长度（该点到原点的距离）。在二维和三维空间中的欧氏距离就是两点之间的实际距离。

利用二维欧式距离的公式计算 $A(x_1,y_1)$ 和 $B(x_2,y_2)$ 两点之间的距离：

$$\rho = \text{sqrt}(\ (x_1-x_2)^2+(y_1-y_2)^2\ ) \qquad (3.86)$$

利用三维欧式距离的公式计算 $A(x_1,y_1,z_1)$ 和 $B(x_2,y_2,z_2)$ 两点之间的距离：

$$\rho = \text{sqrt}(\ (x_1-x_2)^2+(y_1-y_2)^2+(z_1-z_2)^2\ ) \qquad (3.87)$$

（2）k 值的选取

在 KNN 算法中，k 的选取非常重要，KNN 分类的准备率对 k 值很敏感。不同的值有可能带来不同的结果。如果 k 选大了，求出来的 k 最近邻集合可能包含了太多隶属于其他类别的样本点，不具有代表性，最极端的就是 k 取训练集的大小，此时无论输入实例是什么都只是简单地预测它属于在训练实例中最多的类，模型过于简单，忽略了训练实例中大量有用信息。如果 k 选小了，结果对噪声样本点很敏感。在实际中，一般采用交叉验证（一部分样本做训练集，一部分做测试集）或者依靠经验的方法来选取 k 值。k 值初始时取一个比较小的数值，之后不断来调整 k 值的大小来使样本分类最优，最优时的 k 值即为所选值。k 值一般为奇数。有一个经验规则：k 一般低于训练样本数的平方根。

（3）类别的判定

投票决定：少数服从多数，在 k 个近邻中哪个类别的点最多就分为哪类。

加权投票法：根据距离的远近，对近邻的投票进行加权，距离越近，权重越大（权重为距离平方的倒数）。这是考虑到各个对象的相似度有悬殊，不同距离的样本有可能对未知样本产生的影响不同。

（4）k 个邻近样本的选取

在 KKN 算法中，整个样本集中的每一个样本都要与待测样本进行距离的计算，然后在其中取 k 个最近邻。这带来了巨大的距离计算量，也就是"懒惰"算法所带来的计算成本。改进方案有两个：一个是对样本集进行组织与整理，分群分层，尽可能将计算压缩到在接近测试样本邻域的小范围内，避免盲目地与训练样本集中每个样本进行距离计算；另一个就是在原有样本集中挑选出对分类计算有效的样本，使样本总数合理减少，以同时达到既减少计算量又减少存储量的双重效果。KD 树方法采用的是第一个思路，压缩近邻算法采用的是第二个思路。

3. KNN 算法的优缺点

（1）优点

① 简单，易于理解，易于实现，无须估计参数，无须训练。

② 适合对稀有事件进行分类（例如，当流失率很低时（比如低于 0.5%），构造流失预测模型）。

③ 特别适合于多分类问题（multi-modal，对象具有多个类别标注），例如根据基因特征来判断其功能分类，KNN 比 SVM 的表现要好。

（2）缺点

① 懒惰算法，对测试样本分类时的计算量大，内存开销大。

② 可解释性较差，无法给出决策树那样的规则。

③ 该算法在分类时有一个主要的不足，当样本不平衡时，如一个类的样本容量很大而其他类样本容量很小时，有可能导致当输入一个新样本时，该样本的 k 个邻居中大容量类的样本占多数。该算法只计算"最近的"邻居样本，当某一类的样本数量很大时，要么这类样本并不接近目标样本，要么这类样本很靠近目标样本。无论怎样，数量并不能影响运行结果。

4. 改进算法

针对以上算法的不足，算法的改进方向主要分成分类效率和分类效果两方面。

① 分类效率：事先对样本属性进行约简，删除对分类结果影响较小的属性，快速得出待分类样本的类别。该算法比较适用于样本容量比较大的类域的自动分类，而那些样本容量较小的类域采用这种算法比较容易产生误分。

② 分类效果：采用权值的方法（和该样本距离小的邻居权值大）来改进。Han 等人于 2002 年尝试用贪心法针对文件分类做可调整权重的 k 最近邻居法 WAkNN（Weighted Adjusted k Nearest Neighbor），以促进分类效果。Li 等人于 2004 年提出由于不同分类的文件有数量上的差异，因此应该依照训练集合中各种分类的文件数量选取不同数目的最近邻居来参与分类。

3.11 本章小结

从广义上来说，机器学习是一种能够赋予机器学习的能力，以此让它完成直接编程无法完成的功能的方法。从实践的意义上来说，机器学习是一种通过利用数据、训练模型然后使用模型预测的一种方法。

机器学习无疑是当前数据分析领域的一个热点内容。很多人在平时的工作中都或多或少会用到机器学习的算法。从范围上来说，机器学习跟模式识别、统计学习、数据挖掘是类似的，同时，机器学习与其他领域的处理技术结合，形成了计算机视觉、语音识别、自然语言处理等交叉学科。因此，一般说数据挖掘时可以等同于说机器学习。同时，我们平常所说的机器学习应用应该是通用的，不仅仅局限在结构化数据，还有图像、音频等应用。

机器学习的算法很多。很多时候困惑人们的是，很多算法是一类算法，而有些算法又是从其他算法中延伸出来的。

第 4 章 Python 机器学习项目

迈入机器学习与人工智能领域绝非易事。考虑到目前市面上存在着大量可用资源,众多怀有这一抱负的专业人士及爱好者往往发现自己很难建立正确的发展路径。这一领域正不断演变,我们必须紧随时代的步伐。为了应对演进与创新带来的压倒性速度,保持机器学习认知与知识积累的最好方法无疑在于同技术社区开展合作,进而为众多顶尖专家所使用的开源项目及工具提供贡献。

本章用 Python 更新了顶级的 AI 和机器学习项目。TensorFlow 已经成为贡献者 3 位数增长的第一位。Scikit-learn 下降到第二位,但仍有非常大的贡献者群体。

与 2016 年相比,贡献者人数增长快的项目是:

① TensorFlow,增长 169%,从 493 个增加到 1324 个贡献者。
② Deap,增长 86%,从 21 个增加到 39 个贡献者。
③ Chainer,增长 83%,从 84 个增加到 154 个贡献者。
④ Gensim,增长 81%,从 145 个增加到 262 个贡献者。
⑤ Neon,增长 66%,从 47 个增加到 78 个贡献者。
⑥ Nilearn,增长 50%,从 46 个增加到 69 个贡献者。

如图 4.1 所示,2018 年末的更新:大小和贡献者数量成正比,颜色代表贡献者数量的变化,其中红色代表变化越大,蓝色代表变化越小,雪花形状的项目表示可用于深度学习,也可以用于其他机器学习。可以看到像 TensorFlow、Theano 和 Caffe 这样的深度学习项目是最受欢迎的项目之一。

本章主要简要介绍 Python 排名靠前开源项目的功能及应用,并以此为起点开始协作及学习利用 Python 实现机器学习的具体方式。

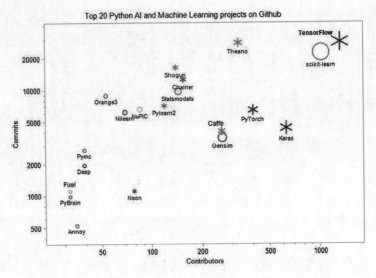

图 4.1　GitHub 上 Top20 AI 和机器学习 Python 开源项目

4.1　SKlearn

SKlearn 是 Scikit-learn 的简称，是一个 Python 库，是专门用于机器学习的模块。文档等资源都可以在官方网站（http://scikit-learn.org/stable/#）里面找到。

4.1.1　SKlearn 包含的机器学习方式

SKlearn 包含的机器学习方式有分类、回归、无监督、数据降维、数据预处理等，包含常见的大部分机器学习方法。关于 SKlearn 的安装，网上教程很多，在此不赘述。建议读者使用 Anaconda，它可用于方便地安装各种库。图 4.2 SKlearn 给出了如何选择正确的方法。

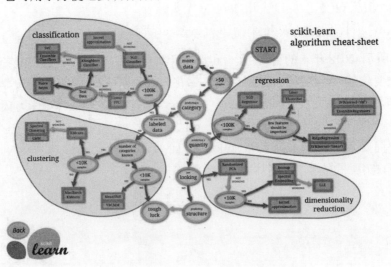

图 4.2　SKlearn 给出如何选择正确的方法

官方网站上有更清晰的图，网址为 http://scikit-learn.org/stable/tutorial/machine_learning_map/index.html。图表对于什么样的问题采用什么样的方法给出了清晰的描述，包括数据量不同的区分。

4.1.2　SKlearn 的强大数据库

数据库网址为 http://scikit-learn.org/stable/modules/classes.html#module-sklearn.datasets，里面包含了很多数据（见图 4.3），可以直接拿来使用。

图 4.3　sklearn.datasets: Datasets

在 SKlearn 官方网站，对于每一个数据集，在后面都给出了使用示例，比如 Boston 房价数据集示例，如图 4.4 所示。

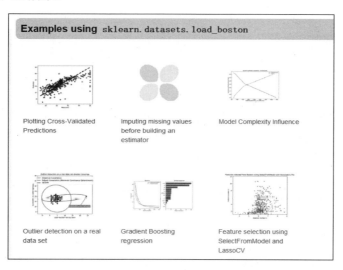

图 4.4　Boston 房价数据集的示例

4.1.3 鸢尾花数据集举例

【例 4.1】 鸢尾花数据集分类（见图 4.5）

使用一个 3 类数据集，绘制不同分类器的分类概率，并使用支持向量分类器对其进行分类，L1 和 L2 使用 One-Vs-Rest 或多项设置对逻辑回归进行惩罚，并使用高斯过程分类。

在默认情况下，Linear SVC 不是一个概率分类器，但是在这个例子中，它有一个内置的校准选项（probability=True）。带有 One-Vs-Rest 的逻辑回归不是开箱即用的多类分类器。因此，它在分离第 2 类和第 3 类时比其他估计值有更多的困难。

```python
#!/usr/bin/env python
# -*- coding: utf-8 -*-
print(__doc__)

# Author: Alexandre Gramfort <alexandre.gramfort@inria.fr>
# License: BSD 3 clause

import matplotlib.pyplot as plt
import numpy as np

from sklearn.metrics import accuracy_score
from sklearn.linear_model import LogisticRegression
from sklearn.svm import SVC
from sklearn.gaussian_process import GaussianProcessClassifier
from sklearn.gaussian_process.kernels import RBF
from sklearn import datasets

iris = datasets.load_iris()
X = iris.data[:, 0:2]  # we only take the first two features for visualization
y = iris.target

n_features = X.shape[1]

C = 10
kernel = 1.0 * RBF([1.0, 1.0])  # for GPC

# Create different classifiers.
classifiers = {
    'L1 logistic': LogisticRegression(C=C, penalty='l1',
                                      solver='saga',
                                      multi_class='multinomial',
                                      max_iter=10000),
    'L2 logistic (Multinomial)': LogisticRegression(C=C, penalty='l2',
                                      solver='saga',
```

```python
                                    multi_class='multinomial',
                                    max_iter=10000),
    'L2 logistic (OvR)': LogisticRegression(C=C, penalty='l2',
                                    solver='saga',
                                    multi_class='ovr',
                                    max_iter=10000),
    'Linear SVC': SVC(kernel='linear', C=C, probability=True,
                    random_state=0),
    'GPC': GaussianProcessClassifier(kernel)
}

n_classifiers = len(classifiers)

plt.figure(figsize=(3 * 2, n_classifiers * 2))
plt.subplots_adjust(bottom=.2, top=.95)

xx = np.linspace(3, 9, 100)
yy = np.linspace(1, 5, 100).T
xx, yy = np.meshgrid(xx, yy)
Xfull = np.c_[xx.ravel(), yy.ravel()]

for index, (name, classifier) in enumerate(classifiers.items()):
    classifier.fit(X, y)

    y_pred = classifier.predict(X)
    accuracy = accuracy_score(y, y_pred)
    print("Accuracy (train) for %s: %0.1f%% " % (name, accuracy * 100))

    # View probabilities:
    probas = classifier.predict_proba(Xfull)
    n_classes = np.unique(y_pred).size
    for k in range(n_classes):
        plt.subplot(n_classifiers, n_classes, index * n_classes + k + 1)
        plt.title("Class %d" % k)
        if k == 0:
            plt.ylabel(name)
        imshow_handle = plt.imshow(probas[:, k].reshape((100, 100)),
                        extent=(3, 9, 1, 5), origin='lower')
        plt.xticks(())
        plt.yticks(())
        idx = (y_pred == k)
        if idx.any():
            plt.scatter(X[idx, 0], X[idx, 1], marker='o', c='w', edgecolor='k')
```

```
ax = plt.axes([0.15, 0.04, 0.7, 0.05])
plt.title("Probability")
plt.colorbar(imshow_handle, cax=ax, orientation='horizontal')

plt.show()
```

输出:

```
Automatically created module for IPython interactive environment
Accuracy (train) for L1 logistic: 82.7%
Accuracy (train) for L2 logistic (Multinomial): 82.7%
Accuracy (train) for L2 logistic (OvR): 80.0%
Accuracy (train) for Linear SVC: 82.0%
Accuracy (train) for GPC: 82.7%
```

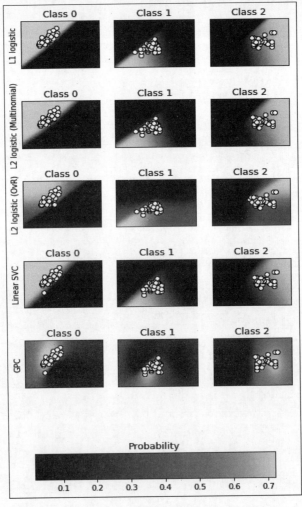

图 4.5　鸢尾花数据集分类

4.1.4　Boston 房价数据集的示例

【例 4.2】 Boston 房价数据集分类（见图 4.6）

该例展示最小协方差行列式的协方差鲁棒估计集中于数据分布的主要模式的能力：位置估计似乎很好，虽然协方差很难估计，因为香蕉形状的分布。不管怎样，可以去掉一些外围的观测值。One-Class SVM 能够捕获真实的数据结构，但其难点在于调整内核带宽参数，以在数据散点矩阵的形状和数据过拟合的风险之间取得良好的折中。在这个例子中（概率=True），带有 One-Vs-Rest7684 逻辑回归不是开箱即用的多类分类器。因此，它在分离第 2 类和第 3 类时比其他估计值有更多的困难。

```python
#!/usr/bin/env python
# -*- coding: utf-8 -*-
print(__doc__)

# Author: Virgile Fritsch <virgile.fritsch@inria.fr>
# License: BSD 3 clause

import numpy as np
from sklearn.covariance import EllipticEnvelope
from sklearn.svm import OneClassSVM
import matplotlib.pyplot as plt
import matplotlib.font_manager
from sklearn.datasets import load_boston

# Get data
X1 = load_boston()['data'][:, [8, 10]]  # two clusters
X2 = load_boston()['data'][:, [5, 12]]  # "banana"-shaped

# Define "classifiers" to be used
classifiers = {
    "Empirical Covariance": EllipticEnvelope(support_fraction=1.,
                                    contamination=0.261),
    "Robust Covariance (Minimum Covariance Determinant)":
    EllipticEnvelope(contamination=0.261),
    "OCSVM": OneClassSVM(nu=0.261, gamma=0.05)}
colors = ['m', 'g', 'b']
legend1 = {}
legend2 = {}

# Learn a frontier for outlier detection with several classifiers
xx1, yy1 = np.meshgrid(np.linspace(-8, 28, 500), np.linspace(3, 40, 500))
xx2, yy2 = np.meshgrid(np.linspace(3, 10, 500), np.linspace(-5, 45, 500))
for i, (clf_name, clf) in enumerate(classifiers.items()):
```

```python
    plt.figure(1)
    clf.fit(X1)
    Z1 = clf.decision_function(np.c_[xx1.ravel(), yy1.ravel()])
    Z1 = Z1.reshape(xx1.shape)
    legend1[clf_name] = plt.contour(
        xx1, yy1, Z1, levels=[0], linewidths=2, colors=colors[i])
    plt.figure(2)
    clf.fit(X2)
    Z2 = clf.decision_function(np.c_[xx2.ravel(), yy2.ravel()])
    Z2 = Z2.reshape(xx2.shape)
    legend2[clf_name] = plt.contour(
        xx2, yy2, Z2, levels=[0], linewidths=2, colors=colors[i])

legend1_values_list = list(legend1.values())
legend1_keys_list = list(legend1.keys())

# Plot the results (= shape of the data points cloud)
plt.figure(1)  # two clusters
plt.title("Outlier detection on a real data set (boston housing)")
plt.scatter(X1[:, 0], X1[:, 1], color='black')
bbox_args = dict(boxstyle="round", fc="0.8")
arrow_args = dict(arrowstyle="->")
plt.annotate("several confounded points", xy=(24, 19),
        xycoords="data", textcoords="data",
        xytext=(13, 10), bbox=bbox_args, arrowprops=arrow_args)
plt.xlim((xx1.min(), xx1.max()))
plt.ylim((yy1.min(), yy1.max()))
plt.legend((legend1_values_list[0].collections[0],
        legend1_values_list[1].collections[0],
        legend1_values_list[2].collections[0]),
        (legend1_keys_list[0], legend1_keys_list[1], legend1_keys_list[2]),
        loc="upper center",
        prop=matplotlib.font_manager.FontProperties(size=12))
plt.ylabel("accessibility to radial highways")
plt.xlabel("pupil-teacher ratio by town")

legend2_values_list = list(legend2.values())
legend2_keys_list = list(legend2.keys())

plt.figure(2)  # "banana" shape
plt.title("Outlier detection on a real data set (boston housing)")
plt.scatter(X2[:, 0], X2[:, 1], color='black')
plt.xlim((xx2.min(), xx2.max()))
```

```
plt.ylim((yy2.min(), yy2.max()))
plt.legend((legend2_values_list[0].collections[0],
            legend2_values_list[1].collections[0],
            legend2_values_list[2].collections[0]),
           (legend2_keys_list[0], legend2_keys_list[1], legend2_keys_list[2]),
            loc="upper center",
            prop=matplotlib.font_manager.FontProperties(size=12))
plt.ylabel("% lower status of the population")
plt.xlabel("average number of rooms per dwelling")

plt.show()
```

输出：

```
Automatically created module for IPython interactive environment
```

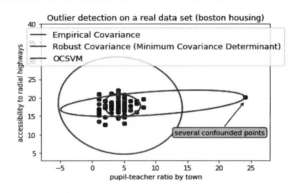

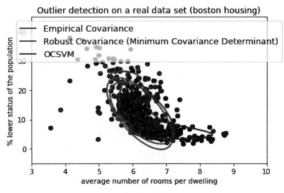

图 4.6　Boston 房价数据集分类

4.2　TensorFlow

TensorFlow 最初是由谷歌机器智能研究机构 Google Brain Team 的研究人员和工程师开发的。该系统的设计是为了便于机器学习的研究，并使其快速、容易地从研究原型过渡到生产系统。GitHub 网址为 https://github.com/tensorflow/tensorflow。

4.2.1 TensorFlow 简介

1. 关于 TensorFlow

TensorFlow™ 是一个采用数据流图（Data Flow Graphs），用于数值计算的开源软件库。图中的节点（Node）表示数学操作，线（Edge）表示在节点间相互联系的多维数据数组，即张量（Tensor）。它灵活的架构使之可以在多种平台上展开计算，例如台式计算机中的一个或多个 CPU（或 GPU）、服务器、移动设备等。TensorFlow 最初由 Google 大脑小组（隶属于 Google 机器智能研究机构）的研究员和工程师们开发，用于机器学习和深度神经网络方面的研究。这个系统的通用性很好，也可广泛用于其他计算领域。

2. 数据流图

数据流图用"节点"和"线"的有向图来描述数学计算。"节点"一般用来表示施加的数学操作，但也可以表示数据输入（Feed In）的起点/输出（Push Out）的终点，或者是读取/写入持久变量（Persistent Variable）的终点。"线"表示"节点"之间的输入/输出关系。这些数据"线"可以运输"大小可动态调整"的多维数据数组，即"张量"（Tensor）。张量从图中流过的直观图像是这个工具取名为"TensorFlow"的原因。一旦输入端的所有张量准备好，节点将被分配到各种计算设备来异步并行地执行运算。图 4.7 为 TensorFlow 数据流图。

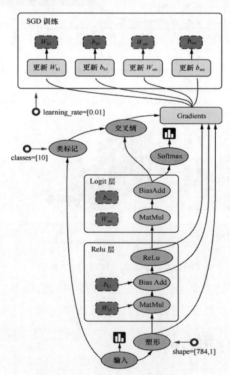

图 4.7　TensorFlow 数据流图

3. TensorFlow 的特征

（1）高度的灵活性

TensorFlow 不是一个严格的"神经网络"库。只要我们可以将计算表示为一个数据流图，就可以使用 TensorFlow。我们来构建图和描写驱动计算的内部循环。TensorFlow 提供了有用的工具来帮助我们组装"子图"（常用于神经网络），当然也可以自己在 TensorFlow 基础上编写"上层库"。定义顺手好用的新复合操作和编写一个 Python 函数一样容易，而且也不用担心性能损耗。当然，万一我们发现找不到想要的底层数据操作，也可以自己编写一点 C++ 代码来丰富底层的操作。

（2）真正的可移植性（Portability）

TensorFlow 在 CPU 和 GPU 上运行，比如说可以运行在台式机、服务器、手机移动设备等。想要在没有特殊硬件的前提下在自己的笔记本上测试并运行一下机器学习的新想法？采用 TensorFlow 就可办到。准备将自己的训练模型在多个 CPU 上规模化运算，又不想修改代码？通过 TensorFlow 也可办到。想要将训练好的模型作为产品的一部分用到手机 App 中？通过 TensorFlow 同样可以实现。改变主意，想要将模型作为云端服务运行在自己的服务器上，或者运行在 Docker 容器里？TensorFlow 不会让我们失望。

（3）将科研和产品联系在一起

过去如果要将科研中的机器学习想法用到产品中就需要大量的代码重写工作。那样的日子一去不复返了！在 Google，科学家用 TensorFlow 尝试新的算法，产品团队则用 TensorFlow 来训练和使用计算模型，并直接提供给在线用户。使用 TensorFlow 可以让应用型研究者将想法迅速运用到产品中，也可以让学术性研究者更直接地彼此分享代码，从而提高科研产出率。

（4）自动求微分

基于梯度的机器学习算法会受益于 TensorFlow 自动求微分的能力。作为 TensorFlow 用户，只需要定义预测模型的结构，将这个结构和目标函数（Objective Function）结合在一起，并添加数据，TensorFlow 将自动计算相关的微分导数。计算某个变量相对于其他变量的导数仅仅是通过扩展图来完成的，所以用户能一直清楚地看到究竟在发生什么。

（5）多语言支持

TensorFlow 有一个合理的 C++ 使用界面，也有一个易用的 Python 使用界面来构建和执行图。既可以直接写 Python/C++ 程序，也可以利用交互式的 IPython 界面来尝试一些想法，它可以帮助我们将笔记、代码、可视化等有条理地归置好。

（6）性能最优化

由于 TensorFlow 给予了线程、队列、异步操作等以最佳的支持，因此 TensorFlow 让我们可以将手边硬件的计算潜能全部发挥出来。我们可以自由地将 TensorFlow 图中的计算元素分配到不同设备上，TensorFlow 可以帮我们管理好这些不同副本。

任何人都可以用 TensorFlow。学生、研究员、爱好者、极客、工程师、开发者、发明家、创业者等都可以在 Apache 2.0 开源协议下使用 TensorFlow。

4.2.2　TensorFlow 的下载与安装

可使用二进制包或者源代码安装 TensorFlow。TensorFlow Python API 依赖 Python 2.7 版本。在 Linux 和 Mac 下最简单的安装方式是使用 pip 安装。为了简化安装步骤，建议使用 virtualenv。

1. 二进制安装

（1）Ubuntu/Linux

```
# 仅使用 CPU 的版本
$ pip install https://storage.googleapis.com/tensorflow/linux/cpu/tensorflow-0.5.0-cp27-none-linux_x86_64.whl
# 开启 GPU 支持的版本 (安装该版本的前提是已经安装了 CUDA sdk)
$ pip install https://storage.googleapis.com/tensorflow/linux/gpu/tensorflow-0.5.0-cp27-none-linux_x86_64.whl
```

（2）Mac OS X

在 OS X 系统上，建议先安装 homebrew，然后执行 brew install python，以便能够使用 homebrew 中的 Python 安装 TensorFlow。另外，还推荐在 virtualenv 中安装 TensorFlow。

```
# 当前版本只支持 CPU
$ pip install https://storage.googleapis.com/tensorflow/mac/tensorflow-0.5.0-py2-none-any.whl
```

2. 基于 Docker 的安装

也可以通过 Docker 运行 TensorFlow。该方式的优点是不用操心软件依赖问题。首先，安装 Docker。一旦 Docker 启动运行，就可以通过命令启动一个容器：

```
$ docker run -it b.gcr.io/tensorflow/tensorflow
```

该命令将启动一个已经安装好 TensorFlow 及相关依赖的容器。

默认的 Docker 镜像只包含启动和运行 TensorFlow 所需依赖库的一个最小集。官方网站额外提供了下面的容器，同样可以通过上述 docker run 命令来安装：

```
b.gcr.io/tensorflow/tensorflow-full
```

镜像中的 TensorFlow 是从源代码完整安装的，包含编译和运行 TensorFlow 所需的全部工具。在该镜像上，可以直接使用源代码进行实验，而不需要再安装上述的任何依赖。

3. 基于 virtualenv 的安装

推荐使用 virtualenv 创建一个隔离的容器来安装 TensorFlow。这是可选的，但是这样做能使排查安装问题变得更容易。

首先，安装所有必备工具：

```
# 在 Linux 上：
$ sudo apt-get install python-pip python-dev python-virtualenv

# 在 Mac 上：
$ sudo easy_install pip    # 如果还没有安装 pip
$ sudo pip install --upgrade virtualenv
```

接下来，建立一个全新的 virtualenv 环境。为了将环境建在 ~/tensorflow 目录下，执行：

```
$ virtualenv --system-site-packages ~/tensorflow
$ cd ~/tensorflow
```

然后，激活 virtualenv：

```
$ source bin/activate      # 如果使用 bash
$ source bin/activate.csh  # 如果使用 csh
(tensorflow)$  # 终端提示符应该发生变化
```

在 virtualenv 内，安装 TensorFlow：

```
(tensorflow)$ pip install --upgrade <$url_to_binary.whl>
```

接下来，使用类似命令运行 TensorFlow 程序：

```
(tensorflow)$ cd tensorflow/models/image/mnist
(tensorflow)$ python convolutional.py
# 当使用完 TensorFlow
(tensorflow)$ deactivate  # 停用 virtualenv
$  # 命令提示符会恢复原样
```

4. 尝试第一个 TensorFlow 程序

（1）启用 GPU 支持（可选）

如果使用 pip 二进制包安装了开启 GPU 支持的 TensorFlow，就必须确保系统里安装了正确的 CUDA SDK 和 CUDNN 版本。你还需要设置 LD_LIBRARY_PATH 和 CUDA_HOME 环境变量。可以考虑将下面的命令添加到 ~/.bash_profile 文件中，这样每次登录后会自动生效。注意，下面的命令假定 CUDA 安装目录为 /usr/local/cuda：

```
export LD_LIBRARY_PATH="$LD_LIBRARY_PATH:/usr/local/cuda/lib64"
export CUDA_HOME=/usr/local/cuda
```

（2）运行 TensorFlow

```
#打开一个 Python 终端
$ python

>>> import tensorflow as tf
```

```
>>> hello = tf.constant('Hello, TensorFlow!')
>>> sess = tf.Session()
>>> print sess.run(hello)
Hello, TensorFlow!
>>> a = tf.constant(10)
>>> b = tf.constant(32)
>>> print sess.run(a+b)
42
```

5. 从源码安装

克隆 TensorFlow 仓库：

```
$ git clone --recurse-submodules https://github.com/tensorflow/tensorflow
```

--recurse-submodules 参数是必须要有的，用于获取 TensorFlow 依赖的 ProtoBuf 库。

（1）Linux 安装

首先安装 Bazel 的依赖，然后使用下列命令下载和编译 Bazel 的源码：

```
$ git clone https://github.com/bazelbuild/bazel.git
$ cd bazel
$ git checkout tags/0.1.0
$ ./compile.sh
```

上面命令中的代码标签为 0.1.0，兼容 TensorFlow 1.10.0。将执行路径 output/bazel 添加到 $PATH 环境变量中。

安装其他依赖：

```
$ sudo apt-get install python-numpy swig python-dev
```

（2）可选：安装 CUDA（在 Linux 上开启 GPU 支持）

为了编译并运行能够使用 GPU 的 TensorFlow，需要先安装 NVIDIA 提供的 CUDA Toolkit 7.0 和 CUDNN 6.5 V2。

TensorFlow 的 GPU 特性只支持 NVidia Compute Capability >= 3.5 的显卡。被支持的显卡包括但不限于：

- NVidia Titan
- NVidia Titan X
- NVidia K20
- NVidia K40

① 下载并安装 CUDA Toolkit 7.0。将工具安装到诸如 /usr/local/cuda 之类的路径。

② 下载并安装 CUDNN Toolkit 6.5。解压并复制 CUDNN 文件到 CUDA Toolkit 7.0 安装路径下。假设 CUDA Toolkit 7.0 安装在/usr/local/cuda，可执行以下命令：

```
tar xvzf cudnn-6.5-linux-x64-v2.tgz
```

```
sudo cp cudnn-6.5-linux-x64-v2/cudnn.h /usr/local/cuda/include
sudo cp cudnn-6.5-linux-x64-v2/libcudnn* /usr/local/cuda/lib64
```

③ 配置 TensorFlow 的 CUDA 选项。

从源码树的根路径执行:

```
$ ./configure
Do you wish to build TensorFlow with GPU support? [y/n] y
GPU support will be enabled for TensorFlow
Please specify the location where CUDA 7.0 toolkit is installed. Refer to
README.md for more details. [default is: /usr/local/cuda]: /usr/local/cuda
Please specify the location where CUDNN 6.5 V2 library is installed. Refer to
README.md for more details. [default is: /usr/local/cuda]: /usr/local/cuda
Setting up Cuda include
Setting up Cuda lib64
Setting up Cuda bin
Setting up Cuda nvvm
Configuration finished
```

这些配置将建立到系统 CUDA 库的符号链接。每当 CUDA 库的路径发生变更时，必须重新执行上述步骤，否则无法调用 Bazel 编译命令。

④ 编译目标程序。

开启 GPU 支持，从源码树的根路径执行:

```
$ bazel build -c opt --config=cuda //tensorflow/cc:tutorials_example_trainer
$ bazel-bin/tensorflow/cc/tutorials_example_trainer --use_gpu
# 大量输出信息，这个例子用 GPU 迭代计算一个 2×2 矩阵的主特征值 (major eigenvalue)
# 最后几行输出和下面的信息类似
000009/000005 lambda = 2.000000 x = [0.894427 -0.447214] y = [1.788854 -0.894427]
000006/000001 lambda = 2.000000 x = [0.894427 -0.447214] y = [1.788854 -0.894427]
000009/000009 lambda = 2.000000 x = [0.894427 -0.447214] y = [1.788854 -0.894427]
```

注意，GPU 支持需通过编译选项 "--config=cuda" 开启。尽管可以在同一个源码树下编译开启 CUDA 支持和禁用 CUDA 支持的版本，还是推荐在切换这两种不同的编译配置时使用 "bazel clean" 清理环境。在执行 Bazel 编译前必须先运行 configure，否则编译会失败并提示错误信息。

4.2.3 TensorFlow 的基本使用

使用 TensorFlow，必须了解 TensorFlow 如下特性:

- 使用图 (Graph) 来表示计算任务。
- 在被称之为会话 (Session) 的上下文 (Context) 中执行图。
- 使用 Tensor 表示数据。

- 通过变量（Variable）维护状态。
- 使用 feed 和 fetch 可以为任意的操作(Arbitrary Operation)赋值或者从其中获取数据。

TensorFlow 是一个编程系统，使用图来表示计算任务。图中的节点被称之为 op（operation 的缩写）。一个 op 获得 0 个或多个 Tensor，执行计算，产生 0 个或多个 Tensor。每个 Tensor 是一个类型化的多维数组。例如，可以将一小组图像集表示为一个四维浮点数数组，这 4 个维度分别是[batch, height, width, channels]。

一个 TensorFlow 图描述了计算的过程。为了进行计算，图必须在会话里被启动。会话将图的 op 分发到诸如 CPU 或 GPU 之类的设备上，同时提供执行 op 的方法。这些方法执行后，将产生的 Tensor 返回。在 Python 语言中，返回的 Tensor 是 numpy ndarray 对象；在 C 和 C++ 语言中，返回的 Tensor 是 tensorflow::Tensor 实例。

1. 计算图

TensorFlow 程序通常被组织成一个构建阶段和一个执行阶段。在构建阶段，op 的执行步骤被描述成一个图。在执行阶段，使用会话执行图中的 op。例如，通常在构建阶段创建一个图来表示和训练神经网络，然后在执行阶段反复执行图中的训练 op。TensorFlow 支持 C、C++、Python 编程语言。目前，TensorFlow 的 Python 库更加易用，提供了大量的辅助函数来简化构建图的工作，这些函数尚未被 C 和 C++ 库支持。3 种语言的会话库（Session Libraries）是一致的。

2. 构建图

构建图的第一步是创建源 op（Source op）。源 op 不需要任何输入，例如常量（Constant）。源 op 的输出被传递给其他 op 做运算。在 Python 库中，op 构造器的返回值代表被构造出的 op 的输出，这些返回值可以传递给其他 op 构造器作为输入。TensorFlow Python 库有一个默认图（Default Graph），op 构造器可以为其增加节点。这个默认图对许多程序来说已经足够用了。

```
import tensorflow as tf
# 创建一个常量 op，产生一个 1×2 矩阵，这个 op 被作为一个节点
# 加到默认图中
#
# 构造器的返回值代表该常量 op 的返回值
matrix1 = tf.constant([[3., 3.]])
# 创建另外一个常量 op，产生一个 2×1 矩阵
matrix2 = tf.constant([[2.],[2.]])
# 创建一个矩阵乘法 matmul op，把 'matrix1' 和 'matrix2' 作为输入
# 返回值 'product' 代表矩阵乘法的结果
product = tf.matmul(matrix1, matrix2)
```

默认图现在有 3 个节点：两个 constant() op 和一个 matmul() op。为了真正进行矩阵相乘运算，并得到矩阵乘法的结果，我们必须在会话里启动这个图。

3. 在一个会话中启动图

构造阶段完成后，才能启动图。启动图的第一步是创建一个 Session 对象，如果无任何创建参数，那么会话构造器将启动默认图。

```
# 启动默认图
sess = tf.Session()
# 调用 sess 的 'run()' 方法来执行矩阵乘法 op, 传入 'product' 作为该方法的参数
# 上面提到, 'product' 代表了矩阵乘法 op 的输出, 传入它是向方法表明, 我们希望取回
# 矩阵乘法 op 的输出
#
# 整个执行过程是自动化的, 会话负责传递 op 所需的全部输入, op 通常是并发执行的
#
# 函数调用 'run(product)' 触发了图中 3 个 op (两个常量 op 和一个矩阵乘法 op) 的执行
#
# 返回值 'result' 是一个 numpy 'ndarray' 对象
result = sess.run(product)
print result
# ==> [[ 12.]]
# 任务完成, 关闭会话
sess.close()
```

Session 对象在使用完后需要关闭以释放资源。除了显式调用 close 外,也可以使用"with"代码块来自动完成关闭动作。

```
with tf.Session() as sess:
    result = sess.run([product])
    print result
```

在具体实现上，TensorFlow 将图形定义转换成分布式执行的操作，以充分利用可用的计算资源（如 CPU 或 GPU）。一般不需要显式指定使用 CPU 还是 GPU，TensorFlow 能自动检测。如果检测到 GPU，TensorFlow 会尽可能地利用找到的第一个 GPU 来执行操作。

如果机器上有超过一个可用的 GPU，那么除第一个外的其他 GPU 默认是不参与计算的。为了让 TensorFlow 使用这些 GPU，你必须将 op 明确指派给它们执行。with...Device 语句用来指派特定的 CPU 或 GPU 执行操作：

```
with tf.Session() as sess:
    with tf.device("/gpu:1"):
        matrix1 = tf.constant([[3., 3.]])
        matrix2 = tf.constant([[2.],[2.]])
        product = tf.matmul(matrix1, matrix2)
        ...
```

设备用字符串进行标识。目前支持的设备包括：

- "/cpu:0": 机器的 CPU。

- "/gpu:0"：机器的第一个 GPU（如果有）。
- "/gpu:1"：机器的第二个 GPU。

……

4. 交互式使用

文档中的 Python 示例使用一个会话 Session 来启动图，并调用 Session.run() 方法执行操作。为了便于使用诸如 IPython 之类的 Python 交互环境，可以使用 InteractiveSession 代替 Session 类，使用 Tensor.eval() 和 Operation.run() 方法代替 Session.run()。这样可以避免使用一个变量来持有会话。

```
# 进入一个交互式 TensorFlow 会话
import tensorflow as tf
sess = tf.InteractiveSession()
x = tf.Variable([1.0, 2.0])
a = tf.constant([3.0, 3.0])
# 使用初始化器 initializer op 的 run() 方法初始化 'x'
x.initializer.run()
# 增加一个减法 sub op，从 'x' 减去 'a'，运行减法 op，输出结果
sub = tf.sub(x, a)
print (sub.eval())
#==> [-2. -1.]
```

5. tensor

TensorFlow 程序使用 tensor 数据结构来代表所有的数据。在计算图中，操作间传递的数据都是 tensor。我们可以把 TensorFlow tensor 看作是一个 n 维的数组或列表。一个 tensor 包含一个静态类型 rank 和一个 shape。

6. 变量

变量维护图执行过程中的状态信息。下面的例子演示如何使用变量实现一个简单的计数器。

【例 4.3】

```
# 创建一个变量，初始化为标量 0
state = tf.Variable(0, name="counter")
# 创建一个 op，作用是使 state 增加 1
one = tf.constant(1)
new_value = tf.add(state, one)
update = tf.assign(state, new_value)
# 启动图后，变量必须先经过初始化 (init) op 初始化，
# 首先必须增加一个初始化 op 到图中
init_op = tf.initialize_all_variables()
# 启动图，运行 op
with tf.Session() as sess:
```

```
# 运行 'init' op
sess.run(init_op)
# 打印 'state' 的初始值
print sess.run(state)
# 运行 op, 更新 'state', 并打印 'state'
for _ in range(3):
  sess.run(update)
  print (sess.run(state))
```

输出:

```
0
1
2
3
```

代码中 assign() 操作是图所描绘的表达式的一部分，正如 add() 操作一样。所以在调用 run() 执行表达式之前，它并不会真正执行赋值操作。通常会将一个统计模型中的参数表示为一组变量。例如，可以将一个神经网络的权重作为某个变量存储在一个 tensor 中。在训练过程中，通过重复运行训练图，更新这个 tensor。

7. fetch

为了取回操作的输出内容，可以在使用 Session 对象的 run() 调用执行图时传入一些 tensor，这些 tensor 会帮助我们取回结果。在之前的例子里，只取回了单个节点 state，但是我们也可以取回多个 tensor:

```
input1 = tf.constant(3.0)
input2 = tf.constant(2.0)
input3 = tf.constant(5.0)
intermed = tf.add(input2, input3)
mul = tf.mul(input1, intermed)
with tf.Session():
  result = sess.run([mul, intermed])
  print result
```

输出:

```
[array([ 21.], dtype=float32), array([ 7.], dtype=float32)]
```

需要获取的多个 tensor 值在 op 的一次运行中一起获得（而不是逐个去获取 tensor）。

8. feed

上述示例在计算图中引入了 tensor，以常量或变量的形式存储。TensorFlow 还提供了 feed 机制，该机制可以临时替代图中任意操作中的 tensor，对图中任何操作提交补丁，直接插入一个 tensor。

feed 使用一个 tensor 值临时替换一个操作的输出结果。可以提供 feed 数据作为 run() 调用的参数。feed 只在调用它的方法内有效，方法结束，feed 就会消失。最常见的用例是将某些特殊的操作指定为 "feed" 操作，标注的方法是使用 tf.placeholder() 为这些操作创建占位符。

```
input1 = tf.placeholder(tf.types.float32)
input2 = tf.placeholder(tf.types.float32)
output = tf.mul(input1, input2)
with tf.Session() as sess:
  print (sess.run([output], feed_dict={input1:[7.], input2:[2.]}))
```

输出：

```
[array([ 14.], dtype=float32)]
```

如果没有正确提供 feed，placeholder() 操作将会产生错误。

4.3 Theano

Theano 是一个 Python 库，专门用于定义、优化、求值数学表达式，它的效率高，适用于多维数组，特别适合做机器学习。一般来说，使用时需要安装 Python 和 NumPy。

1. 搭建环境

安装：

```
pip install theano
```

使用 Theano 来搭建机器学习（深度学习）框架，有以下优点：

- Theano 能够自动计算梯度。
- 只需要两步就能搭建框架、定义函数和计算梯度。

2. 定义函数举例

- 步骤 0　声明使用 Theano：import theano。
- 步骤 1　定义输入：x=theano.tensor.scalar()。
- 步骤 2　定义输出：y=2*x。
- 步骤 3　定义函数：f = theano.function([x],y)。
- 步骤 4　调用函数：print f(-2)。

步骤 1　定义输入变量：

```
a = theano.tensor.scalar()
b =theano.tensor.matrix()
```

简化:

```
import theano.tensor as T
```

步骤 2　定义输出变量和输入变量的关系:

```
    x1=T.matrix()
    x2=T.matrix()
    y1=x1*x2
    y2=T.dot(x1,x2) #矩阵乘法
```

步骤 3　声明函数:

```
    f= theano.function([x],y)
```

函数输入必须是 list 带[]。

【例 4.4】

```
#encoding:utf-8
import theano
import theano.tensor as T
a= T.matrix()
b= T.matrix()
c = a*b
d = T.dot(a,b)
F1= theano.function([a,b],c)
F2= theano.function([a,b],d)
A=[[1,2],[3,4]]
B=[[2,4],[6,8]] #2×2 矩阵
C=[[1,2],[3,4],[5,6]]  #3×2 矩阵
print (F1(A,B))
print (F2(C,B))
```

输出:

```
[[ 2.  8.]
 [18. 32.]]
[[14. 20.]
 [30. 44.]
 [46. 68.]]
```

3. 计算梯度

计算 dy/dx,直接调用 g=T.grad(y,x),其中 y 必须是一个标量(Scalar)。

【例 4.5】标量对标量的导数

```
import theano
import theano.tensor as T
```

```
x= T.scalar('x')
y = x**2
g = T.grad(y,x)
f= theano.function([x],y)
f_prime=theano.function([x],g)
print (f(-2))
print (f_prime(-2))
```

输出:

```
4.0
-4.0
```

【例 4.6】标量对向量的导数

```
import theano
import theano.tensor as T
x1= T.scalar()
x2= T.scalar()
y = x1*x2
g = T.grad(y,[x1,x2])
f= theano.function([x1,x2],y)
f_prime=theano.function([x1,x2],g)
print (f(2,4))
print (f_prime(2,4))
```

输出:

```
8.0
[array(4.), array(2.)]
```

【例 4.7】标量对矩阵的导数

```
#encoding:utf-8
import theano
import theano.tensor as T
A= T.matrix()
B= T.matrix()
C=A*B              #不是矩阵乘法, 是对于位置相乘
D=T.sum(C)
g=T.grad(D,A)      #注意 D 是求和, 所以肯定是一个标量, 但 g 是一个矩阵
y_prime=theano.function([A,B],g)
A=[[1,2],[3,4]]
B=[[2,4],[6,8]]
print (y_prime(A,B))
```

输出:

```
[[2. 4.]
 [6. 8.]]
```

4. 搭建神经网络

【例 4.8】单个神经元（示意图见图 4.8）

假设 w、b 已知，y=neuron(x;w,b)。

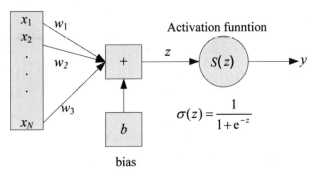

图 4.8　单个神经元示意图

```
import theano
import theano.tensor as T
import random
import numpy as np

x = T.vector()
w = T.vector()
b = T.scalar()

z= T.dot(w,x)+b
y= 1/(1+T.exp(-z))

neuron =theano.function(
    inputs=[x,w,b],
    outputs=[y]
)

w = [-1,1]
b=0
for i in range(100):
    x = [random.random(),random.random()]
    print (x)
    print (neuron(x,w,b))
```

输出：

```
[0.4545852874950511, 0.416681705051839]
```

```
[array(0.49052524)]
[0.46180143032906784, 0.9076234710829244]
[array(0.60964543)]
[0.7419406475925908, 0.5390090280881658]
[array(0.44944048)]
[0.205681374553067, 0.2656315789138177]
[array(0.51498306)]
[0.46846476249157587, 0.19150011356446262]
[array(0.43119809)]
[0.487244567628357, 0.8852345233279647]
[array(0.59820463)]
[0.5435601397655122, 0.4160263528359741]
[array(0.4681597)]
[0.8834284617813434, 0.28844191088092164]
[array(0.35549153)]
[0.1483704019465204, 0.6467017389057291]
[array(0.62206711)]
[0.17262703136935942, 0.7064879457115925]
[array(0.63038316)]
[0.31484521328436, 0.6534365594753911]
[array(0.5838483)]
[0.9134298333395832, 0.2511517245899655]
[array(0.34022805)]
[0.7814480428765854, 0.02220094647031623]
[array(0.31880975)]
[0.6043292454236322, 0.1249563818188254]
[array(0.38240023)]
[0.2515960972404958, 0.11150355935865375]
[array(0.46503403)]
[0.2408718150456035, 0.12609045119634366]
[array(0.47133612)]
[0.6371527519246765, 0.9872516235556419]
[array(0.58664155)]
[0.22949569147558724, 0.726492868621309]
[array(0.6217534)]
[0.08984249940101197, 0.9328984509357924]
[array(0.69910844)]
[0.2697874678484984, 0.36138259972963005]
[array(0.52288279)]
[0.7577699236087145, 0.10409420655869317]
[array(0.3421617)]
[0.1981294010648138, 0.04409242875293928]
[array(0.46156672)]
```

[0.000264166349317807, 0.8773720835360445]
[array(0.70622255)]
[0.884221286966989, 0.8035707101014958]
[array(0.47984828)]
[0.42777732321503137, 0.16561986998279477]
[array(0.43483343)]
[0.7608950920494931, 0.2294710283861886]
[array(0.37018481)]
[0.928443627226761, 0.4888880296306106]
[array(0.39184687)]
[0.7512227548825495, 0.7435654322569621]
[array(0.49808568)]
[0.2328838710182578, 0.870283676687307]
[array(0.65416545)]
[0.28177875846043354, 0.2856562837456367]
[array(0.50096938)]
[0.24458738165843164, 0.8185210141552987]
[array(0.63967034)]
[0.9633035908701494, 0.9940646071029446]
[array(0.50768965)]
[0.48156476580215746, 0.45421870345644855]
[array(0.49316391)]
[0.9031776463740699, 0.9981533297798952]
[array(0.52372609)]
[0.733518197114175, 0.7365208792557031]
[array(0.50075067)]
[0.2303657271011741, 0.3739056453204973]
[array(0.53582349)]
[0.4996222987302158, 0.33014348884550504]
[array(0.45773142)]
[0.34124463485572387, 0.7687497214170605]
[array(0.60527775)]
[0.0456347598116551, 0.49832719573743645]
[array(0.61127919)]
[0.4936802990250455, 0.9447250445751071]
[array(0.6108876)]
[0.5212804576711783, 0.5173532236542339]
[array(0.49901819)]
[0.46319795080823933, 0.785990263406357]
[array(0.579997)]
[0.9806560099241677, 0.532997706294097]
[array(0.38991767)]
[0.29157226655083135, 0.8310452545212633]

```
[array(0.63168981)]
[0.7571438648428029, 0.5251492482944401]
[array(0.44226008)]
[0.20727337978586458, 0.8873912524798562]
[array(0.663765)]
[0.9981856319224749, 0.9743209619364135]
[array(0.49403412)]
[0.4081068807713466, 0.9316831192057239]
[array(0.62798363)]
[0.9725166355364266, 0.2990658067363007]
[array(0.33772457)]
[0.31345922904854284, 0.883849620078016]
[array(0.63885325)]
[0.43460564868957285, 0.5687137789608808]
[array(0.53347687)]
[0.9079158451191726, 0.21874174390519463]
[array(0.33421682)]
[0.4476519355729156, 0.02066843873021673]
[array(0.39484687)]
[0.6646909308318957, 0.9604912358840543]
[array(0.57341555)]
[0.5955277741194636, 0.7940178946176788]
[array(0.54946025)]
[0.5303307857240181, 0.8583548846344214]
[array(0.58127853)]
[0.4587332625356091, 0.23632794111695243]
[array(0.44462673)]
[0.6554421691889762, 0.15108475519129017]
[array(0.37651721)]
[0.6140807159106558, 0.9612157037208469]
[array(0.58592265)]
[0.6238995354574016, 0.04039921228440169]
[array(0.35812757)]
[0.5986819394247734, 0.011019201300952797]
[array(0.35717131)]
[0.8118132336575589, 0.1912318902730611]
[array(0.34964925)]
[0.05876142446997934, 0.4188035424751192]
[array(0.58905063)]
[0.07144803700817415, 0.2244846175605384]
[array(0.53818465)]
[0.6863959888457952, 0.9260964051445529]
[array(0.55963982)]
```

```
[0.7947952416509035, 0.3042910564828961]
[array(0.3797748)]
[0.5805925698214676, 0.788699632610369]
[array(0.55183981)]
[0.06329104095941795, 0.08141611862440457]
[array(0.50453115)]
[0.1312633013134339, 0.5581533698674992]
[array(0.6051308)]
[0.16403799549148823, 0.7651559038823948]
[array(0.64591202)]
[0.1000945151735213, 0.043231398998646475]
[array(0.48578805)]
[0.14265601465730604, 0.28703284719226707]
[array(0.53603164)]
[0.10814917475556873, 0.45021121835975764]
[array(0.58469133)]
[0.7733706689870505, 0.36718508933903793]
[array(0.3998271)]
[0.25003607082303436, 0.3080637179933057]
[array(0.51450284)]
[0.6154137259359883, 0.38499953340666804]
[array(0.44264996)]
[0.7678609706844424, 0.0441640111505347]
[array(0.32657941)]
[0.834174331315116, 0.31765290885487163]
[array(0.373666)]
[0.7845023254025405, 0.48665566879052835]
[array(0.42608397)]
[0.9686158298628277, 0.6291022812004713]
[array(0.41592765)]
[0.19904801524428228, 0.07245052814099462]
[array(0.46839283)]
[0.8171670038783342, 0.5400223625564138]
[array(0.43115394)]
[0.05750982086012679, 0.301172401901069]
[array(0.56061604)]
[0.6143531288128768, 0.7917008573162686]
[array(0.54422109)]
[0.5222149845011839, 0.2959333388918195]
[array(0.44366974)]
[0.3492928672063633, 0.5618292210308188]
[array(0.55293497)]
[0.4943004712059401, 0.03820388872055014]
```

```
[array(0.38791223)]
[0.795757810238379, 0.9953700929310543]
[array(0.54973803)]
[0.185323279190733, 0.00940576938302251]
[array(0.45613369)]
[0.7827125673073853, 0.7864203312183299]
[array(0.50092694)]
[0.5064905570180893, 0.9923777583710366]
[array(0.61913708)]
[0.08438961605228579, 0.9460316889107401]
[array(0.70300362)]
[0.2870123917460403, 0.5828625405109975]
[array(0.57342774)]
[0.8728168433228658, 0.6355654915761423]
[array(0.44096382)]
[0.28249565152652345, 0.04650730850359808]
[array(0.4412752)]
[0.7454456842755076, 0.3883653707301581]
[array(0.41166652)]
[0.22828784163896743, 0.24942250063603033]
[array(0.50528347)]
[0.6625468598557048, 0.20100290201285564]
[array(0.38661962)]
[0.12097667856748151, 0.06163641635101447]
[array(0.48516929)]
[0.4660218958271558, 0.3632424801871662]
[array(0.47432774)]
```

【例 4.9】逻辑回归实现

```
#encoding:utf-8
import numpy
import theano
import theano.tensor as T
rng = numpy.random


N = 400                              # training sample size
feats = 784                          # number of input variables


# generate a dataset: D = (input_values, target_class)
D = (rng.randn(N, feats), rng.randint(size=N, low=0, high=2))
training_steps = 10000


# Declare Theano symbolic variables
```

```python
x = T.dmatrix("x")
y = T.dvector("y")

# initialize the weight vector w randomly
#
# this and the following bias variable b
# are shared so they keep their values
# between training iterations (updates)
w = theano.shared(rng.randn(feats), name="w")

# initialize the bias term
b = theano.shared(0., name="b")

print("Initial model:")
print(w.get_value())
print(b.get_value())

# Construct Theano expression graph
p_1 = 1 / (1 + T.exp(-T.dot(x, w) - b))   # Probability that target = 1
prediction = p_1 > 0.5                    # The prediction thresholded
xent = -y * T.log(p_1) - (1-y) * T.log(1-p_1) # Cross-entropy loss function
cost = xent.mean() + 0.01 * (w ** 2).sum()    # The cost to minimize
gw, gb = T.grad(cost, [w, b])             # Compute the gradient of the cost
                                          # w.r.t weight vector w and
                                          # bias term b
                                          # (we shall return to this in a
                                          # following section of this tutorial)

# Compile
train = theano.function(
        inputs=[x,y],
        outputs=[prediction, xent],
        updates=((w, w - 0.1 * gw), (b, b - 0.1 * gb)))
predict = theano.function(inputs=[x], outputs=prediction)

# Train
for i in range(training_steps):
    pred, err = train(D[0], D[1])

print("Final model:")
print(w.get_value())
print(b.get_value())
print("target values for D:")
```

```
print(D[1])
print("prediction on D:")
print(predict(D[0]))
```

输出：

```
Initial model:
[ 1.83361209e-01 -2.34299784e-01 -1.99289954e-01  3.30225270e-01
  2.61411100e-01  1.37172080e+00  6.95701634e-01 -4.19117941e-01
 -6.54718295e-01  1.35633305e+00 -8.89642317e-01  7.01798007e-01
 -8.68285106e-01  1.91330806e-02 -1.00369799e+00 -4.25885808e-01
  1.03143831e+00 -3.96802778e-01 -2.07665609e+00  8.92347376e-01
 -2.82921846e-01  7.41864460e-01 -1.33104629e+00  1.22280918e-01
  3.00125048e-01 -7.06475225e-01 -9.60393071e-01 -8.88197204e-02
 -8.99422842e-01  1.37253349e+00 -4.35941901e-02  7.15892346e-01
 -1.95087089e+00  1.00023993e+00 -4.23403867e-01  4.79205368e-01
  6.57590332e-01  3.30781032e-01  3.07954028e-01  6.08546020e-01
  4.33178127e-01  1.61045220e-01  3.42090885e-01 -6.90538585e-01
 -3.50014211e-01  1.24447713e+00 -3.77666644e-01  4.00660786e-02
 -4.74436471e-01 -9.94681401e-01 -1.54192381e+00 -8.68488630e-02
  5.04734696e-01 -5.63262747e-01  2.29794348e-02  9.52619633e-01
  5.44655142e-01 -3.08598561e-01 -7.39046297e-01 -6.94254019e-02
  1.43798277e-01  8.65424451e-01  1.43408215e+00 -1.50850349e+00
 -8.35679996e-01 -6.74906594e-01  2.73121617e-01  1.32972351e+00
  9.92923893e-01  2.16422484e-01  1.40753302e+00  2.17668378e-01
 -1.55021066e+00  1.31663202e-01  8.99285362e-01 -1.00031508e-01
 -9.00181868e-01  1.17033187e+00 -7.63227337e-01 -6.28259144e-01
  6.81794855e-01 -8.53342233e-01 -7.11030696e-01 -1.55358864e+00
 -8.01807545e-01  8.53048255e-01  1.40572501e+00  1.66879550e+00
  1.67791797e-01 -5.74513051e-01 -7.28060652e-01 -9.32592433e-01
 -4.02073361e-01 -2.02415267e+00 -2.01675599e-01  6.47807650e-01
  6.30877932e-02  1.98585120e-01 -1.33602906e-01 -1.43297051e+00
 -1.68300713e-01 -3.29085168e-01  1.97503819e-01  2.55354412e-01
  1.38981121e+00 -3.36401270e-01  1.74178339e-01 -7.56962898e-01
 -1.47930265e+00  9.11953241e-01  8.37607360e-01 -1.06164526e+00
 -1.28615159e+00 -1.31692473e-01 -6.76344540e-01  7.61470252e-02
 -8.86301789e-01  1.31356285e+00 -1.04834777e-01 -1.96471155e-01
  7.97055011e-01  1.07599740e+00  2.12320292e+00 -8.33265963e-01
  7.36240500e-01  1.31921965e+00 -1.52028447e+00  8.41935548e-02
  8.59397489e-01  8.67835788e-01  1.40344630e+00 -2.02182991e-01
 -6.00255171e-02  8.75147919e-01 -1.48428457e-01  7.80347678e-01
 -5.21828872e-01 -3.20664722e-01  2.58491381e-01  3.31404383e-01
 -3.57241933e-01  1.80060657e+00  1.65760332e+00 -9.67215944e-02
  3.41239468e-01 -2.10350471e+00  2.10835778e-01 -1.11919845e+00
 -1.67066004e-01 -6.84663302e-02 -1.03629396e+00 -9.49036483e-01
 -1.04188293e+00 -7.98432881e-01 -9.59205196e-03 -2.04046063e-01
 -4.74307933e-01  1.16395454e+00 -2.83178386e-02  1.40082089e+00
 -6.75595050e-01 -3.08596097e-01  8.43847511e-02 -1.66562431e+00
  1.12632323e-01  7.61943428e-01  2.65185181e-02 -1.28365186e-01
```

```
  3.28730482e-01  4.88669971e-01  1.59606733e+00  1.05144635e+00
 -6.79450942e-01 -9.20828885e-01  1.59430355e+00 -1.01331225e+00
 -1.11864942e+00  1.80766089e-01 -7.54826944e-01  6.06708068e-01
  9.08601219e-01  1.50268265e+00  4.93108137e-01 -8.69255710e-01
  1.29760384e+00 -1.14486344e+00 -3.82664395e-01 -7.30736849e-02
  3.84494118e-01  1.20156021e-01 -9.68501643e-01 -8.03123788e-01
  6.79151056e-01 -7.75999269e-01 -3.34095457e-01 -9.05893905e-01
  1.21917866e+00  9.71025606e-01 -9.84081564e-01  1.33771122e+00
 -2.98172905e-01 -1.06693994e+00  6.93368934e-01 -3.55177957e-01
 -2.14346306e-02 -1.49270301e+00  4.74268196e-01 -2.91763695e-02
  4.68611647e-01 -4.69710756e-02  9.45648949e-02  8.39587354e-01
  8.81628814e-01  1.95765096e+00 -2.06912697e+00 -7.54944370e-01
 -1.28736096e+00  1.04623173e-02  7.09240776e-01  5.66968767e-01
 -1.16011579e+00  8.04176716e-01 -1.55371938e+00  1.45921572e+00
  3.28566995e-01 -8.76208851e-01 -3.47270245e-01  8.20942395e-01
 -3.11914331e-01 -2.02373915e-01  9.85055426e-01 -1.33596383e+00
  1.55049714e+00  9.57121537e-01 -8.48978679e-01  5.98289261e-01
 -3.58975350e-01 -5.32577744e-01  6.31115518e-01 -3.23414877e-01
 -3.28149821e-01 -8.50157490e-01 -8.82378580e-01  1.29220195e+00
  3.17928313e-01  6.44869730e-01 -3.44073807e-01 -6.13594295e-01
  5.87005820e-01 -7.63167593e-01 -8.99452730e-01 -1.53331287e+00
 -8.78531247e-01  3.06385534e-01 -1.34693569e+00 -7.83183418e-01
 -4.40887478e-01 -3.91296084e-02  2.78904269e-01  9.59918864e-01
 -9.32972808e-02 -1.24791693e-01  1.14240247e+00  2.87559809e-01
 -4.12109265e-02 -1.08693631e+00  8.35126811e-01 -1.06001760e+00
  3.59997092e-03  4.23288919e-01 -1.93142127e-01  2.96271320e-02
  2.73249374e-01 -4.70173309e-01  3.72696866e-01  5.18472754e-01
  9.84135450e-01 -7.31390597e-01  6.54604352e-01  6.69743450e-01
  2.41646847e+00  5.14657210e-01  1.34182628e+00  9.93496772e-02
  1.34801994e+00 -4.25661205e-01 -4.48712953e-01  1.26966694e+00
  4.28716919e-01  1.77002702e+00  9.73624258e-01  4.31428271e-01
 -1.58344652e+00  9.80532706e-01 -7.53561438e-01 -1.01080529e+00
  3.17742637e-01  6.86038827e-01  9.03093783e-01  2.79559233e-01
 -1.04671460e-01  4.37602015e-01  5.98746033e-01  5.65356853e-01
  1.28340392e+00 -9.75903459e-01 -2.42209386e-01  9.94239949e-01
  7.25186129e-01  3.27652516e-01 -1.55113748e-01  1.54343079e+00
  5.60477964e-02  1.71349972e-01  3.15168596e-01  9.86381520e-01
  1.44403386e-01 -6.24632237e-01  1.52096951e+00  6.28184161e-01
  1.30269517e+00  4.09267057e-01 -8.14009525e-01  9.29842209e-02
  1.33961399e-01  6.97114888e-01 -2.34290292e-01  5.43442018e-01
 -4.80672123e-01  1.84050802e-03 -4.88583473e-01  9.62695058e-01
 -4.23233896e-01  1.15263187e+00 -8.87890522e-01  2.65165247e-01
  1.31448200e-01 -1.00484764e+00  1.14422192e+00  8.68743224e-01
 -1.08864328e+00 -3.16246623e-01 -9.12128061e-01 -5.04045502e-01
 -7.03022515e-01  2.18693091e-01  1.70606720e-01  1.01533919e-01
 -6.50633144e-01 -1.57132466e+00  1.51256838e+00 -1.51980449e+00
 -2.04813495e-01  1.05464696e-01 -1.94072157e+00 -4.53533121e-01
  2.31353669e+00 -1.10221772e+00 -9.10632503e-02  1.57950260e-01
```

```
 2.05816900e+00 -1.13510475e+00 -6.44912574e-01  1.72529678e-01
 1.31571189e+00 -2.51394237e+00 -2.34970427e-01 -9.40869371e-01
 6.06981731e-01  2.78127828e-01  2.80386611e-01  5.86301621e-01
-6.75040434e-01  4.38506493e-01  1.51511814e+00 -8.28478131e-03
-6.16386860e-01 -6.29115783e-01 -3.93687972e-01  2.53331442e-01
 1.13912526e+00  6.62612107e-01  4.85016217e-01  1.31626988e+00
 1.42497469e+00  5.14123251e-01 -8.47001571e-01  5.25037740e-01
-8.05632969e-02  3.03584168e-02 -3.09058988e-01  1.10890971e+00
-1.16095123e+00 -9.92383477e-02 -2.32287531e+00 -9.17157487e-01
 7.09712396e-01  1.17021729e-01  2.50311348e-01 -8.44218876e-01
 8.44538743e-01  7.40469062e-01 -7.21417666e-01 -5.38760431e-01
-4.13266024e-01  2.10323705e+00 -1.59657646e-01 -9.18717552e-01
-1.15644518e+00  3.91309780e-01  1.79220315e+00  8.89609537e-01
-1.76546508e+00  3.80505157e-02  7.46027709e-01 -1.41768163e-01
-8.43261351e-01 -1.72873504e+00  1.23695973e+00 -1.40687383e+00
-5.44736650e-01  1.85726677e+00 -9.35984329e-01 -7.09750098e-01
-5.60941220e-01  2.57114055e-01 -3.14369678e-01 -5.13849043e-01
-7.63932162e-03 -5.12442758e-01  6.18532096e-02 -1.47359191e+00
 7.32214612e-01  8.91291002e-01 -6.42541483e-02  5.48200522e-01
 7.48165627e-01  3.80066435e-01 -2.15948149e-01 -1.93647912e+00
 3.66392038e-01  4.38081046e-01  2.86387962e-01  5.70087338e-01
 5.83806737e-01  1.37150602e+00  5.44003316e-01 -6.90682607e-01
-1.81281355e+00 -1.05237045e+00  1.57479122e+00  8.78351998e-02
 1.48144828e+00 -7.61949275e-01 -3.01954183e-01  1.04157989e+00
 3.92209654e-01 -5.34631519e-01 -2.02919129e-01  2.80000094e-01
 1.74917610e+00 -1.38088317e+00  5.69771885e-01 -7.34086833e-01
-3.59336803e-01 -1.50526249e-01  6.67359139e-01  5.47089506e-01
-2.57349918e-01 -1.52531244e+00 -2.20971497e+00 -1.53263932e+00
 7.08192305e-01 -1.07364451e+00 -3.18173542e-01  2.90944542e-01
-1.05299105e+00 -4.26665919e-01  1.58814321e+00  1.20867721e+00
-5.81652754e-02  4.62198403e-02  1.25815045e+00  3.27112417e-01
-1.40376116e-01  6.33927470e-03 -6.46839918e-01  2.17581243e+00
-9.49443843e-01  5.52033795e-01 -9.76991469e-02 -1.15498405e+00
-1.92938696e+00  1.64538693e-01  1.03779531e-01 -1.01177558e+00
-5.61361822e-02 -6.55972812e-01  1.21336489e+00  1.08207327e+00
 1.04860012e+00 -9.05010243e-01  8.66218173e-01 -5.07057081e-01
-4.85763774e-01  1.06713619e+00 -2.37839637e-01 -7.31290182e-01
-1.65722330e+00 -1.61217989e+00 -2.22396800e+00  4.00516118e-01
-1.59647705e-01  3.23104323e-01  1.35219608e-01 -1.66116666e+00
-1.17041239e+00  6.90182544e-02 -6.44484711e-01  1.92650787e-01
-1.52593939e+00 -1.02029433e+00  1.05794680e+00 -1.75207573e+00
-1.54881207e+00  5.50653732e-01 -4.31022861e-01  3.10075820e-01
-1.23985391e+00  5.37038865e-01  5.75985414e-01  2.41741188e+00
 5.20164700e-01  6.54744296e-01 -1.30725231e-01 -1.12730189e+00
 5.64548821e-01  8.15995528e-01 -1.98158945e-02 -9.62332461e-01
 3.45022713e-01  8.28916047e-01  1.34066813e+00  9.61680381e-01
 7.08573750e-01  3.44553789e-01  5.90408189e-01  3.80377908e-01
-5.11239393e-01 -6.07841698e-01 -3.43054668e-01  1.30656745e-01
```

```
 4.22553961e-01  1.03428891e+00  7.44311719e-01  6.61734717e-01
-1.58564274e+00 -1.55886024e+00 -1.13654126e-01 -1.42067669e+00
 2.09820849e-01  1.23187133e+00 -1.87183632e+00  2.16719764e-01
 1.90795348e+00  8.05402560e-01  9.85796655e-01 -4.87075003e-01
 1.67615484e+00  5.69311190e-01  1.54103888e+00 -7.17034538e-02
-2.21359096e-01  6.64927254e-02 -5.61742498e-02  6.42305512e-01
 1.70283301e+00 -1.34034409e+00  1.49108691e+00 -1.15066913e+00
-3.02516791e-01 -1.07228522e+00 -4.63840711e-03  1.16581427e+00
-1.47555588e+00  1.38632837e+00  4.80938101e-01 -3.88267206e-01
 3.60403631e-01  1.38021860e+00  1.91576812e-01  2.82608033e-02
 2.31946076e+00 -7.07151163e-02  2.41453426e-01 -1.21236350e+00
-1.34047882e+00 -5.40535356e-01 -8.34343075e-01  1.16848410e+00
 2.48880940e+00 -6.61980813e-01  7.96165816e-01  9.60464883e-01
 1.80807282e+00  4.48646907e-01  5.45234571e-01  2.01331884e+00
 2.33064632e-01  4.11392470e-01  6.40052537e-01  1.71610789e+00
-5.82158525e-01  4.95161687e-01 -1.87778694e+00  1.21518079e+00
 1.34759811e+00 -6.89863971e-01  1.52998168e-01  1.15858451e-01
-2.72838754e-01  2.80172242e-01 -1.92251086e+00  7.91821456e-02
-6.92905891e-01 -2.37650493e-01  5.29706035e-01 -6.61289821e-01
 5.79680805e-01  1.25957336e+00  8.52680549e-01 -7.35945616e-02
-1.37541272e-02 -7.75658132e-01 -1.54724589e+00  7.15478145e-01
 8.29157442e-01 -9.52269142e-01 -9.62374811e-01  1.68493426e+00
 7.22458794e-02  2.18154697e+00  7.03612645e-01 -3.24538037e-02
-8.77722197e-01 -1.47808528e-01  4.83255769e-01  3.66618762e-01
-7.61609363e-01  1.05191757e+00  8.58666331e-01  1.37306512e+00
 6.53414099e-01  2.46947560e-01  7.49307875e-01  2.72081148e+00
 6.47840849e-01 -1.67070516e+00  1.10230343e+00  3.93046811e-01
 2.59019667e+00  4.96179928e-01 -3.41857241e-01 -1.18534980e+00
-9.15907050e-01 -1.10164615e+00 -4.13605798e-01  1.04521290e+00
 5.83986432e-01  8.41772514e-01  4.55527072e-02 -9.91558846e-01
-6.46977788e-01  8.04722669e-01  7.20436127e-01 -9.51086472e-01
-1.72668609e+00  1.58309144e+00 -3.32802845e-01 -3.82499927e-01
 1.27021231e+00 -1.28339573e+00  3.96721774e-01 -1.12392301e+00
 8.38180940e-01 -7.49430887e-01  1.92428116e-01  1.53327732e-01
 7.40475184e-01 -5.57322365e-01 -1.39325633e-01  1.34196718e+00
 1.12459904e+00 -4.94359085e-02 -8.15041108e-01 -4.95841775e-01
-5.39063999e-01  1.81368992e+00  4.79839739e-01 -2.40731763e-01
-8.15580738e-01 -2.53489127e-01  8.58267166e-01 -4.87059366e-01
-7.45957751e-01  2.49936179e-01  8.37856030e-01 -9.38194368e-01
 1.40889269e+00  9.85961581e-01 -1.90323516e-02  1.03106244e+00
-7.18934534e-01 -4.57711323e-01  2.00837761e-01 -1.42869307e-01
-2.41140128e-01 -5.07020457e-01 -6.04959585e-01 -1.21062871e-01
 2.38283756e-01 -9.04835946e-01  2.20413491e-01  2.63644157e-01
-1.63659782e+00  1.68204811e-02 -3.54366524e-02 -7.60873125e-01
 1.70473394e-01  1.13700102e+00  8.20160000e-01  2.57900035e-01
-2.69458248e-01  1.13988090e+00 -6.03098187e-01  3.88123678e-01
-2.89919391e-01 -8.96952059e-01 -6.47233856e-01  2.82079609e-01
 1.92173909e-01  1.61642898e-01  4.36492066e-01  1.43806500e+00
```

```
  7.76978232e-01 -4.87718314e-01 -9.68812427e-01 -4.11020159e-01
 -1.13169297e+00 -6.50985310e-01 -6.01219952e-01 -1.23428481e+00
 -9.52741086e-01 -7.54868320e-01  1.03559681e+00 -2.08880376e-01
  2.41470765e-01 -3.98923528e-02 -1.74849464e+00  1.37899185e-01
  6.66576269e-01 -1.51625212e+00  3.64369347e-01  1.68112924e+00
 -1.21984157e+00  5.29479289e-01  1.09102684e+00  3.60966726e-01
  1.15229686e+00 -1.15829675e+00  1.78441898e+00 -1.32464589e+00
 -3.78427875e-01 -1.41908865e+00 -3.97173764e-01 -6.31292068e-01
  1.01795182e+00  1.01865186e+00 -8.76988927e-01  2.06132717e+00
  9.80137030e-01 -1.35297113e+00 -7.24361673e-01  2.02803976e+00]
0.0
Final model:
[-8.59839832e-02  3.16062846e-02 -2.36348321e-01 -6.97120527e-02
 -8.76274216e-02  8.29309921e-03 -4.58852212e-02 -1.39421677e-01
  1.33108610e-03  4.91315131e-02 -6.63035729e-02 -7.64296396e-02
 -3.29548687e-02 -6.02297843e-02 -1.35229421e-01  1.77122680e-02
  1.51214661e-02  1.27530524e-01  1.84018300e-03 -1.34522792e-01
  6.67088042e-02  6.95181570e-02 -9.09385969e-02 -9.80364507e-02
 -8.47597200e-02  4.84978809e-03  3.81954029e-02  1.72288259e-01
  3.38843571e-02 -1.99497566e-02 -5.01228658e-02 -5.31670153e-02
 -3.09126943e-01  5.78651090e-02  1.25939930e-01  6.32919892e-02
  2.51688051e-01  1.33281925e-01  1.08206636e-01  5.63176839e-03
  3.86116997e-02 -1.27181643e-01  2.08633895e-02 -2.77796618e-02
 -7.29343114e-02 -8.13630459e-02  8.77125165e-02  7.08767149e-02
 -4.90463798e-02  1.22046686e-01 -3.08866703e-02  5.76866942e-02
 -1.41784547e-01  9.54010916e-02 -1.52523078e-03  1.50035400e-02
  9.82507829e-02 -2.20720124e-02  1.72407073e-03 -1.60280522e-01
  5.00230958e-02 -1.14021740e-01 -1.48367302e-02  7.78935645e-02
 -1.58662154e-01  1.59548094e-01  2.07096767e-02  3.22953678e-03
  1.40850728e-01 -2.17691554e-01  6.34188265e-02  7.91532711e-02
  1.51899803e-01 -4.91714521e-02 -2.26891921e-03  1.04521097e-01
  1.58542728e-01  3.14235114e-02  6.27373862e-02 -7.55526455e-02
 -7.50323182e-02  1.83817091e-02  5.15450931e-02  1.37160680e-01
 -2.19935006e-02  1.01726264e-01 -4.68742438e-02  1.96836505e-01
 -1.24951591e-01 -8.70511474e-03 -1.82300205e-01  8.82314101e-02
  8.05409806e-02  6.75110609e-02  8.64671592e-02 -1.00883189e-01
  1.52021129e-01 -2.89773368e-01  6.53898245e-02  8.20210306e-02
 -7.22517239e-02  3.36966053e-02  8.66106457e-02 -1.17054989e-01
  3.27595914e-02 -1.09620935e-01  9.42844701e-02 -3.82349844e-02
  1.99484923e-01  1.34446678e-02  9.42913852e-02 -2.93737263e-02
 -1.21668399e-02  1.28928281e-01  1.82835204e-01 -5.53084039e-02
  2.88955242e-02  1.08456122e-01 -3.36433529e-02 -2.04485318e-02
  5.08557614e-02 -8.34571582e-02  1.10732334e-01  8.93047845e-02
 -1.85592258e-02 -8.53852616e-03 -1.26383093e-02 -1.39066145e-01
  1.18282658e-01 -5.63406066e-02  3.09549545e-02 -5.97710293e-02
  2.20250297e-02  1.81395955e-02 -1.21339790e-01  6.38437006e-02
  6.32531485e-02  1.03616237e-01  1.28484830e-01 -2.45663305e-02
 -2.07701298e-02 -1.95976957e-01  1.05577687e-02  4.77790132e-02
```

```
  4.45377890e-02 -8.13162606e-02  2.77251290e-02  8.38657650e-02
  1.23788313e-01 -1.27072235e-01  5.22465105e-03 -4.77642724e-03
 -7.61550641e-02  1.09764818e-01 -3.74470202e-02  8.43127248e-02
 -5.99091746e-02 -1.94214712e-01 -1.77968981e-01 -6.14679082e-02
  1.12359488e-02  9.80499623e-02  8.38331079e-02  5.31149579e-02
 -2.18290512e-02 -2.52480326e-02 -1.80283673e-02  6.44289370e-02
  1.05038528e-01 -7.37759842e-02 -8.67925440e-02  2.59488727e-02
  5.09979609e-02 -1.13566327e-02  1.09293613e-01  1.74941082e-01
  1.84977549e-01  1.20630629e-02 -6.89563074e-02  1.77894762e-02
 -1.45279041e-01  1.06973874e-01  1.74552008e-01  1.98864191e-01
  1.34508148e-01  1.47876561e-02 -1.29520606e-01  1.23844435e-02
  1.39780506e-01 -2.66995260e-02 -2.71712781e-02  1.49078336e-01
  1.37219450e-01  5.28695155e-02 -7.77977460e-02  1.04543421e-02
 -6.39277247e-02  2.87934096e-02  9.94196114e-02 -5.39132284e-03
 -4.22169793e-02 -5.68647685e-02  1.92864734e-01 -6.91361123e-02
 -3.08229699e-02 -1.36957230e-01  7.25808050e-02 -1.91327882e-01
 -1.43597434e-01  2.05506851e-02  1.20766884e-01 -1.26511102e-01
 -1.00685992e-01 -6.32428850e-03  1.34514349e-01 -9.26429509e-02
  1.86967939e-01  8.22461176e-03 -7.47540753e-02 -6.84303138e-03
 -1.46503021e-01 -3.17201324e-02  4.69602090e-02  2.45544336e-02
 -1.11255416e-01  4.92831396e-02  9.08212440e-02  5.07028077e-02
  2.02717940e-02 -9.12753980e-03  8.04657077e-02  1.80785870e-02
  1.04159206e-01  1.16693100e-01 -3.84138348e-02 -1.24586539e-01
  3.64446727e-02 -1.20534889e-01 -3.39753456e-02  2.47225965e-02
 -1.35789895e-01  1.43215107e-01  1.88662546e-01 -4.52268023e-02
  1.62451807e-01  2.03490677e-01 -7.34918687e-02 -1.84883027e-01
  2.15340427e-01 -1.37908137e-02 -1.04498567e-02  3.35776327e-02
 -8.93612386e-03 -1.32986173e-01 -8.23761151e-02 -5.63071662e-02
  1.01872915e-01 -8.06235065e-02 -6.62577125e-02  7.80768993e-02
 -2.54145051e-01 -1.27639810e-01 -4.73630280e-02  2.36165322e-02
  1.17156030e-02 -2.64129039e-02  1.60844450e-02  1.17866234e-01
  9.19179488e-02 -1.12933239e-01  4.40202501e-02  1.29506070e-01
 -1.43953649e-01 -7.67115090e-02  1.39928147e-01 -4.33159044e-02
 -2.10962321e-02  4.88322037e-03 -1.47193011e-01  5.29347109e-02
  4.90443353e-02  1.51513340e-02  6.44848492e-02  4.94980176e-02
  1.01418286e-01  3.38353121e-02 -1.63650863e-01  7.27725078e-02
 -1.81503549e-02  1.94418829e-02  8.63164177e-02 -4.40415708e-03
 -2.42493080e-02  3.33827293e-02 -5.22090988e-02 -1.84420703e-02
 -3.57528316e-02 -1.03343762e-01 -1.16314033e-01 -3.01951017e-02
  1.01093913e-01 -2.38274751e-02  4.99585540e-02 -1.26072524e-01
  3.62888394e-02 -4.32720526e-02  4.68067582e-02  1.65985442e-01
 -3.78332112e-02 -7.33140461e-02  1.73117951e-01 -4.36029657e-02
 -4.03022608e-02  8.08372980e-02  1.27587199e-02 -5.46269888e-02
  6.02189532e-02 -5.82056319e-02  9.45994763e-02  2.15856901e-02
 -8.14525650e-02  9.05865298e-03  2.03214385e-01  1.66502419e-02
 -1.91496885e-02 -2.02765334e-02 -7.22185443e-02 -1.50647442e-01
 -3.47973984e-02  4.80519591e-02 -8.35262375e-04  3.78268868e-02
  5.24740056e-02  1.24044259e-01 -7.24697587e-02 -3.41362013e-02
```

```
  1.24692138e-01 -1.63860387e-02  6.48273035e-02  1.31837499e-01
 -7.17243978e-03  5.62051433e-02  1.24971235e-01 -4.05222475e-02
  1.25024798e-01 -3.05564181e-03 -1.22470342e-01 -5.46361302e-02
 -2.02664200e-02 -4.84673595e-04 -6.97897554e-03  1.49058494e-02
 -4.26618896e-02 -1.43143016e-02  1.20187641e-01 -8.89019349e-02
 -1.12903040e-03 -2.28208683e-02  6.56899758e-02 -2.22840930e-01
  4.05441201e-02  7.57826283e-02 -1.43012987e-01 -3.12416819e-02
 -8.15699326e-03 -1.31809825e-01  6.16370833e-02  4.71378846e-02
 -4.80952469e-02 -7.49984568e-02 -7.98997268e-03  3.88548952e-02
 -8.19503954e-02  1.07620677e-03  6.71876554e-02  1.17755085e-01
  5.83144989e-02  4.08440183e-02  5.25265889e-02  1.63773366e-01
 -4.68816064e-02 -1.86337885e-02 -9.80714433e-02  3.47042566e-02
  2.68161678e-02  1.48847950e-02 -6.80367010e-02  1.26506015e-01
 -2.52014411e-02 -9.02441305e-02  1.46191223e-02  4.27773266e-02
 -6.96032700e-02  9.51222402e-02 -1.06939079e-01  3.47198247e-02
 -8.34494481e-02 -1.39988611e-01  3.58865588e-02  1.38426640e-02
 -1.02048998e-01 -1.97569951e-01 -3.49853252e-04  4.33820800e-02
  2.45738332e-02  2.41792835e-01 -8.94235952e-02 -1.28816704e-01
 -6.88151222e-02  3.98901303e-02  1.52226001e-01  1.17265923e-01
 -4.65975050e-03 -5.21394113e-02 -1.11736280e-01 -1.07984204e-01
  7.12802398e-02 -3.05873465e-02  2.91689192e-03  4.00477499e-03
 -1.45431291e-02  1.87977925e-02 -1.53971958e-01  1.07023438e-01
 -7.03004012e-02 -1.13439453e-01 -2.90769892e-02  8.63519902e-03
 -1.05458589e-01 -4.31234020e-02  1.12226338e-01  3.77235987e-02
  4.00739407e-03  2.35255684e-02  1.87716335e-01  1.22746194e-01
 -9.40400510e-02 -1.00294015e-01  1.21142897e-01  1.60990363e-01
  3.33877658e-02  2.25479428e-02 -6.01523777e-02  6.19465225e-03
  4.59219358e-02  5.40358782e-02 -5.00380317e-02 -1.26200940e-01
 -1.82988627e-02  2.66669278e-02 -1.75975124e-01 -9.39588504e-02
  1.42941805e-01 -1.26358123e-01  7.73692455e-02  2.57420641e-02
 -1.04592820e-01 -6.54012040e-03 -1.65613947e-01  2.15090841e-02
 -1.48180100e-01 -1.05078900e-01 -8.57778680e-02 -1.25021904e-01
  1.85375634e-01 -3.99930839e-02 -7.21067316e-02  6.91146564e-02
  3.57298586e-02 -4.02183648e-02  1.70611890e-01 -1.21784479e-03
  1.60285011e-02 -7.32448450e-02  5.96386319e-02  1.18755419e-01
  1.97295219e-02  3.28667545e-02 -1.23760835e-01 -1.79784018e-01
 -1.87471923e-01  3.69686271e-04  1.22210699e-02 -9.11306651e-02
  1.08708272e-02  3.17264600e-02  6.21532594e-03  5.47533293e-02
 -1.92610130e-01 -2.00461458e-02 -1.14651004e-01 -8.38064388e-02
 -2.48337824e-02 -7.56050169e-02  2.77636920e-02 -2.04288077e-04
 -3.57443239e-02 -1.01983353e-01 -2.41492773e-02  1.27073199e-01
  1.41120762e-01 -4.96193440e-02 -4.33919911e-02  5.38525390e-02
 -1.21998214e-01 -1.92334532e-01 -1.54635235e-01  2.33390678e-02
 -1.24767242e-01  1.58136441e-01  6.20135601e-03 -1.14748599e-01
  2.02039436e-02 -3.62931218e-02 -2.88078186e-02  1.24957806e-03
  4.02651037e-02  4.79126197e-02 -8.44259109e-02  1.31732617e-03
  3.04850104e-02 -1.01803776e-02  9.81822706e-02  1.39893340e-01
  6.14877882e-02 -7.19017843e-02  1.18875497e-01 -7.70738437e-02
```

```
-3.11385577e-03 -1.04876962e-01  9.25780138e-02 -3.72865173e-02
-1.16819773e-01  9.24014697e-02  1.09834112e-01 -5.06255764e-02
 2.22657418e-01  1.87890816e-02  1.60280632e-01  1.57861688e-01
-7.71987101e-02  3.74242298e-03 -7.61715900e-03 -1.80108346e-01
 2.40172591e-01 -1.07677849e-03  1.16886275e-01  1.65353383e-01
 5.74920176e-02 -1.56038480e-02  1.18719743e-01 -5.75284963e-03
-3.79467122e-02  2.71948327e-02  6.09909867e-02  1.28525792e-01
 4.58933158e-02 -1.40073580e-02 -3.27579542e-02  8.23116689e-02
-6.97454232e-02  2.52965716e-02 -1.32199054e-01 -1.40247020e-02
 3.71957838e-02 -1.92388525e-01 -7.17628213e-03  4.06880268e-02
-6.37621916e-02  1.60692340e-01  6.79084264e-02  3.65579705e-02
 2.38941195e-02  8.99284525e-02 -3.89259337e-02  5.98351913e-02
-1.22707230e-01 -5.94434294e-02 -2.44280040e-01  3.66897802e-02
 7.70021509e-02  6.42126435e-02 -7.16210341e-02  2.81697506e-01
-4.85526234e-02 -3.76295288e-02 -3.94600508e-02 -1.99260408e-01
 8.20591675e-02 -4.03412815e-02  9.95130813e-02 -5.18932976e-02
 1.05345532e-02 -3.12644693e-02 -5.83065608e-03  3.25769519e-02
-1.81778971e-01 -6.31865523e-02 -2.97602094e-03 -8.13936420e-02
-1.76140091e-01 -4.95477996e-02  7.25337748e-02 -7.89001221e-02
-1.10757628e-01 -5.03489015e-03  1.22248495e-01 -7.63827323e-03
-1.12731568e-01 -1.33781297e-01 -1.14937645e-01 -1.27043850e-01
 7.23856932e-02  7.43666155e-03  1.08215014e-01 -4.86247909e-02
 4.63229721e-02 -4.26710573e-02  8.21151860e-02 -3.21774172e-02
-5.07798403e-02  1.93550965e-02 -2.13247711e-02  5.84017432e-02
 2.44713678e-02 -6.29772938e-02 -8.51109130e-04 -3.26142845e-02
-5.29673284e-02 -7.45682491e-02 -1.20072851e-01 -8.51343713e-02
-7.93586879e-02 -1.25525551e-01  9.70559069e-02  6.06650529e-02
-2.21908174e-02 -6.04831457e-02  4.01386364e-02 -1.06470683e-01
 9.72075995e-03 -6.36394580e-03  1.23925774e-01  8.12847240e-02
 7.38792723e-02 -8.40865484e-02 -7.69191958e-02 -1.68243431e-02
 1.04165592e-01 -3.10664782e-02 -1.00254580e-01  1.99163039e-01
 3.61983234e-02 -1.51862434e-01  1.48831255e-02  3.49395665e-02
 2.31491403e-01 -1.51851445e-01 -7.23613108e-02 -5.85798307e-02
 1.57192882e-01  6.69123039e-02 -1.16108997e-01  4.31855474e-02
 1.49376787e-02  4.64804907e-02 -1.49443355e-01  6.41001031e-02
-1.98575722e-01 -3.97317508e-02  7.75364562e-04  9.06047747e-02
-1.27529707e-01  8.55974359e-03  6.73814726e-02  9.26616420e-03
 1.05519401e-01  1.01572396e-01 -5.06032884e-02 -1.00455457e-01
-2.13590300e-01  7.89581363e-02 -2.26090252e-01 -7.24635312e-02
 2.26445439e-02 -1.78595437e-02 -5.14679924e-02 -5.93728436e-02
-7.10579590e-02  1.89270307e-02  2.71503151e-02  8.17150321e-02
-2.90359041e-02 -3.78551285e-02  6.77519449e-03 -1.75491153e-02
 8.45608871e-02 -1.22904716e-01 -5.71785908e-02 -1.01147443e-01
 3.42560007e-02 -5.85758249e-02  8.73633787e-02  2.48846078e-01
 9.53093963e-02  9.81471545e-02  2.23562615e-01  1.71398062e-02
 6.28061559e-03  8.01433450e-02  6.37680987e-02 -1.90472746e-01
 9.89803478e-03  6.57919286e-03  7.29825495e-02  3.68994863e-02
-7.58362240e-02 -7.69194282e-02  2.10488014e-02 -1.08609781e-02
```

```
   1.33603919e-01  3.08076670e-02  1.54842595e-02  5.55736071e-02
  -9.82000758e-02  2.03759694e-01  1.31464787e-01 -8.52385205e-02
   5.57006402e-02 -1.24620291e-01  3.92931415e-03  6.95716352e-02
  -2.17734265e-02  1.09753372e-02 -3.88800106e-02 -3.58391309e-02
  -4.94131869e-02  9.42777108e-03 -9.03769767e-03  4.26305090e-04
   7.07064785e-02 -8.16267582e-02 -4.25430600e-02  3.94165055e-02
   2.05736890e-02 -4.75757940e-02  2.60161436e-02 -1.12084887e-01
   4.72951693e-02  4.78899344e-02 -7.33914188e-02 -8.22409149e-03
  -1.63217956e-02 -9.86756939e-03  1.92176180e-02 -3.10549513e-02
   6.13567448e-02 -6.67136793e-02 -5.00860530e-02 -1.55020493e-01
   6.02806588e-02  2.26193855e-02 -2.17768324e-01 -7.72153090e-02
  -7.88782021e-02  1.25699320e-01  6.53427485e-02 -9.47857891e-03
  -4.57297290e-02 -4.84437631e-02  1.03793104e-01 -1.02129033e-01
   2.22387210e-01 -1.05337981e-01  1.30510310e-01 -1.96290556e-01
  -5.73107543e-02 -1.28378975e-01  9.94919010e-02  1.28568129e-01
  -1.43090519e-01  1.61221757e-01 -1.07627424e-01 -8.93025071e-02]
-0.20528642873868336
target values for D:
[0 1 0 1 1 1 1 1 0 0 0 1 0 0 1 1 0 1 1 1 0 0 0 0 0 0 1 1 0 1 1 0 0 0 1 0
 1 0 0 1 0 1 0 0 0 0 0 1 1 0 1 0 0 0 1 1 0 0 0 0 0 0 1 0 1 0 0 1 1 0 0 1
 0 1 1 0 1 1 0 0 1 1 1 1 0 0 1 0 0 1 1 1 0 0 1 1 0 1 0 1 1 1 1 1 0 0 0 0
 1 1 0 0 1 0 1 1 0 1 1 1 1 1 0 0 1 1 1 1 0 1 0 0 1 0 0 1 0 0 1 1 0 1 0 0
 0 0 1 0 1 1 0 1 1 1 1 1 0 1 1 1 1 0 1 1 0 1 1 0 1 0 0 0 1 0 0 0 1
 1 0 0 1 1 1 1 0 1 0 0 0 0 0 0 1 1 1 1 1 1 0 1 0 0 0 1 0 0 1 1 1 0 1 0 0
 0 1 1 1 1 0 0 1 1 0 1 0 0 1 0 1 1 0 1 0 1 1 0 0 0 0 1 0 0 0
 0 1 1 1 1 0 0 1 0 0 1 0 0 0 0 0 1 0 0 1 1 0 0 1 0 1 1 1 1 0 0 0 0 1 1 1
 1 1 0 0 0 0 1 0 0 1 1 0 0 1 0 1 0 0 1 1 0 0 1 1 1 1 1 0 0 1 0 1 0 0 0 1
 0 0 0 0 1 1 0 0 0 0 0 1 0 1 0 0 0 0 0 1 0 0 0 1 0 1 0 0 1 0 0 0 0 1 0 1 1
 1 0 1 0 1 1 0 0 0 1 1 1 1 0 1 0 1 1 0 0 0 0 1 1 1 0 0 1 1 0]
prediction on D:
[False  True False  True  True  True  True  True  True False False False
  True False False  True  True False  True  True  True False False False
 False False False  True  True False  True  True False False False  True
 False  True False False  True False False  True False False False False
 False  True  True False  True False False False  True  True False False
 False False False False  True False  True False False  True  True False
 False  True False  True  True False  True  True False False  True  True
  True  True False False  True False False  True  True  True False False
 False  True  True False  True False  True  True  True  True  True False
 False False False  True  True False False  True False  True  True False
  True  True  True  True  True False False  True  True  True  True  True
 False  True False False  True False False  True False False  True  True
 False  True False False False False  True False  True  True False  True
 False  True  True  True  True False  True  True  True  True False  True
  True False  True  True False  True  True False  True False False False
  True False False False  True  True  True False False False False  True
  True  True  True  True  True  True False  True False False False  True
  True  True  True False  True False False False  True False False  True
```

```
 True  True False  True False False False  True  True  True False False
 True  True False  True False False  True  True  True False  True False
False  True False  True  True False  True False  True  True False False
False False False  True False False False False  True  True  True  True
False False  True False False  True False False False False False  True
False False  True  True False False  True False  True  True  True  True
 True False False False False  True  True  True  True  True False False
False False  True False False  True  True False False  True False  True
False False  True False  True  True  True  True  True  True  True  True
False False  True False  True False False False  True False False False
False  True  True False False False False False  True False  True False
False False False False  True False False False  True False  True False
False  True False False False False  True False  True  True  True False
 True False  True  True False False False  True  True  True  True False
 True False  True  True False False False  True  True  True  True False
False  True  True False]
```

其他一些 Theano 应用可以在 http://deeplearning.net/software/theano/index.html 网站上查看。

4.4 Caffe

Caffe（Convolutional Architecture for Fast Feature Embedding，快速特征嵌入的卷积结构）是一种常用的深度学习框架（卷积神经网络框架），主要应用在视频、图像处理方面的应用上。

Caffe 是一个清晰、可读性高、快速的深度学习框架。作者是贾扬清，加州大学伯克利的 Ph.D.，现就职于 Facebook。Caffe 的官网是 http://caffe.berkeleyvision.org/。采用高效的 C++语言实现，并内置有 Python 和 MATLAB 接口，以供开发人员使用 Python 或 MATLAB 来开发和部署以深度学习为核心算法的应用。Caffe 适用于互联网级别的海量数据处理，包括语音、图片、视频等多媒体数据。Caffe 的高速运算是通过 GPU 来实现的，在 K40 或者 Titan GPU 上每天可处理 4000 万张图片，相当于 1 张图片仅用 2.5 毫秒，速度非常快。

4.4.1 Caffe 框架与运行环境

1. Caffe 框架

Caffe 提供了一个用于训练、测试、微调和开发模型的完整工具包，而且拥有文档完善的例子用于这些工作。同样的，它也是一个对于研究人员和其他开发者进入尖端机器学习的理想起点，这使得它在短时间内就能用于产业开发。Caffe 的特性和优点主要有：

- 模块性：Caffe 本着尽可能模块化的原则，使新的数据格式、网络层和损失函数容易扩展。网络层和损失函数已定义，大量示例展示了这些部分是怎样组成一个识别系统用于不同情况工作的。
- 表示和实现的分离：Caffe 模型的定义已经用 Protocol Buffer 语言写成配置文件。Caffe

支持在任意有向无环图形式的网络构建。根据实例化，Caffe 保留网络需要的内存，并且从主机或者 GPU 底层的位置抽取内存。在 CPU 和 GPU 之间转换只需要调用一个函数。
- 测试范围：每一个在 Caffe 中的单独模块都会进行测试，没有相应测试就不能有新代码加入项目。这样就可以快速改进和重构代码库。
- Python 和 MATLAB 结合：Caffe 提供了 Python 和 MATLAB 相结合的目前研究代码的快速原型和接口。两种语言都用在了构造网络和分类输入中。
- 预训练参考模型：Caffe 提供了参考模型用于视觉工作，包括里程碑式的"Alex Net"、Image Net 模型的变形和 R-CNN 探测模型。

目前，有许多主流的深度学习框架，每个框架都有其优点。Caffe 与其他深度学习开发工具相比，主要有以下两个区别：① Caffe 完全用 C++语言来实现，便于移植，并且无硬件和平台的限制，适用于商业开发和科学研究。② Caffe 提供了许多训练好的模型，通过微调（Fine-Tuning）这些模型,在不用重写大量代码的情况下就可以快速、高效地开发出新的应用，这也是当下软件应用开发的趋势。

2. 编译 Caffe

Caffe 的核心模块有 3 个，分别是 Blobs、Layers 和 Nets。Blobs 用来进行数据存储、数据交互和处理，统一制定了数据内存的接口。Layers 是神经网络的核心，定义了许多层级结构，将 Blobs 视为输入输出。Nets 是一系列 Layers 的集合，并且这些层结构通过连接形成一个网图。

（1）Blobs

Blobs 本质是一个 N 维向量，用来存储数据信息，包括图片、深度网络进行前向传输时的数据和反向求梯度过程时的梯度数据等。对于图像数据来说，Blobs 通常是一个 4 维向量，其格式为（Number，Channel，Height，Width）。其中，Channel 表示图像的通道数，若图像是单通道的灰度图，则 Channel=1；若是 3 通道的 RGB 图像，则 Channel=3。Height 和 Width 分别表示图像的高度和宽度。Number 表示图像批块（Batch），批处理可以使神经网络有更大的吞吐量。

（2）Layers

Layers 是神经网络的核心，Caffe 设计实现了许多层结构，包括卷积、池化、损失等层结构，利用这些层结构可以实现绝大部分的神经网络模型。Layers 将下层的数据输出作为输入，进而通过内部运算输出。Layers 层的定义和使用一般需要 3 个步骤：① 建立层，包括建立连接关系和初始化其中一些变量参数；② 前向传输过程，给定输入并计算出相应的输出；③ 反向传播过程，进行反向梯度的计算，并把梯度保存在层结构中。

（3）Nets

Nets 是由 Layers 组成的，定义了输入、输出、网络各层，并将各层连接成一个有向无环图（DAG），由此定义一个网络。一个典型的网络应该有数据输入，并且以一个代价函数作为输出，针对不同的任务（例如分类和重构），应选择不同的代价函数。

（4）Solver

定义针对 Nets 网络模型的求解方法，记录网络的训练过程，保存网络模型参数，中断并恢复网络的训练过程。自定义 Solver 能够实现不同的网络求解方式。

3. Caffe 的硬件环境

（1）CPU 的选择

Caffe 支持 CPU 和 GPU 训练。如果采用 CPU 训练，CPU 支持的线程越多越好，因为 Caffe 本身显性地使用两个线程：一个线程用来读取数据，另一个线程用来执行前向传播和反向传播。如果采用 GPU 训练，则大量运算由 GPU 完成，CPU 只运行 Caffe 的两个线程，因此即使选用更多的 CPU 也无法大幅度加速训练，训练时效取决于 GPU。

（2）GPU 的选择

因为 Caffe 只支持 CUDA（Computer Unified Device Architecture）库，而 CUDA 库是 NVIDIA 显卡专用的，所以选择 Caffe 作为深度学习框架一定要选 NVIDIA 显卡。如果电脑使用两个不同显卡的版本，那么训练速度是两张低速卡一起训练的速度。

（3）内存的选择

选择支持双通道的内存以及高频率的内存有利于训练，GPU 训练下，内存频率不是重要影响因素。

（4）硬盘选择

Caffe 采用单独线程异步方式从硬盘中顺序读取数据，需要根据实际情况看是否考虑固态硬盘（SSD），硬盘容量和数据集密切相关。

4. Caffe 的依赖库

- Boost 库：一个可移植、提供源代码的 C++库，作为标准库的后备，是 C++标准化进程的开发引擎之一。Caffe 采用 C++作为主开发语言，其中大量的代码依赖于 Boost 库。
- GFlags 库：Google 一个开源的处理命令行参数库，使用 C++开发。Caffe 库采用 GFlags 库开发 Caffe 的命令行。
- GLog 库：一个应用程序的日志库，提供基于 C++风格的流日志 API，Caffe 运行时的日志依赖于 GLog 库。
- LevelDB 库：Google 实现的一个非常高效的 Key-Value 数据库，单进程服务，性能非常高，是 Caffe 支持的两种数据库之一。
- LMDB 库：是一个超级小、超级快的 Key-Value 数据存储服务，使用内存映射文件，因此在读取数据的性能时跟内存数据库一样，其大小受限于虚拟地址空间的大小，是 Caffe 支持的两种数据库之一。
- ProtoBuf 库：Google Protocol Buffer，一种轻便高效的结构化数据存储格式，可用于结构化数据的串行化（序列化），适合做数据存储或 RPC 数据交换格式；也可用于通信协议、数据存储等领域的语言无关、平台无关、可扩展的序列化结构数据格式。Caffe 使用起来非常方便，很大程度上是因为采用.proto 文件作为用户的输入接口。用户通过编写.proto 文件定义网络模型和 Solver。按序排列时二进制字符串尺寸最小，高效序列化，易读的文本格式与二进制版本兼容，可用多种语言实现高效的接口，尤其是 C++和 Python。这些优势造就了 Caffe 模型的灵活性与扩展性。

- HDF5 库：Hierarchical Data File，一种高效存储和分发科学数据的新型数据格式，可存储不同类型的图像和数码数据的文件格式，可在不同的机器上进行传输，同时还有统一处理这种文件格式的函数库。Caffe 支持 HDF5 格式。
- snappy 库：一个 C++库，用来压缩和解压缩的开发包，旨在提供高速压缩速度和合理的压缩率。Caffe 在数据处理时依赖于 snappy 库。

5. Caffe 的接口

（1）Caffe Python 接口_Pycaffe

- caffe.Net：主要接口，负责导入数据、校验数据、计算模型。
- caffe.Classifier：用于图像分类。
- caffe.Detector：用于图像检测。
- caffe.SGDSolver：是暴露在外的 Solver 的接口。
- caffe.io：处理输入输出、数据预处理。
- caffe.draw：可视化 net 的结构。
- Caffe blobs：以 numpy ndarrays 的形式表示，方便而且高效。

（2）Cafe MATLAB 接口_matcaffe

- caffe train：用于模型学习。
- caffe test：用于测试运行模型的得分，并且用百分比表示网络输出的最终结果。
- caffe time：用来检测系统性能和测量模型相对执行时间，通过逐层计时与同步来执行模型检测。

6. Caffe 模型基本组成

（1）预处理图像的 leveldb 构建

- 输入：一批图像和标注（2 和 3）。
- 输出：leveldb（4）。

指令里包含如下信息：

- convert_imageset（构建 leveldb 的可运行程序）。
- train/（此目录存放处理的 jpg 或者其他格式的图像）。
- label.txt（图像文件名及其标注信息）。

- 输出的 leveldb 文件夹的名字。
- CPU/GPU（指定是在 CPU 上还是在 GPU 上运行）。

（2）CNN 网络配置文件

- Imagenet_solver.prototxt（包含全局参数配置的文件）。
- Imagenet.prototxt（包含训练网络配置的文件）。
- Imagenet_val.prototxt（包含测试网络配置的文件）。

4.4.2 网络模型

网络模型可定义网络的每一层。例如，图 4.9 就是用 Caffe 中/python/draw_net.py 画出的 siamese 模型。

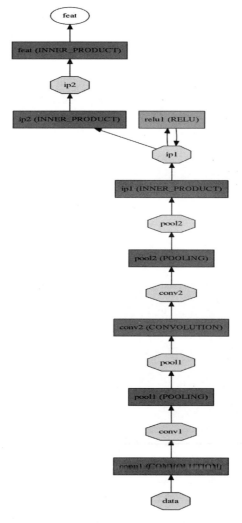

图 4.9　用 Caffe 中/python/draw_net.py 画出的 siamese 模型

(1) 以 LeNet 网络模型层为例来解释

DATA 一般包括训练数据和测试数据层两种类型。

① 训练数据层：

```
layer {
 name: "mnist"
 type: "Data"
 top: "data"
 top: "label"
 include {
   phase: TRAIN
 }
 transform_param {
   scale: 0.00390625
 }
 data_param {
   source: "examples/mnist/mnist_train_lmdb"
   batch_size: 64
   backend: LMDB
 }
}
```

② 测试数据层：

```
layer {
 name: "mnist"
 type: "Data"
 top: "data"
 top: "label"
 include {
   phase: TEST
 }
 transform_param {
   scale: 0.00390625
 }
 data_param {
   source: "examples/mnist/mnist_test_lmdb"
   batch_size: 100
   backend: LMDB
 }
}
```

③ 卷积层（CONVOLUATION）。blobs_lr:1 、blobs_lr:2 分别表示 weight 及 bias 更新时的学习率，这里权重的学习率为 solver.prototxt 文件中定义的学习率，bias 的学习率是权重学

习率的 2 倍，这样一般会得到很好的收敛速度。num_output 表示滤波的个数，kernelsize 表示滤波的大小，stride 表示步长，weight_filter 表示滤波的类型。

```
layer {
  name: "conv1"
  type: "Convolution"
  bottom: "data"
  top: "conv1"
  param {
    lr_mult: 1  //weight 学习率
  }
  param {
    lr_mult: 2  //bias 学习率，一般为 weight 的两倍
  }
  convolution_param {
    num_output: 20   //滤波器个数
    kernel_size: 5
    stride: 1   //步长
    weight_filler {
      type: "xavier"
    }
    bias_filler {
      type: "constant"
    }
  }
}
```

④ 池化层（POOLING）：

```
layer {
  name: "pool1"
  type: "Pooling"
  bottom: "conv1"
  top: "pool1"
  pooling_param {
    pool: MAX
    kernel_size: 2
    stride: 2
  }
}
```

⑤ INNER_PRODUCT（表示全连接）：

```
layer {
  name: "ip1"
```

```
type: "InnerProduct"
bottom: "pool2"
top: "ip1"
param {
  lr_mult: 1
}
param {
  lr_mult: 2
}
inner_product_param {
  num_output: 500
  weight_filler {
    type: "xavier"
  }
  bias_filler {
    type: "constant"
  }
}
}
```

⑥ RELU（激活函数，非线性变化层）：

```
layer {
  name: "relu1"
  type: "ReLU"
  bottom: "ip1"
  top: "ip1"
}
```

⑦ SOFTMAX：

```
layer {
  name: "loss"
  type: "SoftmaxWithLoss"
  bottom: "ip2"
  bottom: "label"
  top: "loss"
}
```

⑧ 参数配置文件。***_solver.prototxt 文件定义一些模型训练过程中需要的参数，比较学习率、权重衰减系数、迭代次数、使用 GPU 还是 CPU 等。

```
# The train/test net protocol buffer definition
net: "examples/mnist/lenet_train_test.prototxt"

# test_iter specifies how many forward passes the test should carry out.
```

```
# In the case of MNIST, we have test batch size 100 and 100 test iterations,
# covering the full 10,000 testing images.
test_iter: 100

# Carry out testing every 500 training iterations.
test_interval: 500

# The base learning rate, momentum and the weight decay of the network.
base_lr: 0.01
momentum: 0.9
weight_decay: 0.0005

# The learning rate policy
lr_policy: "inv"
gamma: 0.0001
power: 0.75

# Display every 100 iterations
display: 100

# The maximum number of iterations
max_iter: 10000

# snapshot intermediate results
snapshot: 5000
snapshot_prefix: "examples/mnist/lenet"

# solver mode: CPU or GPU
solver_mode: GPU
device_id: 0    #在cmdcaffe接口下，GPU序号从0开始，如果有一个GPU，则device_id:0
```

（2）使用 Caffe 训练模型步骤

① 准备数据。在 Caffe 中使用数据来对机器学习算法进行训练时，首先需要了解基本数据组成。不论使用何种框架进行 CNNs 训练，共有 3 种数据集：

- Training Set：用于训练网络。
- Validation Set：用于训练时测试网络准确率。
- Test Set：用于测试网络训练完成后的最终正确率。

② 重建 lmdb/leveldb 文件，Caffe 支持 3 种数据格式输入：images，levelda，lmdb。Caffe 生成的数据分为 2 种格式：lmdb 和 leveldb。

- 它们都是键-值对（Key-Value Pair）嵌入式数据库管理系统编程库。

- 虽然 lmdb 的内存消耗是 leveldb 的 1.1 倍，但是 lmdb 的速度比 leveldb 快 10%至 15%，更重要的是 lmdb 允许多种训练模型同时读取同一组数据集。因此 lmdb 取代 leveldb 成为 Caffe 默认的数据集生成格式。

③ 定义 name.prototxt、name_solver.prototxt 文件。
④ 训练模型。

（3）Caffe 中比较有用且基础的接口（cmdcaffe）

在使用 cmdcaffe 时，需要默认切换到 Caffe_Root 文件夹下。已经训练好的 caffe 模型可在 git 的 caffe 项目中下载，比较经典的模型有 AlexNet.caffemodel、LeNet.caffemodel 和 RCnn.caffemodel，其他的可以在 Caffe 的 Git 官网上面下载。

① 训练模型，以 mnist 为例子：

```
./build/tools/caffe train --solver=examples/mnist/lenet_solver.prototxt
```

Caffe 官网上给的例子不能直接执行，需要利用上述命令使用 tools 下的 Caffe 接口，因为 Caffe 默认都需要从根目录下面执行文件。

② 观察各个阶段的运行时间：

```
./build/tools/caffe time
--model=models/bvlc_reference_caffenet/train_val.prototxt
```

③ 使用已有模型提取特征：

```
./build/tools/extract_features.bin
models/bvlc_reference_caffenet/bvlc_reference_caffenet.caffemodel
examples/_temp/imagenet_val.prototxt conv5 examples/_temp/features 10
```

conv5 表示提取第五个卷积层的特征，examples/_temp/features 表示存放结果的目录（这里的目录需要提前构建好）。

④ 使用自己的数据集对已经训练好的模型进行 fine-tuning 的操作（使用 cmdcaffe 接口来进行）：

```
./build/tools/caffe train -solver models/finetune_flickr_style/solver.prototxt
-weights models/bvlc_reference_caffenet/bvlc_reference_caffenet.caffemodel -gpu 0
[option] 2>&1 | tee log.txt
```

⑤ draw_net.py 可以根据.prototxt 文件将模式用图示的方法表示出来：

```
./python/draw_net.py ./examples/siamese/mnist_siamese.prototxt  ./examples/siamese/mnist_siamese.png
```

使用该接口进行网络的绘制示例化，第一个参数为模型文件，第二个参数为所绘模型图的保存地址。

4.5　Gensim

Gensim 是一个免费 Python 库，能够从文档中有效地自动抽取语义主题。Gensim 中的算法包括 LSA（Latent Semantic Analysis）、LDA（Latent Dirichlet Allocation）、RP（Random Projections），通过在一个训练文档语料库中检查词汇统计联合出现模式，可以用来发掘文档语义结构，这些算法属于无监督学习，可以处理原始的、非结构化的文本（Plain Text）。

4.5.1　Gensim 特性与核心概念

1. Gensim 特性

- 内存独立，对于训练语料来说，没必要在任何时间将整个语料都驻留在 RAM 中。
- 有效实现了许多流行的向量空间算法，包括 TF-IDF、分布式 LSA、分布式 LDA 以及 RP，并且很容易添加新算法。
- 对流行的数据格式进行 I/O 封装和转换。
- 在其语义表达中，可以进行相似查询。
- Gensim 的创建原因是缺乏简单地实现主题建模的可扩展软件框架。

2. Gensim 核心概念

- 语料（Corpus）：一组原始文本的集合，用于无监督地训练文本主题的隐藏层结构。语料中不需要人工标注的附加信息。在 Gensim 中，Corpus 通常是一个可迭代的对象（比如列表）。每一次迭代返回一个可用于表达文本对象的稀疏向量。
- 向量（Vector）：由一组文本特征构成的列表，是一段文本在 Gensim 中的内部表达。
- 稀疏向量（Sparse Vector）：通常，我们可以略去向量中多余的 0 元素。此时，向量中的每一个元素都是一个(key, value)的元组。
- 模型（Model）：一个抽象的术语，定义了两个向量空间的变换（从文本的一种向量表达变换为另一种向量表达）。

4.5.2　训练语料的预处理

Gensim 是使用 Python 语言开发的，为了减少安装中的烦琐，直接使用 Anaconda 工具进行集中安装，输入"pip install genism"即可，这里不再赘述。

训练语料的预处理指的是将文档中原始的字符文本转换成 Gensim 模型所能理解的稀疏向量的过程。

通常，要处理的原生语料是一堆文档的集合，每一篇文档又是一些原生字符的集合。在交给 Gensim 的模型训练之前，需要将这些原生字符解析成 Gensim 能处理的稀疏向量的格式。由于语言和应用的多样性，需要先对原始文本进行分词、去除停用词等操作，得到每一篇文档的特征列表。例如，在词袋模型中，文档的特征就是其包含的单词：

```
texts = [['human', 'interface', 'computer'],
['survey', 'user', 'computer', 'system', 'response', 'time'],
['eps', 'user', 'interface', 'system'],
['system', 'human', 'system', 'eps'],
['user', 'response', 'time'],
['trees'],
['graph', 'trees'],
['graph', 'minors', 'trees'],
['graph', 'minors', 'survey']]
```

其中，语料的每一个元素对应一篇文档。

接下来，可以调用 Gensim 提供的 API 建立语料特征（此处是单词）的索引字典，并将文本特征的原始表达转化成词袋模型对应的稀疏向量的表达。依然以词袋模型为例：

```
from gensim import corpora
dictionary = corpora.Dictionary(texts)
corpus = [dictionary.doc2bow(text) for text in texts]
print corpus[0] # [(0, 1), (1, 1), (2, 1)]
```

到这里，训练语料的预处理工作就完成了，得到了语料中每一篇文档对应的稀疏向量（这里是 bow 向量）。向量的每一个元素代表一个单词在这篇文档中出现的次数。值得注意的是，虽然词袋模型是很多主题模型的基本假设，但是这里介绍的 doc2bow 函数并不是将文本转化成稀疏向量的唯一途径。

最后，出于内存优化的考虑，Gensim 支持文档的流式处理。我们需要做的只是将上面的列表封装成一个 Python 迭代器，每一次迭代都返回一个稀疏向量即可。

```
class MyCorpus(object):
    def __iter__(self):
        for line in open('mycorpus.txt'):
            # assume there's one document per line, tokens separated by whitespace
            yield dictionary.doc2bow(line.lower().split())
```

4.5.3 主题向量的变换

对文本向量的变换是 Gensim 的核心。通过挖掘语料中隐藏的语义结构特征，最终可以变换出一个简洁高效的文本向量。

在 Gensim 中，每一个向量变换的操作都对应着一个主题模型，例如上面提到的对应着词袋模型的 doc2bow 变换。每一个模型又都是一个标准的 Python 对象。下面以 TF-IDF 模型为例介绍 Gensim 模型的一般使用方法。

首先是模型对象的初始化。通常 Gensim 模型都接受一段训练语料（注意，在 Gensim 中，语料对应着一个稀疏向量的迭代器）作为初始化的参数。显然，越复杂的模型需要配置的参数越多。

```
from gensim import models
tfidf = models.TfidfModel(corpus)
```

其中，corpus 是一个返回 bow 向量的迭代器。这两行代码将完成对 corpus 中出现的每一个特征的 IDF 值的统计工作。

接下来，可以调用这个模型将任意一段语料（依然是 bow 向量的迭代器）转化成 ifidf 向量的迭代器。需要注意的是，这里的 bow 向量必须与训练语料的 bow 向量共享同一个特征字典（共享同一个向量空间）。

```
doc_bow = [(0, 1), (1, 1)]
print tfidf[doc_bow] # [(0, 0.70710678), (1, 0.70710678)]
```

同样是出于内存的考虑，model[corpus]方法返回的是一个迭代器。如果要多次访问 model[corpus]的返回结果，可以先将结果向量序列化到磁盘上，也可以将训练好的模型持久化到磁盘上，以便下一次使用：

```
tfidf.save("./model.tfidf")
tfidf = models.TfidfModel.load("./model.tfidf")
```

Gensim 内置了多种主题模型的向量变换，包括 LDA、LSI、RP、HDP 等。这些模型通常以 bow 向量或 tfidf 向量的语料为输入，生成相应的主题向量。所有的模型都支持流式计算。

4.5.4 文档相似度的计算

在得到每一篇文档对应的主题向量后，就可以计算文档之间的相似度，进而完成文本聚类、信息检索之类的任务。在 Gensim 中，也提供了这一类任务的 API 接口。以信息检索为例，对于一篇待检索的 query，目标是从文本集合中检索出主题相似度最高的文档。

首先，需要将待检索的 query 和文本放在同一个向量空间里进行表达（以 LSI 向量空间为例）：

```
# 构造 LSI 模型并将待检索的 query 和文本转化为 LSI 主题向量
# 转换之前的 corpus 和 query 均是 BOW 向量
lsi_model = models.LsiModel(corpus, id2word=dictionary,num_topics=2)
documents = lsi_model[corpus]
query_vec = lsi_model[query]
```

接下来，用待检索的文档向量初始化一个相似度计算的对象：

```
index = similarities.MatrixSimilarity(documents)
```

也可以通过 save()和 load()方法持久化这个相似度矩阵：

```
index.save('/tmp/test.index')
index = similarities.MatrixSimilarity.load('/tmp/test.index')
```

如果待检索的目标文档过多，使用 similarities.MatrixSimilarity 类往往会带来内存不够用的问题。此时，可以改用 similarities.Similarity 类。二者的接口基本保持一致。最后，借助 index 对象计算任意一段 query 和所有文档的余弦相似度：

```
sims = index[query_vec]
#返回一个元组类型的迭代器：(idx, sim)
```

TF-IDF 是一种统计方法，用以评估一个字词对于一个文件集或一个语料库其中一份文件的重要程度。字词的重要性随着它在文件中出现的次数成正比增加，但同时会随着它在语料库中出现的频率成反比下降。TF-IDF 加权的各种形式常被搜索引擎应用，作为文件与用户查询之间相关程度的度量或评级。

【例 4.10】

```
# -*- coding: utf-8 -*-
"""
Created on Wed Mar 20 09:32:50 2019
@author: DLG
"""

from gensim import corpora
from collections import defaultdict
documents = ["Human machine interface for lab abc computer applications",
             "A survey of user opinion of computer system response time",
             "The EPS user interface management system",
             "System and human system engineering testing of EPS",
             "Relation of user perceived response time to error measurement",
             "The generation of random binary unordered trees",
             "The intersection graph of paths in trees",
             "Graph minors IV Widths of trees and well quasi ordering",
             "Graph minors A survey"]

# 去掉停用词
stoplist = set('for a of the and to in'.split())
texts = [[word for word in document.lower().split() if word not in stoplist]
         for document in documents]

# 去掉只出现一次的单词
frequency = defaultdict(int)
for text in texts:
    for token in text:
        frequency[token] += 1
texts = [[token for token in text if frequency[token] > 1]
         for text in texts]
```

```python
dictionary = corpora.Dictionary(texts)    # 生成词典

# 将文档存入字典，字典有很多功能，比如
# diction.token2id 存放的是单词-id key-value 对
# diction.dfs 存放的是单词的出现频率
dictionary.save('D:\Anaconda3\workspace\deerwester.dict')  # store the
dictionary, for future reference
corpus = [dictionary.doc2bow(text) for text in texts]
corpora.MmCorpus.serialize('D:\Anaconda3\workspace\deerwester.mm', corpus)  #
store to disk, for later use
import os
from gensim import corpora, models, similarities
from pprint import pprint
from matplotlib import pyplot as plt
import logging

# logging.basicConfig(format='%(asctime)s : %(levelname)s : %(message)s',
level=logging.INFO)

def PrintDictionary(dictionary):
    token2id = dictionary.token2id
    dfs = dictionary.dfs
    token_info = {}
    for word in token2id:
        token_info[word] = dict(
            word = word,
            id = token2id[word],
            freq = dfs[token2id[word]]
        )
    token_items = token_info.values()
    token_items = sorted(token_items, key = lambda x:x['id'])
    print('The info of dictionary: ')
    pprint(token_items)
    print('--------------------------')

def Show2dCorpora(corpus):
    nodes = list(corpus)
    ax0 = [x[0][1] for x in nodes] # 绘制各个doc代表的点
    ax1 = [x[1][1] for x in nodes]
    # print(ax0)
    # print(ax1)
    plt.plot(ax0,ax1,'o')
    plt.show()
```

```python
if (os.path.exists("/tmp/deerwester.dict")):
    dictionary = corpora.Dictionary.load('/tmp/deerwester.dict')
    corpus = corpora.MmCorpus('/tmp/deerwester.mm')
    print("Used files generated from first tutorial")
else:
    print("Please run first tutorial to generate data set")

PrintDictionary(dictionary)

# 尝试将 corpus(bow 形式) 转化成 TF-IDF 形式
# step 1 -- initialize a model 将文档由按照词频表示转变为按照 TF-IDF 格式来表示
tfidf_model = models.TfidfModel(corpus)
doc_bow = [(0, 1), (1, 1),[4,3]]
doc_tfidf = tfidf_model[doc_bow]

# 将整个 corpus 转为 TF-IDF 格式
corpus_tfidf = tfidf_model[corpus]
# pprint(list(corpus_tfidf))
# pprint(list(corpus))

## LSI 模型 ******************************************************
# 转化为 LSI 模型，可用作聚类或分类
lsi_model = models.LsiModel(corpus_tfidf, id2word=dictionary, num_topics=2)
corpus_lsi = lsi_model[corpus_tfidf]
nodes = list(corpus_lsi)
# pprint(nodes)
lsi_model.print_topics(2)            # 打印各 topic 的含义

# ax0 = [x[0][1] for x in nodes]     # 绘制各个 doc 代表的点
# ax1 = [x[1][1] for x in nodes]
# print(ax0)
# print(ax1)
# plt.plot(ax0,ax1,'o')
# plt.show()

lsi_model.save('D:\Anaconda3\workspace\model.lsi') # same for tfidf, lda, ...
lsi_model = models.LsiModel.load('D:\Anaconda3\workspace\model.lsi')
# ******************************************************

## LDA 模型 ******************************************************
lda_model = models.LdaModel(corpus_tfidf, id2word=dictionary, num_topics=2)
corpus_lda = lda_model[corpus_tfidf]
```

```python
Show2dCorpora(corpus_lsi)
# nodes = list(corpus_lda)
# pprint(list(corpus_lda))

# 此外，还有Random Projections 和Hierarchical Dirichlet Process等模型
corpus_simi_matrix = similarities.MatrixSimilarity(corpus_lsi)
# 计算一个新的文本与既有文本的相关度
test_text = "Human computer interaction".split()
test_bow = dictionary.doc2bow(test_text)
test_tfidf = tfidf_model[test_bow]
test_lsi = lsi_model[test_tfidf]
test_simi = corpus_simi_matrix[test_lsi]
print(list(enumerate(test_simi)))
```

输出：

```
runfile('D:/Anaconda3/workspace/untitled1.py', wdir='D:/Anaconda3/workspace')
2019-03-20 14:57:11,548 : INFO : adding document #0 to Dictionary(0 unique tokens: [])
2019-03-20 14:57:11,550 : INFO : built Dictionary(12 unique tokens: ['computer', 'human', 'interface', 'response', 'survey']...) from 9 documents (total 29 corpus positions)
2019-03-20 14:57:11,551 : INFO : saving Dictionary object under D:\Anaconda3\workspace\deerwester.dict, separately None
2019-03-20 14:57:11,552 : INFO : saved D:\Anaconda3\workspace\deerwester.dict
2019-03-20 14:57:11,554 : INFO : storing corpus in Matrix Market format to D:\Anaconda3\workspace\deerwester.mm
2019-03-20 14:57:11,559 : INFO : saving sparse matrix to D:\Anaconda3\workspace\deerwester.mm
2019-03-20 14:57:11,560 : INFO : PROGRESS: saving document #0
2019-03-20 14:57:11,561 : INFO : saved 9x12 matrix, density=25.926% (28/108)
2019-03-20 14:57:11,563 : INFO : saving MmCorpus index to D:\Anaconda3\workspace\deerwester.mm.index
2019-03-20 14:57:11,565 : INFO : collecting document frequencies
2019-03-20 14:57:11,566 : INFO : PROGRESS: processing document #0
2019-03-20 14:57:11,566 : INFO : calculating IDF weights for 9 documents and 11 features (28 matrix non-zeros)
2019-03-20 14:57:11,570 : INFO : using serial LSI version on this node
2019-03-20 14:57:11,571 : INFO : updating model with new documents
2019-03-20 14:57:11,572 : INFO : preparing a new chunk of documents
2019-03-20 14:57:11,573 : INFO : using 100 extra samples and 2 power iterations
2019-03-20 14:57:11,573 : INFO : 1st phase: constructing (12, 102) action matrix
2019-03-20 14:57:11,578 : INFO : orthonormalizing (12, 102) action matrix
2019-03-20 14:57:11,609 : INFO : 2nd phase: running dense svd on (12, 9) matrix
```

```
2019-03-20 14:57:11,620 : INFO : computing the final decomposition
2019-03-20 14:57:11,621 : INFO : keeping 2 factors (discarding 47.565% of energy spectrum)
2019-03-20 14:57:11,621 : INFO : processed documents up to #9
2019-03-20 14:57:11,622 : INFO : topic #0(1.594): 0.703*"trees" + 0.538*"graph" + 0.402*"minors" + 0.187*"survey" + 0.061*"system" + 0.060*"response" + 0.060*"time" + 0.058*"user" + 0.049*"computer" + 0.035*"interface"
2019-03-20 14:57:11,623 : INFO : topic #1(1.476): -0.460*"system" + -0.373*"user" + -0.332*"eps" + -0.328*"interface" + -0.320*"time" + -0.320*"response" + -0.293*"computer" + -0.280*"human" + -0.171*"survey" + 0.161*"trees"
2019-03-20 14:57:11,625 : INFO : topic #0(1.594): 0.703*"trees" + 0.538*"graph" + 0.402*"minors" + 0.187*"survey" + 0.061*"system" + 0.060*"response" + 0.060*"time" + 0.058*"user" + 0.049*"computer" + 0.035*"interface"
2019-03-20 14:57:11,627 : INFO : topic #1(1.476): -0.460*"system" + -0.373*"user" + -0.332*"eps" + -0.328*"interface" + -0.320*"time" + -0.320*"response" + -0.293*"computer" + -0.280*"human" + -0.171*"survey" + 0.161*"trees"
2019-03-20 14:57:11,628 : INFO : saving Projection object under D:\Anaconda3\workspace\model.lsi.projection, separately None
2019-03-20 14:57:11,630 : INFO : saved D:\Anaconda3\workspace\model.lsi.projection
2019-03-20 14:57:11,632 : INFO : saving LsiModel object under D:\Anaconda3\workspace\model.lsi, separately None
2019-03-20 14:57:11,648 : INFO : not storing attribute projection
2019-03-20 14:57:11,649 : INFO : not storing attribute dispatcher
2019-03-20 14:57:11,651 : INFO : saved D:\Anaconda3\workspace\model.lsi
2019-03-20 14:57:11,651 : INFO : loading LsiModel object from D:\Anaconda3\workspace\model.lsi
2019-03-20 14:57:11,652 : INFO : loading id2word recursively from D:\Anaconda3\workspace\model.lsi.id2word.* with mmap=None
2019-03-20 14:57:11,653 : INFO : setting ignored attribute projection to None
2019-03-20 14:57:11,653 : INFO : setting ignored attribute dispatcher to None
2019-03-20 14:57:11,654 : INFO : loaded D:\Anaconda3\workspace\model.lsi
2019-03-20 14:57:11,654 : INFO : loading LsiModel object from D:\Anaconda3\workspace\model.lsi.projection
2019-03-20 14:57:11,655 : INFO : loaded D:\Anaconda3\workspace\model.lsi.projection
2019-03-20 14:57:11,656 : INFO : using symmetric alpha at 0.5
2019-03-20 14:57:11,656 : INFO : using symmetric eta at 0.5
2019-03-20 14:57:11,657 : INFO : using serial LDA version on this node
2019-03-20 14:57:11,659 : INFO : running online (single-pass) LDA training, 2 topics, 1 passes over the supplied corpus of 9 documents, updating model once every 9 documents, evaluating perplexity every 9 documents, iterating 50x with a convergence threshold of 0.001000
2019-03-20 14:57:11,660 : WARNING : too few updates, training might not converge;
```

```
consider increasing the number of passes or iterations to improve accuracy
2019-03-20 14:57:11,696 : INFO : -3.597 per-word bound, 12.1 perplexity estimate
based on a held-out corpus of 9 documents with 15 words
2019-03-20 14:57:11,697 : INFO : PROGRESS: pass 0, at document #9/9
2019-03-20 14:57:11,702 : INFO : topic #0 (0.500): 0.104*"trees" + 0.099*"system"
+ 0.093*"interface" + 0.091*"user" + 0.089*"time" + 0.087*"human" + 0.086*"eps"
+ 0.085*"response" + 0.080*"computer" + 0.066*"graph"
2019-03-20 14:57:11,703 : INFO : topic #1 (0.500): 0.136*"graph" + 0.134*"trees"
+ 0.114*"minors" + 0.098*"survey" + 0.078*"system" + 0.068*"user" +
0.066*"computer" + 0.064*"response" + 0.062*"eps" + 0.061*"human"
2019-03-20 14:57:11,704 : INFO : topic diff=0.354399, rho=1.000000
Please run first tutorial to generate data set
The info of dictionary:
[{'freq': 2, 'id': 0, 'word': 'computer'},
 {'freq': 2, 'id': 1, 'word': 'human'},
 {'freq': 2, 'id': 2, 'word': 'interface'},
 {'freq': 2, 'id': 3, 'word': 'response'},
 {'freq': 2, 'id': 4, 'word': 'survey'},
 {'freq': 3, 'id': 5, 'word': 'system'},
 {'freq': 2, 'id': 6, 'word': 'time'},
 {'freq': 3, 'id': 7, 'word': 'user'},
 {'freq': 2, 'id': 8, 'word': 'eps'},
 {'freq': 3, 'id': 9, 'word': 'trees'},
 {'freq': 3, 'id': 10, 'word': 'graph'},
 {'freq': 2, 'id': 11, 'word': 'minors'}]
------------------------
2019-03-20 14:57:11,983 : WARNING : scanning corpus to determine the number of
features (consider setting 'num_features' explicitly)
2019-03-20 14:57:11,998 : INFO : creating matrix with 9 documents and 2 features
[(0, 0.9991645), (1, 0.9963216), (2, 0.9990505), (3, 0.99886364), (4, 0.99996823),
(5, -0.058117405), (6, -0.02158928), (7, 0.0135240555), (8, 0.25163394)]
```

可以看到，这里除了 corpus 以外，还多了 num_topic 的选项。这是指的是潜在主题（topic）的数目，也等于转成 LSI 模型以后每个文档对应的向量长度。转化以后的向量在各项的值即为该文档在该潜在主题的权重。因此，LSI 和 LDA 的结果也可以看作该文档的文档向量，用于后续的分类、聚类等算法。值得注意的是，id2word 是所有模型都有的选项，可以指定使用的词典。从图 4.10 可以很清楚地看到，9 个文档可以看成两类，分别是前 5 行和后 4 行。与向量的相似度计算方式一样，采用余弦方法计算得到。一般来讲，使用 LSI 模型得到的向量进行计算效果比较好。得到的结果为[(0, 0.9991645), (1, 0.9963216), (2, 0.9990505), (3, 0.99886364), (4, 0.99996823), (5, -0.058117405), (6, -0.02158928), (7, 0.0135240555), (8, 0.25163394)]，显然属于第一类。

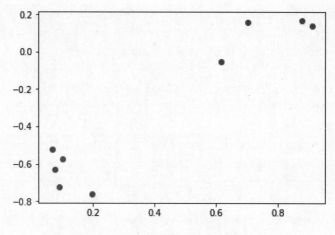

图 4.10　num_topics=2 图直观的显示

4.6　Pylearn2

Pylearn2 是一个基于 Theano 的机器学习库，大部分功能都是基于 Theano 顶层实现的。这意味着用户可以用数学表达式去编写 Pylearn2 插件（新模型、算法等）。Theano 不仅会帮助用户优化这些表达式，并且会将这些表达式编译到 CPU 或者 GPU 中。

1. Pylearn2 功能特性

- 研究人员可以添加他们所需要的功能，避免提前设置过多自上而下的计划（容易导致用户难以使用）。
- 一个实现高效科学实验的机器学习工具箱。
- LISA 实验室发布的所有模型/算法都应该在 Pylearn2 中具有引用说明。
- Pylearn2 可能会引用其他库，如 Scikit-learn。
- Pylearn2 与 SKlearn 的不同之处在于，Pylearn2 旨在提供极大的灵活性，研究人员可以自行实现任何事情；SKlearn 旨在作为一个"黑箱"，即使用户对算法底层实现没有了解也可以输出实验结果。
- 包括矢量、图像、视频等数据集接口。
- 一个针对普通 MLP/RBM/SDA/卷积实验所有需要内容的小型框架。
- 易于重复使用 Pylearn2 的子组件。
- 使用 Pylearn2 库的一个子组件不会强制用户使用或学习使用所有其他子组件，用户可以自行选择。
- 支持跨平台序列化学习模型。
- 十分简单易用（蒙特利尔大学的 IFT6266）。

2. Pylearn2 下载与安装

目前没有 PyPI 安装包，所以 Pylearn2 不能用 pip 命令来进行安装。

（1）用户必须使用如下命令从 GitHub 上下载：

```
$ git clone git://github.com/lisa-lab/pylearn2.git
```

为了让 Pylearn2 能够在用户安装的 Python 中使用，要在 Pylearn2 顶层目录中执行如下命令：

```
python setup.py develop
```

（2）配置环境变量。一般用 .bash_profile，也可以放在 .bashrc 里：

```
export PYLEARN2_DATA_PATH=/path/to/data
```

（3）进入 Pylearn2：

```
$ python setup.py build
$ python setup.py install --prefix=$PYTHON_HOME
```

（4）测试。参考 http://deeplearning.net/software/pylearn2/tutorial/index.html#tutorial，首先下载数据集 cifar-10: http://www.cs.toronto.edu/~kriz/cifar.html。

```
$ wget http://www.cs.toronto.edu/~kriz/cifar-10-python.tar.gz
$ tar -zxvf cifar-10*
```

① 创建数据集：

```
$ python make_dataset.py
```

② 训练模型：

```
$ train.py cifar_grbm_smd.yaml
```

③ 检验模型：

```
$ show_weights.py cifar_grbm_smd.pkl
$ plot_monitor.py cifar_grbm_smd.pkl
```

4.7 Shogun

Shogun 是一个机器学习工具箱，由 Soeren Sonnenburg 和 Gunnar Raetsch 创建，重点是大尺度上的内核学习方法，特别是支持向量机（SVM，Support Vector Machines）的学习工具箱。它提供了一个通用的连接到几个不同的 SVM 实现方式中的 SVM 对象接口，目前发展最先进的 LIBSVM 和 SVMlight 也位于其中，每个 SVM 都可以与各种内核相结合。工具箱不仅为常用的内核程序（如线性、多项式、高斯和 S 型核函数）提供了高效的实现途径，还自带了一些近期的字符串内核函数，例如 Fischer、TOP、Spectrum、加权度内核与移位，后来有效的

LINADD 优化内核函数也已经实现。

此外，Shogun 还提供了使用自定义预计算内核工作的自由，其中一个重要特征就是可以通过多个子内核的加权线性组合来构造的组合核，每个子内核无须工作在同一个域中。通过使用多内核学习可知最优子内核的加权。

目前 Shogun 可以解决 SVM 2 类的分类和回归问题。此外，Shogun 也添加了像线性判别分析（LDA）、线性规划（LPM）、（内核）感知等大量线性方法和一些用于训练隐马尔可夫模型的算法。

GitHub 项目地址为 https://github.com/shogun-toolbox/shogun。

4.8 Chainer

基于 Python 的 Chainer 是用 Python 开发的，允许在运行时检查和自定义 Python 中的所有代码和可理解的 Python 消息。

大多数现有的深度学习框架都是基于 Define-and-Run 的方案。也就是说，首先要有一个预先被定义的网络结构，然后用户才能给这个网络训练数据。因为所有的网络都是在前向/后向传播计算完成之前就定义好的了，所以所有的逻辑必须以数据的形式嵌入网络结构中去。然而，Chainer 采纳了 Define-by-Run 的方案，比如网络是通过实际的前向计算在运行中定义的。更确切地说，Chainer 存储计算的历史而不是编程逻辑。这个策略能够充分利用 Python 中编程逻辑的力量。定义方案是 Chainer 的核心概念。总而言之，Define-and-Run 的方案是结构领着数据走，有了结构才能够通过提供数据来训练网络；Define-by-Run 的方案是数据领着结构走，有了数据参数的定义才有网络的概念。

1. 网络搭建

（1）前向传播 / 反向传播的计算

```
x=Variable(np.array([[1,2,3],[4,5,6]],dtype=np.float32))
y=x**2-2*x+1
y.grad=np.ones((2,3),dtype=np.float32)
y.backward()
x.grad
```

从这个例子就可以看出来，并没有预先定义一个结构，而是有了 x 和 y 的变量，通过调用 y.backward() 来实现反向传播的计算。

（2）links

link 只是一个能容纳参数的对象。一个经常使用的 link 就是 linear link，它代表数学表达式 $f(x) = Wx + b$。

```
f=L.Linear(3,2)
```

这个代表的是一个输入为三维、输出为二维的线性方程。此外，大部分的方程和 link 都只接受 mini-batch 的输入，也就是说，第一维被默认为 batch 的维度，所以在上面的例子中输入必须有(N,3)的形状。所有的参数都以属性的形式存放。在上面的例子中，W 和 b 存放在 f 的属性中，可以通过下面的语句完成：

```
>>>f.W.data
array([[ 1.01847613, 0.23103087, 0.56507462],[ 1.29378033, 1.07823515,
-0.56423163]], dtype=float32)
>>>f.b.data
array([ 0., 0.], dtype=float32)
```

默认下，W 和 b 都是随机初始化。

合起来，一个完整的例子如下：

```
>>>x=Variable(np.array([[1,2,3],[4,5,6]],dtype=np.float32))
>>>y=f(x)
>>>y.data
array([[ 3.1757617 , 1.75755572],[ 8.61950684, 7.18090773]], dtype=float32)
```

梯度这个参数通过调用 backward()方法来实现。需要注意的是，这个梯度是被累加起来的，所以每次调用之前需要对梯度进行清零操作。

```
>>>f.cleargrads()
```

最终的例子是：

```
>>>y.grad=np.ones((2,2),dtype=np.float32)
>>>y.backward()
>>>f.W.gradarray([[ 5., 7., 9.],[ 5., 7., 9.]], dtype=float32)
>>>f.b.gradarray([ 2., 2.], dtype=float32)
```

（3）用 chain 来写一个模块

大部分的神经网络都包含许多个 link。比如，一个多层感知器由多个线性层组成。接下来就写一个这样的例子：

```
>>>classMyChain(Chain):
    ...def__init__(self):
      ...super(MyChain,self).__init__(
          ...l1=L.Linear(4,3),
          ...l2=L.Linear(3,2),
          ...)
      ...def__call__(self,x):
         ...h=self.l1(x)
         ...returnself.l2(h)
```

(4) 优化器

```
>>>model=MyChain()
>>>optimizer=optimizers.SGD()
>>>optimizer.use_cleargrads()
>>>optimizer.setup(model)
```

setup() 方法只是为优化器提供一个 link。若改梯度衰减值之类的则需要调用：

```
>>>optimizer.add_hook(chainer.optimizer.WeightDecay(0.0005))
```

2. Chainer 安装环境

若想其运行在 GPU 上，则需在电脑上安装 NVIDIA CUDA / cuDNN 环境，并且要安装与 CUDA 对应版本的 CuPy 包。CuPy 是 CUDA 上与 NumPy 兼容的多维数组的实现。CuPy 由 cupy.ndarray 构成，是多维数组类的核心，里面有很多函数，同时支持 numpy.ndarray 的接口。

（1）安装 CuPy

```
(For CUDA 8.0)
pip install cupy-cuda80

(For CUDA 9.0)
pip install cupy-cuda90

(For CUDA 9.1)
pip install cupy-cuda91

(For CUDA 9.2)
pip install cupy-cuda92
```

（2）卸载 Chainer

```
pip uninstall chainer
```

3. 利用 Chainer 实践 Mnist 例子（见图 4.11）

【例 4.11】

```
# -*- coding: utf-8 -*-
"""
Created on Fri Mar 22 18:08:12 2019
@author: DLG
"""
#!/usr/bin/env python
from __future__ import print_function
import numpy as np
import chainer
from chainer import backend, backends
```

```python
from chainer.backends import cuda
from chainer import Function, report, training, utils, Variable
from chainer import datasets, iterators, optimizers, serializers
from chainer import Link, Chain, ChainList
import chainer.functions as F
import chainer.links as L
import matplotlib.pyplot as plt
from chainer.datasets import mnist
from chainer import iterators
from chainer.dataset import concat_examples
from chainer.backends.cuda import to_cpu

train, test = mnist.get_mnist(withlabel=True, ndim=1)
x, t = train[0]
plt.imshow(x.reshape(28, 28), cmap="gray")
plt.savefig("5.png")

batchsize = 128

train_iter = iterators.SerialIterator(train, batchsize)
test_iter = iterators.SerialIterator(test, batchsize, repeat=False, shuffle=False)

# 定义训练模型
class MyNetwork(Chain):

    def __init__(self, n_mid_units=100, n_out=10):
        super(MyNetwork, self).__init__()
        with self.init_scope():
            self.l1 = L.Linear(None, n_mid_units)
            self.l2 = L.Linear(n_mid_units, n_mid_units)
            self.l3 = L.Linear(n_mid_units, n_out)

    def forward(self, x):
        h = F.relu(self.l1(x))
        h = F.relu(self.l2(h))
        return self.l3(h)

model = MyNetwork()

gpu_id = -1  # Set to 0 if you use GPU
if gpu_id >= 0:
    model.to_gpu(gpu_id)
```

```python
# 定义优化器
optimizer = optimizers.MomentumSGD(lr=0.01, momentum=0.9)
optimizer.setup(model)

# 开始训练模型
max_epoch = 20
while train_iter.epoch < max_epoch:

    # ---------- One iteration of the training loop ----------
    train_batch = train_iter.next()
    image_train, target_train = concat_examples(train_batch, gpu_id)

    # Calculate the prediction of the network
    prediction_train = model(image_train)

    # Calculate the loss with softmax_cross_entropy
    loss = F.softmax_cross_entropy(prediction_train, target_train)

    # Calculate the gradients in the network
    model.cleargrads()
    loss.backward()

    # Update all the trainable parameters
    optimizer.update()
    # --------------------- until here ---------------------

    # Check the validation accuracy of prediction after every epoch
    if train_iter.is_new_epoch:  # If this iteration is the final iteration of the current epoch

        # Display the training loss
        print('epoch:{:02d} train_loss:{:.04f} '.format(
            train_iter.epoch, float(to_cpu(loss.data))), end='')

        test_losses = []
        test_accuracies = []
        while True:
            test_batch = test_iter.next()
            image_test, target_test = concat_examples(test_batch, gpu_id)

            # Forward the test data
```

```python
            prediction_test = model(image_test)

            # Calculate the loss
            loss_test = F.softmax_cross_entropy(prediction_test, target_test)
            test_losses.append(to_cpu(loss_test.data))

            # Calculate the accuracy
            accuracy = F.accuracy(prediction_test, target_test)
            accuracy.to_cpu()
            test_accuracies.append(accuracy.data)

            if test_iter.is_new_epoch:
                test_iter.epoch = 0
                test_iter.current_position = 0
                test_iter.is_new_epoch = False
                test_iter._pushed_position = None
                break

        print('val_loss:{:.04f} val_accuracy:{:.04f}'.format(
            np.mean(test_losses), np.mean(test_accuracies)))

# 保存最佳模型
serializers.save_npz('train_mnist.model', model)

# 利用保存的模型对新数据进行预测
model = MyNetwork()
serializers.load_npz('train_mnist.model', model)

x, t = test[0]
plt.imshow(x.reshape(28, 28), cmap='gray')
plt.savefig('7.png')
print('label: ', t)

# 预测
print(x.shape, end=' -> ')
x = x[None, ...]
print(x.shape)

y = model(x)
y = y.data
pred_label = y.argmax(axis=1)
print("predicted label:", pred_label[0])
```

结果输出：

```
In [1]:runfile('D:/python 语言与大数据/python_work/chainer_mnist.py',
wdir='D:/python 语言与大数据/python_work')
Downloading from http://yann.lecun.com/exdb/mnist/train-images-idx3-ubyte.gz...
Downloading from http://yann.lecun.com/exdb/mnist/train-labels-idx1-ubyte.gz...
Downloading from http://yann.lecun.com/exdb/mnist/t10k-images-idx3-ubyte.gz...
Downloading from http://yann.lecun.com/exdb/mnist/t10k-labels-idx1-ubyte.gz...
epoch:01 train_loss:0.2147 val_loss:0.2709 val_accuracy:0.9223
epoch:02 train_loss:0.1862 val_loss:0.1909 val_accuracy:0.9458
epoch:03 train_loss:0.2352 val_loss:0.1570 val_accuracy:0.9533
epoch:04 train_loss:0.1216 val_loss:0.1300 val_accuracy:0.9614
epoch:05 train_loss:0.0745 val_loss:0.1176 val_accuracy:0.9643
epoch:06 train_loss:0.1498 val_loss:0.1076 val_accuracy:0.9687
epoch:07 train_loss:0.0538 val_loss:0.0973 val_accuracy:0.9694
epoch:08 train_loss:0.1035 val_loss:0.0943 val_accuracy:0.9712
epoch:09 train_loss:0.0463 val_loss:0.0895 val_accuracy:0.9723
epoch:10 train_loss:0.0378 val_loss:0.0928 val_accuracy:0.9723
epoch:11 train_loss:0.1048 val_loss:0.0830 val_accuracy:0.9752
epoch:12 train_loss:0.0662 val_loss:0.0813 val_accuracy:0.9763
epoch:13 train_loss:0.0232 val_loss:0.0799 val_accuracy:0.9752
epoch:14 train_loss:0.0411 val_loss:0.0778 val_accuracy:0.9771
epoch:15 train_loss:0.0210 val_loss:0.0801 val_accuracy:0.9761
epoch:16 train_loss:0.0582 val_loss:0.0761 val_accuracy:0.9764
epoch:17 train_loss:0.0395 val_loss:0.0744 val_accuracy:0.9766
epoch:18 train_loss:0.0200 val_loss:0.0732 val_accuracy:0.9770
epoch:19 train_loss:0.0153 val_loss:0.0733 val_accuracy:0.9774
epoch:20 train_loss:0.0387 val_loss:0.0721 val_accuracy:0.9779
label: 7
(784,) -> (1, 784)
predicted label: 7
```

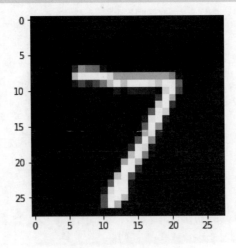

图 4.11　Chainer 实践 Mnist

如果已配好 Chainer 的 GPU 环境，上述代码同样可以在 GPU 上运行，只要将代码中的 gpu_id = -1 设为 gpu_id = 0 即可。

4.9 NuPIC

1. NuPIC 简介

说到 AI，就会想到深度学习、TensorFlow 等。目前深层神经网络占据了 AI 半壁江山。从本质上来说，深度学习所表述的神经网络其实源自于一个"类脑"的想法——通过模仿人类大脑神经元的相互连接，结合权重、反向传播等数学知识所构成。但是深度学习的神经网络运作仍与人类大脑真实运作情况相去甚远。

NuPIC 是一个在 GitHub 上开源了的 AI 算法平台，相比于深度学习，其更为接近人类大脑的运行结构。其算法的理论依据就是纯粹的生物神经学知识，类似突触连接与分解、神经元、多个脑皮层的交互、动作电位等。工程实现也基于此。

NuPIC 是由 Jeff Hawkins 等人创造的算法，其目的是生产出更为接近真实人脑运作的 AI。其理论依据主要是人脑中处理高级认知功能的新皮质部分的运作原理。目前，NuPIC 还远远无法模拟整个人脑。不过，这已经是机器学习上的重大进步了。

NuPIC 的运作比较类似于人的大脑，接受外部输入（时间序列），学习其模式，不断纠错，然后在这个过程中形成经验，进行预测。因此，NuPIC 可用于预测以及异常检测，适用面非常广（目前认为，只要是时间序列都可以）。内部结构决定了其有很强的记忆和学习功能。NuPIC 最特别的一点，也是其相对于其他深度学习算法最大的不同，那就是 NuPIC 模型一旦建好，就能够学习完全不同的多个输入模式，并对它们同时作出预测和异常检测。这一点类似于人脑，大脑不同部位负责不一样的学习任务。深度学习的算法做不到这一点，一个算法面对一个同样的输入模式。这只是深度学习与 NuPIC 的不同，并无优劣。

2. NuPIC 资源

GitHub：https://github.com/numenta/nupic。

国内资料：AIS 论坛，http://www.51hei.com/mcu/3970.html。

NuPIC API 教程：http://nupic.docs.numenta.org/1.0.3/index.html。

3. 总体框架

NuPIC 的核心其实就是一个算法，即 HTM（Hierarchical Temporal Memory）算法。在 HTM 算法中，有区域/层级的概念。IITM 算法的框架大致如图 4.12 所示。

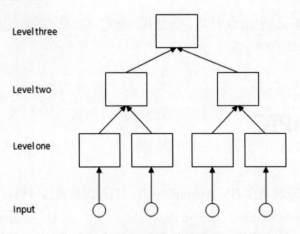

图 4.12 HTM 结构示意图

图 4.12 是一个 HTM 多层网络的框架图。最下面的是外部输入,算法接受的是被编码器编码过的时间序列。最下面的箭头往上就是一级级的区域,也称层级。最上面的箭头所指内容就是输出。上下层之间都有前馈和反馈。所以,HTM 网络也是神经网络的一种,看起来很像卷积网络,但实际上内部原理完全不一样。这张框架图可以用人的大脑皮层运作来描述:每一层的层级就类似于人脑中的每层脑皮质,然后每一层里面包含了无数的神经元以及联系神经元的突触。每一层级的结构大概,如图 4.13 所示。

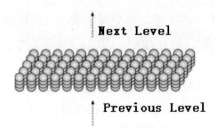

图 4.13 HTM 层级结构

图 4.13 是 HTM 区域的一部分。HTM 区域由许多细胞组成,细胞以成列的二维平面形式组织在一起。HTM 区域及它的柱状结构等同于新大脑皮层区域中的一层神经细胞。

- Cell:细胞,图 4.13 中圆圆的球状。
- 柱状区域:图 4.13 中 4 个 Cell 组合而成的一个列。
- 区域:层级,一个二维平面的柱状区域。

4. 神经学的概念

在 HTM 中,细胞有两种树突:近端树突和远端树突。

- 近端树突会接收外部输入序列的前馈输入。
- 远端树突会连接到同层区域的细胞输入。

(1) 树突区域

HTM 细胞有相当真实的（因此也是复杂的）树突模型。根据理论，每个 HTM 细胞有一个接近细胞核的树突和许多末端树突，称之为近端树突和远端树突。靠近端树突接受前馈输入，远端树突接受附近的细胞横向输入。有一类抑制神经元制约着整个柱状区域的细胞对相似前馈输入的响应。为了简化，用一个柱状区域共享的树突来代替每个细胞上的近核树突。

一个细胞有很多个树突区域。

(2) 突触

突触是树突与轴突相连的部分，因此突触是连接到树突区域的。在 HTM 中，一个细胞有非常多的突触。对于突触，有两个概念：潜在突触和活跃突触。一个树突区域对应这一堆潜在突触，对于这些突触，若"连通值"超过了阈值，则会变成活跃突触。连通值是对每个潜在突触设置的标量。一个突触的连通值是指树突与轴突之间连通性的程度。从生物学上讲，这一程度从完全不连接开始，到刚开始形成一个突触但仍未连接，再到满足连接的最低要求，最后满足最大连接。连通值是一个关于突触的标量，值域为 0.0 到 1.0。学习的过程包含增加或减少突触的连通值。当一个突触的连通值超过阈值时，它就是连通的并且权值为 1。当一个突触的连通值小于阈值，它就是不连通的且权值为 0，因此突触权值是二元性的。

5. HTM 工作流程

对于 HTM 的识别、学习都分为两个流程：首先，对输入进行稀疏离散表征，完成 Spatial Pooling（SP，空间池化），学习阶段会对突触的权值进行更新；其次，基于 Spatial Pooler 的结果，进行横向信息传递、预测，完成 Temporal Pooling（TP，时间池化），学习阶段也会对突触的权值进行更新。Temporal pooling 的输出作为更高层的输入，重复刚才的过程。识别和学习没有明显界限，在识别阶段将学习的部分功能关闭即可。

需要特别指出的是，每个列（Column）如果被激活（之后会解释什么是激活状态），那么它能够表征模式的部分含义，对于一个输入模式，就被一组稀疏的活跃 column 进行有效表征。但是我们知道，"ABCD"与"EBCF"中的模式 B 与 C 是不一样的，那么在 HTM 中如何实现不同上下文的表征呢？在 HTM 中，每个列拥有很多细胞，在不同的上下文时，激活的细胞是不一样的，这样就达到了能够表征不同上下文中相同内容的目的。

稀疏离散表征是 HTM 的重要基础，序列记忆等都基于稀疏离散表征。稀疏离散表征是将自然界语言（如图像、文本、音频等）转换为二进制序列，而且是稀疏的。稀疏离散表征的每一个活跃（值为1）的 bit 都能表达模式的部分含义，但是仅仅一个 bit 又是不够的，只有整体才能表达一个完整的模式。

图 4.14 深色的细胞就是一个表征（是对低层输入的表征），无色的代表没有激活，可见这是一个离散稀疏分布的。这里只是区域中的一小部分。每个柱状区域的细胞从输入的唯一对应子集收到激活信号。有强烈活性抑制的柱状区域活化程度较低。结果就产生了输入的稀疏离散表征。图 4.14 中用灰色表示活跃的柱状区域。

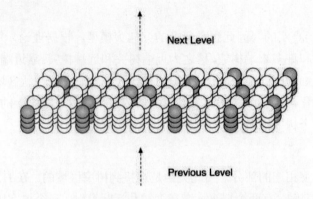

图 4.14　一个 HTM 区域包含着由细胞组成的柱状区域

对于被编码器（Encoder）编码过的信息，已经是一段很长的二进制编码，比如里面有 60% 的 1、40% 的 1，从一定程度上可以说这是稀疏的，当输入到区域后，这段二进制编码会改变区域中的突触连通情况，从而细胞被激活，这样区域就有了一个新的表征。这种表征会比原始输入更为稀疏，比如只有 2% 的 1。

6. 空间池（Spatial Pooler）

促进不活跃的柱状区域，抑制周围的柱状区域来维持稀疏度，确定最小的输入阈值，维持一个庞大的潜在突触池，基于突触的贡献来进行增减权值。

当接收到输入信号后，HTM 中的相应区域要进行空间池化，主要分为以下几步：

首先，对于区域中的每个列（Column），都有一个感受野（Receptive Field），用来接收输入中的子集，其树突上的活跃突触（权值大于一定阈值，初始化时权值在阈值附近进行随机取值）将与输入的 bits 连接，如果连接到活跃 bits（1）的突触数目大于一定阈值，就认为该列可以作为活跃列的备选。

然后，为了达到稀疏表征的目的，不希望太多的列能够激活，所以在一定的抑制半径（通过列的平均 Receptive Field 计算）内，只有前 n（如 $n=10$，当然也可以用总数的百分比）才能被激活。将所有满足条件的列激活，得到的就是输入的稀疏离散表征。

在学习阶段需要更新权值，一般希望特定的突触对于特定的输入具有响应，这样达到不同模式具有不同稀疏表征的效果，所以使得活跃列的潜在突触中所有连接活跃 bits（1）的突触权值自增，而连接不活跃 bits（0）的突触权值自减。其他列的突触权值不变。

当然，在空间池化中，有许多细节需要注意。例如，HTM 希望所有的列都要被用来进行一定模式的表征，所以那些因为覆盖值（Overlap，就是连接活跃 bits 的活跃突触数目）长期不够与那些因为覆盖值长期不能进入抑制半径内前列的列突触权值进行 boosting，即增加其权值。

实现空间池（Spatial Pooler）的代码如下：

```
sp = SpatialPooler(
 # How large the input encoding will be.
 inputDimensions=(encodingWidth),
```

```
# How many mini-columns will be in the Spatial Pooler.
columnDimensions=(2048),
# What percent of the columns's receptive field is available for potential
# synapses?
potentialPct=0.85,
# This means that the input space has no topology.
globalInhibition=True,
localAreaDensity=-1.0,
# Roughly 2%, giving that there is only one inhibition area because we have
# turned on globalInhibition (40 / 2048 = 0.0195)
numActiveColumnsPerInhArea=40.0,
# How quickly synapses grow and degrade.
synPermInactiveDec=0.005,
synPermActiveInc=0.04,
synPermConnected=0.1,
# boostStrength controls the strength of boosting. Boosting encourages
# efficient usage of SP columns.
boostStrength=3.0,
# Random number generator seed.
seed=1956,
# Determines if inputs at the beginning and end of an input dimension should
# be considered neighbors when mapping columns to inputs.
wrapAround=False
)
```

7. 时间池（Temporal Pooler）

当输入用稀疏离散表征后，得到活跃的列（Column），然后进行时间池化（Temporal Pooling），主要分为以下几步：

首先，要对列里面的细胞进行激活。细胞的激活分为两种情况。第一种，区域之前没有做出预测，那么对于空间池得到的激活列中的所有细胞进行激活。第二种，区域之前已经做出了预测，那么在每个活跃列中判断是否有细胞在前一时刻被正确预测：如果有，仅仅激活该细胞，说明符合当前上下文环境；如果没有，那么将这一列的所有细胞进行激活，说明还不清楚上下文环境是什么，所以所有上下文环境都有可能。其他没有被正确预测或者没有再激活列中的细胞保持或者变成不活跃。

其次，要进行预测。对于没有被激活的细胞，观察其连接的树突，如果该树突上连接活跃细胞的活跃突触数目大于一定数目，就认为该树突被激活，然后使其连接的细胞被激活，当一个细胞存在多个树突被激活时，进行 OR 运算。

在学习阶段，为了防止细胞过多地活跃突触就从而进行过多预测，HTM 一般希望每个活跃列中只有一个细胞进行学习。在被正确预测了的细胞中，首先计算该细胞通过活跃突触连接的之前处于学习状态的细胞数目，当其大于一定阈值时进入学习状态；对于没有被正确预测的细胞，选取连接活跃突触数目最多的细胞进入学习状态。对进入学习状态的细胞，将其活跃突触权值自增，其他突触权值自减。对于一些之前被预测、当前没有被预测的（没有激活的）细胞，说明之前的预测有问题，所有该细胞连接的突触自减。对于被正确预测的细胞，希望观察它们是否作出了正确的预测，对于它们权值的强化更新，所以先存在队列中，待到前进一定的时间步长（Time Step）再进行处理。可以看出，如果一个细胞被正确预测，HTM 会强化这种前后细胞的转移关系，强化记忆序列关系。

实现时间池（Temporal Pooler）的代码如下：

```
tm = TemporalMemory(
 # Must be the same dimensions as the SP
 columnDimensions=(2048, ),
 # How many cells in each mini-column.
 cellsPerColumn=32,
 # A segment is active if it has >= activationThreshold connected synapses
 # that are active due to infActiveState
 activationThreshold=16,
 initialPermanence=0.21,
 connectedPermanence=0.5,
 # Minimum number of active synapses for a segment to be considered during
 # search for the best-matching segments.
 minThreshold=12,
 # The max number of synapses added to a segment during learning
 maxNewSynapseCount=20,
 permanenceIncrement=0.1,
 permanenceDecrement=0.1,
 predictedSegmentDecrement=0.0,
 maxSegmentsPerCell=128,
 maxSynapsesPerSegment=32,
 seed=1960
)
```

8. Web 网页数据预测

【例 4.12】 Web 网页数据预测

NuPIC 示例展示了如何使用带有 HTMPredictionModel 的 SDRCategoryEncoder 来分析 Web 站点流量数据，方法是从描述为 Web 页面类别序列的用户会话中提取时间模式。

数据集特性：

- Number of users: 989,818
- Average number of visits per user: 5.7
- Number of categories: 17
- Number of URLs per category: 10 to 5,000

源代码可以在 GitHub 网站下载：https://github.com/numenta/nupic/tree/master/examples/

prediction/category_prediction。

该例在 Python 2.7 环境运行，压缩包解压后运行 webdata.py 源代码指令：

```
python webdata.py
```

webdata.py 的源代码如下：

```
# -*- coding: utf-8 -*-
# ----------------------------------------------------------------------
# Numenta Platform for Intelligent Computing (NuPIC)
# Copyright (C) 2013-2018, Numenta, Inc.  Unless you have an agreement
# with Numenta, Inc., for a separate license for this software code, the
# following terms and conditions apply:
#
# This program is free software: you can redistribute it and/or modify
# it under the terms of the GNU Affero Public License version 3 as
# published by the Free Software Foundation.
#
# This program is distributed in the hope that it will be useful,
# but WITHOUT ANY WARRANTY; without even the implied warranty of
# MERCHANTABILITY or FITNESS FOR A PARTICULAR PURPOSE.
# See the GNU Affero Public License for more details.
#
# You should have received a copy of the GNU Affero Public License
# along with this program.  If not, see http://www.gnu.org/licenses.
#
# http://numenta.org/licenses/
# ----------------------------------------------------------------------
"""
This example shows how to use the `SDRCategoryEncoder` with `HTMPredictionModel`
to analyze web site traffic data by extracting temporal patterns from user
sessions described as a sequences of web page categories.
We will use the [MSNBC.com Anonymous Web Data][1] data set provided by
[UCI Machine Learning Repository][2] to predict the next page the user is more
likely to click. In this data set each page is assigned a category and the user
behavior is recorded as navigating from one page to another.
Dataset characteristics:
  - Number of users: 989,818
  - Average number of visits per user: 5.7
  - Number of categories: 17
  - Number of URLs per category: 10 to 5,000
See [dataset][1] description for more information.
References:
 1. https://archive.ics.uci.edu/ml/datasets/MSNBC.com+Anonymous+Web+Data
```

```
    2. Lichman, M. (2013). UCI Machine Learning Repository
[http://archive.ics.uci.edu/ml].
    Irvine, CA: University of California, School of Information and Computer Science
"""
import os
import random
import sys
import zipfile
from operator import itemgetter

import numpy as np
import prettytable
from prettytable import PrettyTable

from nupic.frameworks.opf.model_factory import ModelFactory

# List of page categories used in the dataset
PAGE_CATEGORIES = [
  "frontpage", "news", "tech", "local", "opinion", "on-air", "misc", "weather",
  "msn-news", "health", "living", "business", "msn-sports", "sports", "summary",
  "bbs", "travel"
]

# Configure the sensor/input region using the "SDRCategoryEncoder" to encode
# the page category into SDRs suitable for processing directly by the TM
SENSOR_PARAMS = {
  "verbosity": 0,
  "encoders": {
    "page": {
      "fieldname": "page",
      "name": "page",
      "type": "SDRCategoryEncoder",
      # The output of this encoder will be passed directly to the TM region,
      # therefore the number of bits should match TM's "inputWidth" parameter
      "n": 1024,
      # Use ~2% sparsity
      "w": 21
    },
  },
}

# Configure the temporal memory to learn a sequence of page SDRs and make
# predictions on the next page of the sequence.
```

```python
TM_PARAMS = {
  "seed": 1960,
  # Use "nupic.bindings.algorithms.TemporalMemoryCPP" algorithm
  "temporalImp": "tm_cpp",
  # Should match the encoder output
  "inputWidth": 1024,
  "columnCount": 1024,
  # Use 1 cell per column for first order prediction.
  # Use more cells per column for variable order predictions.
  "cellsPerColumn": 1,
}

# Configure the output region with a classifier used to decode TM SDRs back
# into pages
CL_PARAMS = {
  "implementation": "cpp",
  "regionName": "SDRClassifierRegion",
  # alpha parameter controls how fast the classifier learns/forgets. Higher
  # values make it adapt faster and forget older patterns faster.
  "alpha": 0.001,
  "steps": 1,
}

# Create a simple HTM network that will receive the current page as input, pass
# the encoded page SDR to the temporal memory to learn the sequences and
# interpret the output SDRs from the temporary memory using the SDRClassifier
# whose output will be a list of predicted next pages and their probabilities.
#
#   page => [encoder] => [TM] => [classifier] => prediction
#
MODEL_PARAMS = {
  "version": 1,
  "model": "HTMPrediction",
  "modelParams": {
    "inferenceType": "TemporalMultiStep",

    "sensorParams": SENSOR_PARAMS,

    # The purpose of the spatial pooler is to create a stable representation of
    # the input SDRs. In our case the category encoder output is already a
    # stable representation of the category therefore adding the spatial pooler
    # to this network would not help and could potentially slow down the
    # learning process
```

```python
    "spEnable": False,
    "spParams": {},

    "tmEnable": True,
    "tmParams": TM_PARAMS,

    "clParams": CL_PARAMS,
  },
}

# Learn page sequences from the first 10,000 user sessions.
# We chose 10,000 because it gives results that are good enough for this example
# Use more records for learning to improve the prediction accuracy
LEARNING_RECORDS = 10000

def computeAccuracy(model, size, top):
  """
  Compute prediction accuracy by checking if the next page in the sequence is
  within the top N predictions calculated by the model
  Args:
    model: HTM model
    size: Sample size
    top: top N predictions to use
  Returns: Probability the next page in the sequence is within the top N
        predicted pages
  """
  accuracy = []

  # Load MSNBC web data file
  filename = os.path.join(os.path.dirname(__file__), "msnbc990928.zip")
  with zipfile.ZipFile(filename) as archive:
    with archive.open("msnbc990928.seq") as datafile:
      # Skip header lines (first 7 lines)
      for _ in xrange(7):
        next(datafile)

      # Skip learning data and compute accuracy using only new sessions
      for _ in xrange(LEARNING_RECORDS):
        next(datafile)

      # Compute prediction accuracy by checking if the next page in the sequence
      # is within the top N predictions calculated by the model
      for _ in xrange(size):
```

```python
      pages = readUserSession(datafile)
      model.resetSequenceStates()
      for i in xrange(len(pages) - 1):
        result = model.run({"page": pages[i]})
        inferences = result.inferences["multiStepPredictions"][1]

        # Get top N predictions for the next page
        predicted = sorted(inferences.items(), key=itemgetter(1),
reverse=True)[:top]

        # Check if the next page is within the predicted pages
        accuracy.append(1 if pages[i + 1] in zip(*predicted)[0] else 0)

  return np.mean(accuracy)

def readUserSession(datafile):
  """
  Reads the user session record from the file's cursor position
  Args:
    datafile: Data file whose cursor points at the beginning of the record
  Returns:
    list of pages in the order clicked by the user
  """
  for line in datafile:
    pages = line.split()
    total = len(pages)
    # Select user sessions with 2 or more pages
    if total < 2:
      continue

    # Exclude outliers by removing extreme long sessions
    if total > 500:
      continue

    return [PAGE_CATEGORIES[int(i) - 1] for i in pages]
  return []

def main():
  # Create HTM prediction model and enable inference on the page field
  model = ModelFactory.create(MODEL_PARAMS)
  model.enableInference({"predictedField": "page"})

  # Use the model encoder to display the encoded SDRs the model will learn
```

```python
sdr_table = PrettyTable(field_names=["Page Category",
                                     "Encoded SDR (on bit indices)"],
                        sortby="Page Category")
sdr_table.align = "l"

encoder = model._getEncoder()
sdrout = np.zeros(encoder.getWidth(), dtype=np.bool)

for page in PAGE_CATEGORIES:
  encoder.encodeIntoArray({"page": page}, sdrout)
  sdr_table.add_row([page, sdrout.nonzero()[0]])

print "The following table shows the encoded SDRs for every page " \
      "category in the dataset"
print sdr_table

# At this point our model is configured and ready to learn the user sessions
# Extract the learning data from MSNBC archive and stream it to the model
filename = os.path.join(os.path.dirname(__file__), "msnbc990928.zip")
with zipfile.ZipFile(filename) as archive:
  with archive.open("msnbc990928.seq") as datafile:
    # Skip header lines (first 7 lines)
    for _ in xrange(7):
      next(datafile)

    print
    print "Start learning page sequences using the first {} user " \
          "sessions".format(LEARNING_RECORDS)
    model.enableLearning()
    for count in xrange(LEARNING_RECORDS):
      # Learn each user session as a single sequence
      session = readUserSession(datafile)
      model.resetSequenceStates()
      for page in session:
        model.run({"page": page})

      # Simple progress status
      sys.stdout.write("\rLearned {} Sessions".format(count + 1))
      sys.stdout.flush()

    print "\nFinished learning"
    model.disableLearning()
```

```python
    # Use the new HTM model to predict next user session
    # The test data starts right after the learning data
    print
    print "Start Inference using a new user session from the dataset"
    prediction_table = PrettyTable(field_names=["Page", "Prediction"],
                        hrules=prettytable.ALL)
    prediction_table.align["Prediction"] = "l"

    # Infer one page of the sequence at the time
    model.resetSequenceStates()
    session = readUserSession(datafile)
    for page in session:
      result = model.run({"page": page})
      inferences = result.inferences["multiStepPredictions"][1]

      # Print predictions ordered by probabilities
      predicted = sorted(inferences.items(),
                  key=itemgetter(1),
                  reverse=True)
      prediction_table.add_row([page, zip(*predicted)[0]])

    print "User Session to Predict: ", session
    print prediction_table

  print
  print "Compute prediction accuracy by checking if the next page in the " \
      "sequence is within the predicted pages calculated by the model:"
  accuracy = computeAccuracy(model, 100, 1)
  print " - Prediction Accuracy:", accuracy
  accuracy = computeAccuracy(model, 100, 3)
  print " - Accuracy Predicting Top 3 Pages:", accuracy

if __name__ == "__main__":
  random.seed(1)
  np.random.seed(1)
  main()
```

抽样输出：

```
The following table shows the encoded SDRs for every page category in the dataset
+----------------+--------------------------------------------------
----------------+
| Page Category | Encoded SDR (on bit indices)
|
```

```
+---------------+----------------------------------------------------------------+
| bbs           | [ 19  26 115 171 293 364 390 442 470 477 550 598 624 670 705 719 744 748
|               |  788 850 956]
| business      | [ 48 104 144 162 213 280 305 355 376 403 435 628 694 724 780 850 854 870
|               |  891 930 955]
| frontpage     | [  4   7  35  37  48  91 118 143 155 313 339 410 560 627 736 762 795 864
|               |  885 889 966]
| health        | [ 50  67 124 209 214 229 288 337 380 402 437 474 566 584 614
|               |  661 754 840 846 894 1008]
| living        | [195 198 209 219 261 317 332 348 353 369 371 375 399 495 501 556 595 758
|               |  799 813 920]
| local         | [  3  48 221 275 284 457 466 516 574 626 645 688 699 761 855 867 899 925
|               |  942 987 997]
| misc          | [ 40  61  90 106 127 179 202 208 217 373 417 523 577 580 722 751 865 925
|               |  926 928 938]
| msn-news      | [ 29  71  72  74 149 241 261 263 276 365 465 528 529 575 577
|               |  661 781 799 830 980 1019]
| msn-sports    | [119 138 150 164 197 263 391 454 510 581 589 614 661 700 724 742 809 886
|               |  889 978 989]
| news          | [ 18  44  71 109 191 322 333 337 375 402 447 587 653 660 794
|               |  837 853 913 936 954 1019]
| on-air        | [ 27  80 134 158 187 199 214 286 374 439 445 484 490
```

```
590  670  |
|              |  771  823  934  952  965 1014]
|
| opinion      | [163 165 216 241 251 260 307 336 382 449 493 540 607 668 679 717
736  866  |
|              |  888  902  981]
|
| sports       | [ 20  39  65 141 147 230 232 248 332 361 467 476 689
847  851  |
|              |  862  866  889  936  958 1010]
|
| summary      | [ 32  34 106 206 302 340 414 564 566 568 596 619 645 657 761 813
879  888  |
|              |  897  944  997]
|
| tech         | [108 276 327 372 411 431 479 577 592 606 650 690 747 756 763 913
936  949  |
|              |  961  981  983]
|
| travel       | [149 164 179 239 316 319 365 427 437 470 632 729 739 748 787 818
821  824  |
|              |  834  906  919]
|
| weather      | [  9  12  21  38  45 146 203 205 284 400 471 506 520 532 595 613
621  639  |
|              |  805  970  987]
|
+--------------+-------------------------------------------------------------
----------------+

Start Learning page sequences using the first 10000 user sessions
Learned 10000 Sessions
Finished Learning

Start Inference using a new user session from the dataset
User Session to Predict: ['on-air', 'misc', 'misc', 'misc', 'on-air', 'misc',
'misc', 'misc', 'on-air', 'on-air', 'on-air', 'on-air', 'tech', 'msn-news', 'tech',
'msn-news', 'local', 'tech', 'local', 'local', 'local', 'local', 'local', 'local']
+----------+------------------------------------------------------------
------------------+
|  Page    | Prediction
|
+----------+------------------------------------------------------------
```

```
--------------------+
| on-air  | ('on-air', 'misc', 'frontpage', 'news', 'summary', 'msn-news',
'weather', 'local')     |
+---------+----------------------------------------------------------------
--------------------+
| misc    | ('misc', 'frontpage', 'on-air', 'local', 'msn-news', 'msn-sports',
'news', 'sports')  |
+---------+----------------------------------------------------------------
--------------------+
| misc    | ('misc', 'frontpage', 'on-air', 'local', 'msn-news', 'msn-sports',
'news', 'sports')  |
+---------+----------------------------------------------------------------
--------------------+
| misc    | ('misc', 'frontpage', 'on-air', 'local', 'msn-news', 'msn-sports',
'news', 'sports')  |
+---------+----------------------------------------------------------------
--------------------+
| on-air  | ('on-air', 'misc', 'frontpage', 'news', 'summary', 'msn-news',
'weather', 'local')     |
+---------+----------------------------------------------------------------
--------------------+
| misc    | ('misc', 'frontpage', 'on-air', 'local', 'msn-news', 'msn-sports',
'news', 'sports')  |
+---------+----------------------------------------------------------------
--------------------+
| misc    | ('misc', 'frontpage', 'on-air', 'local', 'msn-news', 'msn-sports',
'news', 'sports')  |
+---------+----------------------------------------------------------------
--------------------+
| misc    | ('misc', 'frontpage', 'on-air', 'local', 'msn-news', 'msn-sports',
'news', 'sports')  |
+---------+----------------------------------------------------------------
--------------------+
| on-air  | ('on-air', 'misc', 'frontpage', 'news', 'summary', 'msn-news',
'weather', 'local')     |
+---------+----------------------------------------------------------------
--------------------+
| on-air  | ('on-air', 'misc', 'frontpage', 'news', 'summary', 'msn-news',
'weather', 'local')     |
+---------+----------------------------------------------------------------
--------------------+
| on-air  | ('on-air', 'misc', 'frontpage', 'news', 'summary', 'msn-news',
'weather', 'local')     |
```

```
+---------+----------------------------------------------------------------------------------+
| on-air  | ('on-air', 'misc', 'frontpage', 'news', 'summary', 'msn-news', 'weather', 'local') |
+---------+----------------------------------------------------------------------------------+
|  tech   | ('tech', 'frontpage', 'news', 'msn-news', 'on-air', 'business', 'local', 'sports') |
+---------+----------------------------------------------------------------------------------+
| msn-news| ('msn-news', 'frontpage', 'local', 'weather', 'misc', 'on-air', 'msn-sports', 'tech') |
+---------+----------------------------------------------------------------------------------+
|  tech   | ('tech', 'frontpage', 'news', 'msn-news', 'on-air', 'business', 'local', 'sports') |
+---------+----------------------------------------------------------------------------------+
| msn-news| ('msn-news', 'frontpage', 'local', 'weather', 'misc', 'on-air', 'msn-sports', 'tech') |
+---------+----------------------------------------------------------------------------------+
| local   | ('local', 'frontpage', 'misc', 'news', 'msn-news', 'on-air', 'weather', 'sports') |
+---------+----------------------------------------------------------------------------------+
|  tech   | ('tech', 'frontpage', 'news', 'msn-news', 'on-air', 'business', 'local', 'sports') |
+---------+----------------------------------------------------------------------------------+
| local   | ('local', 'frontpage', 'misc', 'news', 'msn-news', 'on-air', 'weather', 'sports') |
+---------+----------------------------------------------------------------------------------+
| local   | ('local', 'frontpage', 'misc', 'news', 'msn-news', 'on-air', 'weather', 'sports') |
+---------+----------------------------------------------------------------------------------+
| local   | ('local', 'frontpage', 'misc', 'news', 'msn-news', 'on-air', 'weather', 'sports') |
+---------+----------------------------------------------------------------------------------+
| local   | ('local', 'frontpage', 'misc', 'news', 'msn-news', 'on-air', 'weather',
```

```
'sports')        |
+----------+-----------------------------------------------------------
--------------------+
| local    | ('local', 'frontpage', 'misc', 'news', 'msn-news', 'on-air', 'weather',
'sports')        |
+----------+-----------------------------------------------------------
--------------------+
| local    | ('local', 'frontpage', 'misc', 'news', 'msn-news', 'on-air', 'weather',
'sports')        |
+----------+-----------------------------------------------------------
--------------------+

Compute prediction accuracy by checking if the next page in the sequence is within
the predicted pages calculated by the model:
- Prediction Accuracy: 0.614173228346
- Accuracy Predicting Top 3 Pages: 0.825196850394
```

4.10 Neon

Neon 是 Nervana Systems 公司一个基于 Python 的深度学习库。它易于使用且具有超高的性能。在某些基准测试中，由 Python 和 Sass 开发的 Neon 的测试成绩甚至要优于 Caffeine、Torch 和谷歌的 TensorFlow。

Neon 是英特尔参考的深度学习框架，致力于在所有硬件上实现最佳性能，为易于使用和可扩展性而设计。

Neon 进行深度学习支持常用层：卷积 Convolution、RNN、LSTM、GRU、BatchNorm 等。训练模型包含预先训练的权重和例子脚本，为开始的模型，包括 VGG、Reinforcement Learning、Deep Residual Networks、Deep Residual Networks、Image Captioning、Sentiment Analysis 等。

Neon 可切换硬件后端：编写一次代码，然后部署到 CPU、GPU 或 Nervana 硬件上。对于快速迭代和模型探索，Neon 的性能是深度学习库中最快的（cuDNNv4 的 2 倍速度）。

Intel Nervana 使用 Neon 在多个领域解决客户的问题。查看最新版本的新特性。Neon v2.0.0+ 通过启用 Intel Math 内核库（MKL）对 CPU 进行了优化，从而获得了更好的性能。使用的 MKL 的 DNN（深度神经网络）组件是免费提供的，并作为安装的一部分自动下载。

1. 安装

Neon 运行在 Python 2.7 或 Python 3.4+上，支持 Linux 和 Mac OS X 机器。

（1）安装环境需求

安装前，请确保设备有以下软件包的最新版本（不同的系统名称显示）：

```
Ubuntu                  OSX             Description
python-pip              pip                             Tool to install python
dependencies
python-virtualenv (*)   virtualenv (*)                  Allows creation of isolated
environments ((*)
libhdf5-dev             h5py            Enables loading of hdf5 formats
libyaml-dev pyaml                       Parses YAML format inputs
pkg-config              pkg-config              Retrieves information about
installed libraries
```

对于 GPU 用户，记得添加 CUDA 路径。例如，在 Ubuntu 上：

```
export PATH="/usr/local/cuda/bin:"$PATH
export LD_LIBRARY_PATH="/usr/local/cuda/lib64:/usr/local/cuda/lib:/usr/local/lib:"$LD_LIBRARY_PATH
```

GPU 用户在 Mac OS X 上：

```
export PATH="/usr/local/cuda/bin:"$PATH
export DYLD_LIBRARY_PATH="/usr/local/cuda/lib:"$DYLD_LIBRARY_PATH
```

（2）安装

建议在虚拟环境中安装 Neon，以确保环境是自包含的。如果使用 Anaconda Python 发行版，请参阅 Anaconda 安装部分。否则，要以这种方式设置 Neon，请运行以下命令：

```
git clone https://github.com/NervanaSystems/neon.git
cd neon; git checkout latest; make
```

上面的代码检查了最新的稳定版本（例如带有标记的版本 v2.6.0）并构建 Neon。或者，可以检出并构建最新的主分支：

```
git clone https://github.com/NervanaSystems/neon.git
cd neon; make
```

这将在 neon/中安装文件，并将在默认路径中使用 Python 版本。neon 将自动下载已发布的 MKLML 库，该库支持 MKL。

要激活虚拟环境，输入：

```
. .venv/bin/activate
```

将看到反映激活环境的提示更改。要启动 Neon 并运行 MNIST 多层感知器示例（深度学习的"Hello World"），请输入：

```
examples/mnist_mlp.py
```

注意，由于 Neon v2.1 的上述功能相当于显式地添加-b mkl，以便在 Intel CPU 上获得更好的性能。换句话说，mkl 后端是默认后端。

```
examples/mnist_mlp.py -b mkl
```

完成时,记得关闭环境:

```
deactivate
```

（3）虚拟环境安装

virtualenv 是一个 Python 工具,它将不同项目所需的依赖项和包保存在不同的环境中。默认情况下,安装在 neon/.venv 目录中创建 Python 可执行文件的副本。要了解更多关于虚拟环境的信息,请访问 http://docs.pythonguide.org/en/latest/dev/virtualenvs/。

（4）全系统的安装

如果不希望使用新的虚拟环境,就可以在系统范围内安装 Neon:

```
git clone https://github.com/NervanaSystems/neon.git
cd neon && make sysinstall
```

（5）Pip install

Neon v2.4.0 及以后版本可以通过名为 nervananeon 的 pypi 安装 pip。

```
pip install nervananeon
```

（6）Anaconda install

如果已经安装并配置了 Python 的 Anaconda 发行版,请遵循以下步骤。

首先,为 Neon 配置并激活一个新的 conda 环境:

```
conda create --name neon pip
source activate neon
```

现在克隆并运行一个系统范围的安装。由于安装是在 conda 环境中进行的,因此依赖项将安装在环境文件夹中。

```
git clone https://github.com/NervanaSystems/neon.git
cd neon && make sysinstall
```

完成后,关闭环境:

```
source deactivate
```

2. 使用脚本运行一个示例

```
python examples/mnist_mlp.py
```

（1）从命令行选择后端引擎

默认选择 GPU 后端,所以上面的命令等价于在系统上找到兼容的 GPU 资源:

```
python examples/mnist_mlp.py -b gpu
```

当 GPU 不可用时,默认选择优化后的 CPU（MKL）后端为 neon v2.1.0,这意味着上面的命令相当于:

```
python examples/mnist_mlp.py -b mkl
```

若有兴趣比较默认的 mkl 后端与未优化的 CPU 后端，可使用以下命令：

```
python examples/mnist_mlp.py -b cpu
```

（2）使用 yaml 文件运行一个示例

```
neon examples/mnist_mlp.yaml
```

要在 yaml 文件中选择特定的后端，可添加或修改包含后端 mkl 以启用 mkl 后端，或包含后端 CPU 以启用 CPU 后端。如果 GPU 可用，那么默认选择 GPU 后端。

【例 4.13】mnist_mlp.py（源码：https://github.com/NervanaSystems/neon/blob/master/examples/mnist_mlp.py）

```
#!/usr/bin/env python
#
# ******************************************************************************
# Copyright 2014-2018 Intel Corporation
#
# Licensed under the Apache License, Version 2.0 (the "License");
# you may not use this file except in compliance with the License.
# You may obtain a copy of the License at
#
#     http://www.apache.org/licenses/LICENSE-2.0
#
# Unless required by applicable law or agreed to in writing, software
# distributed under the License is distributed on an "AS IS" BASIS,
# WITHOUT WARRANTIES OR CONDITIONS OF ANY KIND, either express or implied.
# See the License for the specific language governing permissions and
# limitations under the License.
#
# ******************************************************************************
"""
Train a small multi-layer perceptron with fully connected layers on MNIST data.
This example has some command line arguments that enable different neon features.
Examples:
    python examples/mnist_mlp.py -b gpu -e 10
        Run the example for 10 epochs using the NervanaGPU backend
    python examples/mnist_mlp.py --eval_freq 1
        After each training epoch, process the validation/test data
        set through the model and display the cost.
    python examples/mnist_mlp.py --serialize 1 -s checkpoint.pkl
        After every iteration of training, dump the model to a pickle
        file named "checkpoint.pkl". Changing the serialize parameter
```

```
            changes the frequency at which the model is saved.
       python examples/mnist_mlp.py --model_file checkpoint.pkl
            Before starting to train the model, set the model state to
            the values stored in the checkpoint file named checkpoint.pkl.
"""

from neon.callbacks.callbacks import Callbacks
from neon.data import MNIST
from neon.initializers import Gaussian
from neon.layers import GeneralizedCost, Affine
from neon.models import Model
from neon.optimizers import GradientDescentMomentum
from neon.transforms import Rectlin, Logistic, CrossEntropyBinary, Misclassification
from neon.util.argparser import NeonArgparser
from neon import logger as neon_logger


# parse the command line arguments
parser = NeonArgparser(__doc__)

args = parser.parse_args()

# load up the mnist data set
dataset = MNIST(path=args.data_dir)
train_set = dataset.train_iter
valid_set = dataset.valid_iter

# setup weight initialization function
init_norm = Gaussian(loc=0.0, scale=0.01)

# setup model layers
layers = [Affine(nout=100, init=init_norm, activation=Rectlin()),
          Affine(nout=10, init=init_norm, activation=Logistic(shortcut=True))]

# setup cost function as CrossEntropy
cost = GeneralizedCost(costfunc=CrossEntropyBinary())

# setup optimizer
optimizer = GradientDescentMomentum(
    0.1, momentum_coef=0.9, stochastic_round=args.rounding)

# initialize model object
mlp = Model(layers=layers)
```

```
# configure callbacks
callbacks = Callbacks(mlp, eval_set=valid_set, **args.callback_args)

# run fit
mlp.fit(train_set, optimizer=optimizer,
        num_epochs=args.epochs, cost=cost, callbacks=callbacks)
error_rate = mlp.eval(valid_set, metric=Misclassification())
neon_logger.display('Misclassification error - %.1f%%' % (error_rate * 100))
```

4.11 Nilearn

Nilearn 将机器学习、模式识别、多变量分析等技术应用于神经影像数据的应用中，能完成多体素模式分析（Mutli-Voxel Pattern Analysis，MVPA）、解码、模型预测、构造功能连接、脑区分割、构造连接体等功能，一般用于处理功能磁共振图像（FMRI）、静息状态（Resting-State）或者基于体素的形态学分析（VBM）。对于机器学习专家来说，Nilearn 的价值体现在特定领域特定工程的构造上，也就是将神经影像数据表达成为非常适合于统计学习的特征矩阵。

1. 安装 nilearn 库

建议安装 Anaconda（一个 Python 环境），它满足 Nilearn 所需要的所有第三方库。

进入 CMD（win+r），输入命令：

```
pip install nilearn
```

打开 Python 编辑器，导入 Nilearn：

```
import nilearn
```

2. 机器学习对于神经影像的重要性

- 监督学习的脑图像能预测临床评分或者治疗反应。
- 能够度量泛化评分。
- 能够用于多维多变量统计学分析。
- 能应用数据驱动来探索大脑。

3. Nilearn 大脑图像（FMRI）时间序列做研究分析

最原始的磁共振图像（FMRI）数据是四维的，包含三维的空间信息和一维的时间。在实际应用中，更多的是利用大脑图像时间序列进行研究分析，因为无法直接使用 FMRI 数据进行相关研究。在使用数据之前，需要对原始数据进行一些数据预处理和变换。

（1）Mask

在所有的分析之中，第一步所做的事都是把四维 FMRI 数据转换为二维矩阵，这个过程称为 Mask。通俗的理解就是提取能利用的特征。通过 Mask 得到的二维矩阵包含一维的时间和一维的特征，也就是将 FMRI 数据中每一个时间片上的特征提取出来，再组在一起就是一个二维矩阵，如图 4.15 所示。

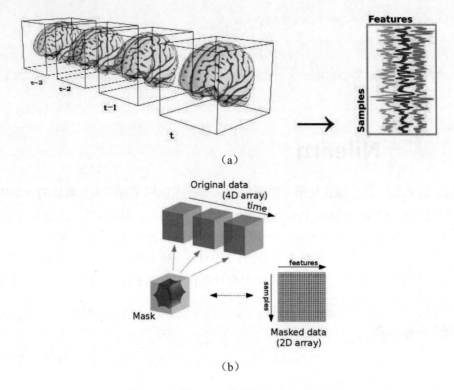

图 4.15 四维 FMRI 数据转换为二维矩阵

在 Nilearn 库中，提供了两个函数计算 Mask：

- nilearn.masking.compute_background_mask
- nilearn.masking.compute_epi_mask

（2）时间序列

在 Mask 之后，接下来要做的任务就是提取时间序列。说到时间序列，不得不说的是图谱。前面 Mask 之后的特征实在是太多，怎样将这些特征与已有的经过验证的图谱对应起来。这时就涉及一个重采样。

在 Nilearn 中，重采样函数为 resample_to_img。重新采样之后就可以得到相应的时间序列了，再把时间序列转换为相关矩阵，并画出图像。

在 Nilearn 库中，提供了两种从 FMRI 数据中提取时间序列的方法：一种基于脑分区（Time-series from a brain parcellation or "MaxProb" atlas），一种基于概率图谱（Time-series from a probabilistic atlas）。（参考文章：Varoquaux and Craddock 2013 年发表在 NeuroImage 上的 *Learning and comparing functional connectomes across subjects*。

4. Glass brain 在 Nilearn 中绘制

【例 4.14】 实验在 Jupyter Notebook 中完成。

```
//Retrieve data from Internet
In[1]:
from nilearn import datasets
motor_images = datasets.fetch_neurovault_motor_task()
stat_img = motor_images.images[0]
d:\Anaconda3\lib\site-packages\h5py\__init__.py:36: FutureWarning: Conversion of
the second argument of issubdtype from 'float' to 'np.floating' is deprecated. In
future, it will be treated as 'np.float64 == np.dtype(float).type'.
  from ._conv import register_converters as _register_converters
Dataset created in C:\Users\DLG/nilearn_data/neurovault
//Glass brain plotting: whole brain sagittal cuts
In[2]:
from nilearn import plotting
plotting.plot_glass_brain(stat_img, threshold=3)
d:\Anaconda3\lib\importlib\_bootstrap.py:219: ImportWarning: can't resolve
package from __spec__ or __package__, falling back on __name__ and __path__
  return f(*args, **kwds)
d:\Anaconda3\lib\site-packages\sklearn\decomposition\nmf.py:972:
DeprecationWarning: invalid escape sequence \s
  """
d:\Anaconda3\lib\site-packages\scipy\ndimage\measurements.py:272:
DeprecationWarning: In future, it will be an error for 'np.bool_' scalars to be
interpreted as an index
  return _nd_image.find_objects(input, max_label)
Out[2]: <nilearn.plotting.displays.OrthoProjector at 0x27bd94d0d68>

//Glass brain plotting: black background （见图 4.16）
In[3]:
plotting.plot_glass_brain(
    stat_img, title='plot_glass_brain',
    black_bg=True, display_mode='xz', threshold=3)
Out[3]: <nilearn.plotting.displays.XZProjector at 0x27bd9d894a8>
```

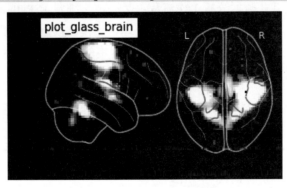

图 4.16　Nilearn 绘制玻璃脑图

```
//Glass brain plotting: Hemispheric sagittal cuts （见图 4.17）
```

```
In[4]:
plotting.plot_glass_brain(stat_img,
                  title='plot_glass_brain with display_mode="lyrz"',
                  display_mode='lyrz', threshold=3)

plotting.show()
d:\Anaconda3\lib\site-packages\scipy\ndimage\measurements.py:272:
DeprecationWarning: In future, it will be an error for 'np.bool_' scalars to be
interpreted as an index
  return _nd_image.find_objects(input, max_label)
```

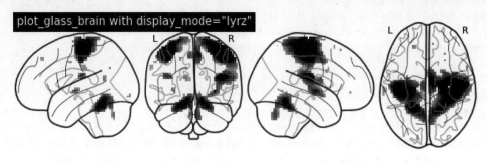

图 4.17 Nilearn 绘制显示模式为 "lyrz" 的玻璃脑图

该项目的源码地址为 http://nilearn.github.io/auto_examples/01_plotting/plot_demo_glass_brain.html#sphx-glr-auto-examples-01-plotting-plot-demo-glass-brain-py。

4.12 Orange3

1. Orange3 数据挖掘工具的介绍

官方网址为 https://orange.biolab.si/。

Orange3 是一个面向新手和专家的开源机器学习和数据可视化工具，带有很多用于数据挖掘或机器学习模型的交互式数据分析工作流程；另外，它绑定了 Python 语言进行脚本开发。包含一系列数据挖掘流程的组件，比如数据预处理、建模、模型评估以及可视化。

数据预处理主要包括数据合并（将两个不同数据集的指定特征合并为同一个数据集）、数据采样、数据异常点去除以及相关性检验（协方差）、rank 等。

- 模型主要包括 CN2 规则归纳、k 近邻、决策树、随机森林、支持向量机、线性回归、逻辑回归、朴素贝叶斯、Adaboost、神经网络和随机梯度下降等。
- 无监督模型有距离矩阵、t-SNE、层次聚类、K-means、Louvain 聚类、PCA 和 MDS 等。

另外，还支持文本分析、词云可视化等。

- 模型评估主要有交叉检验、混淆矩阵、ROC 曲线和 lift 曲线等。

2. Orange3 安装

在官方网站下载 Orange3 并安装（没有其他复杂操作，直接根据提示单击"下一步"按钮即可，若想改变安装位置可自定义修改）。Orange 自带最新或最近版本的 Python 环境，对于复杂的数据集，若 Orange 自带组件处理不便时，通常先将文件数据连接至<Python Script>，通过 Python 处理后再转换成 Orange.Tabel()形式进行后续操作，具体的使用 Python 进行处理的方法及代码可参考官方文档：

```
http://docs.orange.biolab.si/3/data-mining-library/tutorial/data.html
```

另外，该文档还包括一系列使用 Python（Orange）进行数据挖掘的编程教程。想要深入了解并掌握 Orange Python 进行数据挖掘，需要花时间仔细阅读并实践，这里不再赘述。当然，如果已经安装了 Python3 以上版本，你也可以直接执行 pip install Orange3，使用 Python shell 进行编程（前提是你已经对 Orange 的数据结构及相关函数比较熟悉），官方文档为：

```
http://docs.orange.biolab.si/3/data-mining-library/#tutorial
```

在 Python 集成环境 Anaconda3 中启动 Orange3 指令：

```
Python -m Orange.canvas
```

3. Orange3 使用

（1）添加附加组件：在 Options 中找到 Add-ons（见图 4.18），单击后会弹出如图 4.19 所示的对话框。

图 4.18　Add-ons 选项

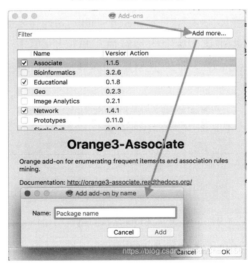

图 4.19　Orange 添加组件包

其中，Filter 栏是 Orange 自带的部分组件（打对勾的为已下载，可根据需要自行下载）；如果想安装其他组件，可以单击 Add more 按钮，输入要下载的包名（类似 Python pycharm 添加 Python 包的步骤）。

（2）对于 Orange 的简单操作，官方文档 https://orange.biolab.si/getting-started/ 提供了许多数据挖掘分析的例子。在打开 Orange3 软件后，会弹出如图 4.20 所示的面板，单击 Examples 图标即可查看。

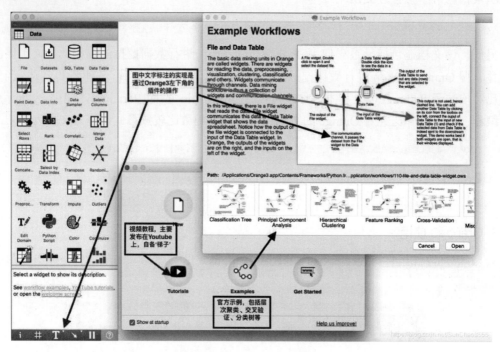

图 4.20　Orange3 界面导航

（3）Orange 数据库操作如图 4.21 所示。SQL Table 为连接数据库的组件，对于首次安装 Orange 的用户，点击后右侧栏中该图标会出现红色三角符号，单击三角符号会提示错误"please install a backend to use this widget"，即缺少 SQL 的编译器（Orange 只支持 PostgreSQL 和 SQL Server 两种数据库）。

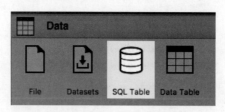

图 4.21　SQL Table 连接数据库的组件

Orange 连接 PostgreSQL 数据库的操作说明如下：
首先，需要下载 PostgreSQL 的配置文件 psycopg2：
https://blog.biolab.si/2018/02/16/how-to-enable-sql-widget-in-orange/

这里为 Python3.6 版本，也可以到 https://pypi.org/搜索 psycopg2，找到对应版本下载。

- Mac OS 下载：https://pypi.python.org/packages/8c/a5/0e61d6f4a140a6e06a9ba40266c4b49123d834f1f97fe9a5ae0b6e45112b/psycopg2-2.7.4-cp36-cp36m- macosx_10_6_intel.macosx_10_9_intel.macosx_10_9_x86_64.macosx_10_10_intel.macosx_10_10_x86_64.whl#md5=1f2b2137c65dc50c16b341774cd822eb。
- Windows 下载：https://pypi.python.org/packages/f9/77/e29b792740ddec37a2d49431efa6c707cf3869c0cc7f28c7411bb6e96d91/psycopg2-2.7.4-cp36-cp36m- win_amd64.whl#md5=119eb3ab86ea8486ab10ef4ea3f67f15。
- Linux 下载：https://pypi.python.org/packages/92/15/92b5c363243376ce9cb879bbec561bba196694eb663a6937b4cb967e230e/psycopg2-2.7.4-cp36-cp36m-manylinux1_x86_64.whl#md5=8288ce1eedf0b70e5f1d8c982fad5a41。

下载完成后，打开上述 Add-ons，将该.whl 文件拖至组件栏中，就会看到 Psycopg 已经安装，此时 SQL widget 依然不能使用，因为还未安装 PostGreSQL 数据库，下载地址为 https://www.postgresql.org/。下载对应系统的版本并安装，安装的过程与 MySQL 类似，需要编辑用户名和密码（务必记住）。另外，对于 PostgreSQL 的界面化管理工具，可选用最近版本的 Navicat Premium。PostgreSQL 是一款强大的开源数据库，想要详细了解可查阅 http://www.postgresqltutorial.com/。

反之，如果只想简单地用作连接 Orange 的插件，可在 Navicat Premium 中直接将 MySQL 数据库中的表直接拖曳并复制到 PostgreSQL 数据库中。注：直接将不属于 PostgreSQL 的数据库文件导入 PostgreSQL 会报错（比如"ERROR:unrecognized configuration parameter "foreign_key_checks", Time: 0.0"，因为不同数据库存储数据的格式和结构不同）。

4.13 PyMC 与 PyMC3

PyMC 是一个实现贝叶斯统计模型和马尔可夫链蒙塔卡洛采样工具拟合算法的 Python 库。PyMC 的灵活性及可扩展性使得它能够适用于解决各种问题。除了包含核心采样功能，PyMC 还包含了统计输出、绘图、拟合优度检验和收敛性诊断等方法。

1. PyMC 特性

PyMC 使得贝叶斯分析尽可能更加容易。以下是一些 PyMC 库的特性：

- 用马尔可夫链蒙特卡洛算法和其他算法来拟合贝叶斯统计分析模型。
- 包含了大范围的常用统计分布。
- 尽可能使用 NumPy 的一些功能。
- 包括一个高斯建模过程的模块。
- 采样循环可以被暂停和手动调整，或者保存和重新启动。

- 创建包括表格和图表的摘要说明。
- 算法跟踪记录可以保存为纯文本、pickles、SQLite 或 MySQL 数据库文档或 HDF5 文档。
- 提供了一些收敛性诊断方法。
- 可扩展性：引入自定义的步骤方法和非常规的概率分布。
- MCMC 循环可以嵌入在较大的程序中，结果可以使用 Python 进行分析。

2. PyMC 安装

PyMC 可以运行在 Mac OS X、Linux 和 Windows 系统中。安装一些其他预装库可以更大程度地提高 PyMC 的性能和功能。

（1）PyMC 的运行要求一些预装库的安装及配置

- Python（2.6 及以上版本）。
- NumPy（1.6 版本及以上）。
- Matplotlib（1.0 版本及以上）。
- SciPy（可选）。
- pyTables（可选）。
- pydot（可选）。
- IPython（可选）。
- nose（可选）。

（2）使用 EasyInstall 安装

安装 PyMC 最简单的方式是在终端输入以下代码：

```
easy_install pymc
```

（3）使用预编译二进制文件进行安装

- 从 PyPI 下载安装器。
- 双击可执行安装包，按照向导进行安装。

（4）编译源码安装

用户可以从 GitHub 网站中下载源代码并解压。

（5）从 GitHub 上安装

用户可以在 GitHub 中查找 PyMC，并执行：

```
git clone git://github.com/pymc-devs/pymc.git
```

历史版本在 /tags 目录中可以找到。

（6）执行测试套件

PyMC 中包含了一个测试用例，用来确保代码中的关键组件能够正常运行。在运行这个测试之前，用户需要保证 nose 已经在本地安装好，在 Python 编译器中执行以下代码：

```
import pymc
pymc.test()
```

3. PyMC 使用

在文件中定义模型，并命名为 mymodel.py。

```python
# Import relevant modules
import pymc
import numpy as np

# Some data
n = 5*np.ones(4,dtype=int)
x = np.array([-.86,-.3,-.05,.73])

# Priors on unknown parameters
alpha = pymc.Normal('alpha',mu=0,tau=.01)
beta = pymc.Normal('beta',mu=0,tau=.01)

# Arbitrary deterministic function of parameters
@pymc.deterministic
def theta(a=alpha, b=beta):
    """theta = logit^{-1}(a+b)"""
    return pymc.invlogit(a+b*x)

# Binomial likelihood for data
d = pymc.Binomial('d', n=n, p=theta, value=np.array([0.,1.,3.,5.]),\
                  observed=True)
```

在 Python 编译器或者相同目录下的其他文件中调用 mymodel.py：

```python
import pymc
import mymodel

S = pymc.MCMC(mymodel, db='pickle')
S.sample(iter=10000, burn=5000, thin=2)
pymc.Matplot.plot(S)
```

这个例子会产生 10 000 个后验样本。这个样本会存储在 Python 序列化数据库中。

4. PyMC3

概率编程允许在用户自定义的概率模型上进行自动贝叶斯推断。新的 MCMC（Markov Chain Monte Carlo）采样方法允许在复杂模型上进行推断。这类 MCMC 采样方法被称为 HMC（Hamliltinian Monte Carlo），但是其推断需要的梯度信息有时是不能获得的。PyMC3 是一个用 Python 编写的开源的概率编程框架，使用 Theano 通过变分推理进行梯度计算，并使用 C 实现加速运算。不同于其他概率编程语言，PyMC3 允许使用 Python 代码来定义模型。这种没有作用域限制的语言极大地方便了模型定义和直接交互。

（1）PyMC3 介绍

PyMC3 具有先进的下一代 MCMC 采样算法，如 No-U-Turn Sampler（NUTS；Hoffman，2014）和 Hamiltonian Monte Carlo 自整定变体（HMC；Duane，1987）。这类采样算法在高维和复杂的后验分布上具有良好的效果，允许对复杂模型进行拟合而不需要对拟合算法有特殊的了解。NUTS 和 HMC 算法从似然函数中获得梯度信息，因此其收敛速度比传统采样方法快很多，特别是针对大模型。NUTS 也具有集合自整定过程，因此使用者不需要了解算法细节。

（2）PyMC3 安装

终端安装：

```
pip install PyMC3
```

Git 安装：

```
pip install --process-dependency-links git+https://github.com/pymc-devs/pymc3
```

【例 4.15】模拟观测数据（见图 4.22）

使用 NumPy 的随机函数 random 模块来产生模拟数据，再使用 PyMC3 尝试恢复相应的参数。

```
import numpy as np
import matplotlib.pyplot as plt
from mpl_toolkits.mplot3d import Axes3D

np.random.seed(123)

alpha=1
sigma=1
beta =[1, 2.5]

N=100

X1=np.random.randn(N)
X2=np.random.randn(N)

Y=alpha + beta[0]*X1 + beta[1]*X2 + np.random.randn(N)*sigma

%matplotlib inline
fig1,ax1 = plt.subplots(1, 2, figsize=(10,4));
ax1[0].scatter(X1, Y);ax1[0].set_xlabel('X1');ax1[0].set_ylabel('Y');
ax1[1].scatter(X2, Y);ax1[1].set_xlabel('X2');ax1[1].set_ylabel('Y');
```

```
fig2 = plt.figure(2);
ax2 = Axes3D(fig2);
ax2.scatter(X1,X2,Y);
ax2.set_xlabel('X1');
ax2.set_ylabel('X2');
ax2.set_zlabel('Y');
```

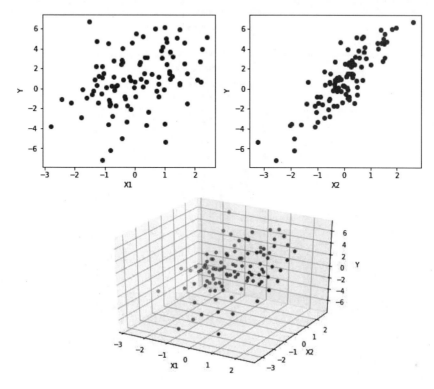

图 4.22 PyMC3 模拟观测数据

4.14 PyBrain

Python 的 PyBrain 模块是比较好用的神经网络建模工具包，对监督学习的数据有开源的处理模块，数据集的建立、网络的训练也十分便捷。当然 PyBrain 不仅仅适用于监督学习的数据，还能够快速构建多种神经网络。

PyBrain 的模块 API：

http://wiki.github.com/pybrain/pybrain/guidelines

PyBrain 的 wiki 指导文档：

http://groups.google.com/group/pybrain

1. PyBrain 特性

PyBrain 的概念是将一系列的数据处理的算法封装到被称之为 Module 的模块中。一个最小的 Module 通常包含基于机器学习算法的可调整的参数集合。Modules 包含一个输入和输出的 buffer，外加误差 buffer 用于存在误差反向传播的场景。

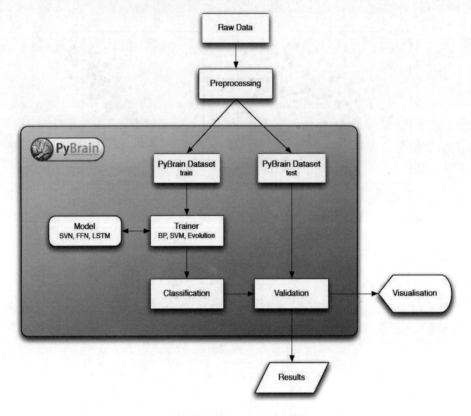

图 4.23　PyBrain 的原理图

Modules 被嵌入到 Network 类中，并且使用 Connection 对象进行连接，其中可能包含一系列可调整的参数，比如连接的权重。而 Network 类本身又是一个 Module，因此可以基于此构建多层网络结构。库中有快捷的方式构造最常用的网络结构，但原则上这个系统允许嵌入最随机的连接方式来形成一个无环图。

网络中的参数通过 Trainer 进行调节，从 Dataset 中学习到最优化的参数。有的增强方式的实验是通过相关的最优化的目标构造模拟环境进行参数学习。

2. 安装 PyBrain 模块

PyBrain 模块的安装首先依赖 NumPy 和 SciPy，确保在安装之前，已经安装了这两个模块。在终端输入安装指令：

```
pip install pybrain
```

或者

```
pip install git+https://github.com/pybrain/pybrain.git
```

后面可以加@参数安装指定的版本：

```
pip install git+https://github.com/pybrain/pybrain.git@@0.3.3
```

3. 创建神经网络

通过 PyBrain 创建神经网络很简单，使用 buildNetwork 方法即可：

```
>>> from pybrain.tools.shortcuts import buildNetwork
>>> net = buildNetwork(2, 3, 1)
```

3 个参数分别表示 2 个输入神经元、3 个隐藏层神经元、1 个输出层神经元。创建后，默认会以随机数初始化这个神经网络，通过.activate()方法可调用：

```
>>> net.activate([2, 1])
array([-0.98646726])
```

activate 方法可接受列表、元组、数组作为输入。在 PyBrain 中，创建完成一个神经网络后每一层默认都有名字。

```
>>> net['in']
<LinearLayer 'in'>
>>> net['hidden0']
<SigmoidLayer 'hidden0'>
>>> net['out']
<LinearLayer 'out'>
```

隐藏层后面有一个数字，以区分不同隐藏层。隐藏层默认是通过 Sigmoid 函数构建的。

```
>>> from pybrain.structure import TanhLayer
>>> net = buildNetwork(2, 3, 1, hiddenclass=TanhLayer)
>>> net['hidden0']
<TanhLayer 'hidden0'>
```

也可以为输出层指定不同类别：

```
>>> from pybrain.structure import SoftmaxLayer
>>> net = buildNetwork(2, 3, 2, hiddenclass=TanhLayer, outclass=SoftmaxLayer)
>>> net.activate((2, 3))
array([ 0.6656323,  0.3343677])
```

还可以设置偏置层：

```
>>> net = buildNetwork(2, 3, 1, bias=True)
>>> net['bias']
<BiasUnit 'bias'>
```

4. PyBrain 示例

【例 4.16】

（1）建立神经网络

在 PyBrain 中，网络由互相关联的模块和连接组成。可以将网络视为有向无环图，其中节点为模块，边缘为连接，这使得 PyBrain 非常灵活。当然，这种结构不是在所有情况下都是必需的。因此，有一种创建网络的简单方法，即 buildNetwork 快捷方式，此调用返回一个具有 2 个输入、3 个隐藏和 1 个输出神经元的网络。在 PyBrain 中，这些图层是 Module 对象，它们已经连接到 FullConnection 对象。

```
#!usr/bin/env python
#_*_coding:utf-8_*_

from pybrain.tools.shortcuts import buildNetwork
from pybrain.structure import TanhLayer
from pybrain.structure import SoftmaxLayer
from pybrain.datasets import SupervisedDataSet
#构建一个神经网络，简单地进行激活并打印各层的名称信息
net = buildNetwork(2,3,1)
print (net.activate([2,1]))
print ("net['in'] = " , net['in'])
print ("net['hidden0'] = " , net['hidden0'])
print ("net['out'] = " , net['out'])

#自定义复杂网络，把神经网络的默认隐藏层参数设置为 Tanh 函数而不是 Sigmoid 函数
#from pybrain.structure import TanhLayer
net = buildNetwork(2,3,1,hiddenclass = TanhLayer)
print ("net['hidden0_1'] = " , net['hidden0'])

#自定义复杂网络，修改输出层的类型
#from pybrain.structure import SoftmaxLayer
net = buildNetwork(2,3,2,hiddenclass = TanhLayer,outclass = SoftmaxLayer)
print (net.activate((2,3)))

net = buildNetwork(2,3,1,bias = True)
print (net['bias'])
```

输出：

```
[0.44803573]
net['in'] = <LinearLayer 'in'>
net['hidden0'] = <SigmoidLayer 'hidden0'>
net['out'] = <LinearLayer 'out'>
```

```
net['hidden0_1'] = <TanhLayer 'hidden0'>
[0.98681316 0.01318684]
<BiasUnit 'bias'>
```

（2）构造数据集

SupervisedDataSet 类用于标准监督学习。它支持输入和目标值，我们必须在对象创建时指定它们的大小。这里我们生成了一个支持二维输入和一维标注信息的数据集。

```
#!usr/bin/env python
#_*_coding:utf-8_*_
from pybrain.datasets import SupervisedDataSet
from pybrain.datasets import UnsupervisedDataSet

ds = SupervisedDataSet(2,1)

ds.addSample((0,0), (0,))
ds.addSample((0,1), (1,))
ds.addSample((1,0), (1,))
ds.addSample((1,1), (0,))
print ('检查数据集的长度')
print (len (ds))
print('用 for 循环迭代的方式访问数据集')
for inpt,target in ds:
    print (inpt,target)
print ('直接访问输入字段和标注字段的数组')
print (ds['input'])
print (ds['target'])
print ('清除数据集')
#ds.clear()
#print ds['input']
#print ds['target']
```

输出：

```
检查数据集的长度
4
用 for 循环迭代的方式访问数据集
[0. 0.] [0.]
[0. 1.] [1.]
[1. 0.] [1.]
[1. 1.] [0.]
直接访问输入字段和标注字段的数组
[[0. 0.]
 [0. 1.]
 [1. 0.]
```

```
 [1. 1.]]
[[0.]
 [1.]
 [1.]
 [0.]]
```
清除数据集

（3）在数据集上训练神经网络

为了调整监督学习中的模块参数，PyBrain 内置一个训练器。训练器的参数分别是一个神经网络模型和一个数据集，训练神经网络模型以适应数据集中的数据。在这里使用 BackpropTrainer（误差反向传播训练器）。单独调用 train()方法只能训练一个完整时期的网络，并返回一个误差值。如果要训练网络直到网络收敛，还有一种方法，即 trainUntilConvergence()，方法名字很好记，"训练直到收敛"，会返回一大堆数据，包含每个训练周期的误差元组。而且会发现误差数组的每个元素都是逐渐减小的，说明网络逐渐收敛。

```
#!usr/bin/env python
#_*_coding:utf-8_*_

#引入建立神经网络所需的相关模块
from pybrain.tools.shortcuts import buildNetwork
from pybrain.structure import TanhLayer

#引入建立数据集所需的相关模块
from pybrain.datasets import SupervisedDataSet

#引入BackpropTrainer反向训练器
from pybrain.supervised.trainers import BackpropTrainer

#引入之前建好的建立数据集中的module

import test_pybrain_2   #test_pybrain_2是module名
'''
ds = SupervisedDataSet(2,1)
ds.addSample((0,0), (0,))
ds.addSample((0,1), (1,))
ds.addSample((1,0), (1,))
ds.addSample((1,1), (0,))
'''
net = buildNetwork(2,3,1,bias = True,hiddenclass=TanhLayer)
trainer = BackpropTrainer(net,test_pybrain_2.ds)    #应用之前建立的ds数据集
#通过调用train()方法来对网络进行训练
print (trainer.train())
#通过调用trainUntilConvertgence()方法对网络训练直到收敛
```

```
print (trainer.trainUntilConvergence())
```

输出:

```
检查数据集的长度
4
用 for 循环迭代的方式访问数据集
[0. 0.] [0.]
[0. 1.] [1.]
[1. 0.] [1.]
[1. 1.] [0.]
直接访问输入字段和标注字段的数组
[[0. 0.]
 [0. 1.]
 [1. 0.]
 [1. 1.]]
[[0.]
 [1.]
 [1.]
 [0.]]
清除数据集
0.8527734007791132
([], [0.008565398313670416])
```

以上是利用 PyBrain 建立简单的神经网络、建立监督数据集、训练神经网络的基本操作与模块。

4.15 Fuel

Fuel 是一个数据管道框架(Data Pipeline Framework),为机器学习模型提供所需的数据。Blocks 和 Pylearn2 这两个神经网络库都有机会使用 Fuel。

Fuel 为机器学习模型提供了它们需要学习的数据:

- 接口到公共数据集,如 MNIST、CIFAR-10(图像数据集)、谷歌的十亿字(文本)等。
- 能够以多种方式对数据进行迭代,例如在带有洗牌/连续的示例 mini-batches 中。
- 一种预处理程序的管道,它允许实时编辑数据,例如通过添加噪声、从句子中提取 n 个格、从图像中提取补丁等。
- 确保整个管道可通过 pickle 序列化,这是能够设置检查点和恢复长时间运行的实验的必要条件。为此,严重依赖 picklable_itertools 库。

Fuel 主要是通过 Blocks 开发应用的,是一个 Theano 工具包,可用于训练神经网络,项目地址为 https://github.com/mila-iqia/fuel。

【例4.17】test_adult.py（源码地址为 https://github.com/mila-iqia/fuel/blob/master/tests/test_adult.py）

```python
#!usr/bin/env python
#_*_coding:utf-8_*_
from importlib import import_module
from unittest.case import SkipTest
import numpy
from numpy.testing import assert_raises, assert_equal, assert_allclose
from fuel.datasets import Adult
from fuel.utils import find_in_data_path
from fuel import config

def skip_if_not_available(modules=None, datasets=None, configurations=None):
    """Raises a SkipTest exception when requirements are not met.
    Parameters
    ----------
    modules : list
        A list of strings of module names. If one of the modules fails to
        import, the test will be skipped.
    datasets : list
        A list of strings of folder names. If the data path is not
        configured, or the folder does not exist, the test is skipped.
    configurations : list
        A list of strings of configuration names. If this configuration
        is not set and does not have a default, the test will be skipped.
    """
    if modules is None:
        modules = []
    if datasets is None:
        datasets = []
    if configurations is None:
        configurations = []
    for module in modules:
        try:
            import_module(module)
        except Exception:
            raise SkipTest
    if datasets and not hasattr(config, 'data_path'):
        raise SkipTest
    for dataset in datasets:
        try:
```

```python
            find_in_data_path(dataset)
        except IOError:
            raise SkipTest
    for configuration in configurations:
        if not hasattr(config, configuration):
            raise SkipTest


def test_adult_test():
    skip_if_not_available(datasets=['adult.hdf5'])

    dataset = Adult(('test',), load_in_memory=False)
    handle = dataset.open()
    data, labels = dataset.get_data(handle, slice(0, 10))

    assert data.shape == (10, 104)
    assert labels.shape == (10, 1)
    known = numpy.array(
        [25., 38., 28., 44., 34., 63., 24., 55., 65., 36.])
    assert_allclose(data[:, 0], known)
    assert dataset.num_examples == 15060
    dataset.close(handle)

    dataset = Adult(('train',), load_in_memory=False)
    handle = dataset.open()
    data, labels = dataset.get_data(handle, slice(0, 10))

    assert data.shape == (10, 104)
    assert labels.shape == (10, 1)
    known = numpy.array(
        [39., 50., 38., 53., 28., 37., 49., 52., 31., 42.])
    assert_allclose(data[:, 0], known)
    assert dataset.num_examples == 30162
    dataset.close(handle)


def test_adult_axes():
    skip_if_not_available(datasets=['adult.hdf5'])

    dataset = Adult(('test',), load_in_memory=False)
    assert_equal(dataset.axis_labels['features'],
                 ('batch', 'feature'))

    dataset = Adult(('train',), load_in_memory=False)
```

```
    assert_equal(dataset.axis_labels['features'],
                ('batch', 'feature'))

def test_adult_invalid_split():
    skip_if_not_available(datasets=['adult.hdf5'])

    assert_raises(ValueError, Adult, ('dummy',))
```

4.16 PyMVPA

PyMVPA（Multivariate Pattern Analysis in Python）是为大数据集提供统计学习分析的Python工具包，提供了一个灵活可扩展的框架。它提供的功能有分类、回归、特征选择、数据导入导出、可视化等。

项目主页如下：

- http://www.pymvpa.org/
- https://github.com/PyMVPA/PyMVPA

【例 4.18】

```
#!/usr/bin/env python
# emacs: -*- mode: python; py-indent-offset: 4; indent-tabs-mode: nil -*-
# vi: set ft=python sts=4 ts=4 sw=4 et:
### ### ### ### ### ### ### ### ### ### ### ### ### ### ### ### ### ### ### ### ##
#
#   See COPYING file distributed along with the PyMVPA package for the
#   copyright and license terms.
#
### ### ### ### ### ### ### ### ### ### ### ### ### ### ### ### ### ### ### ### ##
"""Convenience functions to generate/update datasets for regression testing
"""

__docformat__ = 'restructuredtext'

from os.path import join as pathjoin

import hashlib
import mvpa2

from mvpa2 import pymvpa_dataroot, externals

def get_testing_fmri_dataset_filename():
    """Generate path to the testing filename based on mvpa2/nibabel versions
```

```python
    """
    # explicitly so we do not anyhow depend on dict ordering
    versions_hash = hashlib.md5(
        "_".join(["%s:%s" % (k, externals.versions[k])
                 for k in sorted(externals.versions)])
    ).hexdigest()[:6]

    filename = 'mvpa-%s_nibabel-%s-%s.hdf5' % (
        mvpa2.__version__,
        externals.versions['nibabel'],
        versions_hash)

    return pathjoin(pymvpa_dataroot, 'testing', 'fmri_dataset', filename)
get_testing_fmri_dataset_filename.__test__ = False


def generate_testing_fmri_dataset(filename=None):
    """Helper to generate a dataset for regression testing of mvpa2/nibabel
    Parameters
    ----------
    filename : str
      Filename of a dataset file to store. If not provided, it is composed
      using :func: 'get_testing_fmri_dataset_filename'
    Returns
    -------
    Dataset, string
       Generated dataset, filename to the HDF5 where it was stored
    """
    import mvpa2
    from mvpa2.base.hdf5 import h5save
    from mvpa2.datasets.sources import load_example_fmri_dataset
    # Load our sample dataset
    ds_full = load_example_fmri_dataset(name='1slice', literal=False)
    # Subselect a small "ROI"
    ds = ds_full[20:23, 10:14]
    # collect all versions/dependencies for possible need to troubleshoot later
    ds.a['wtf'] = mvpa2.wtf()
    ds.a['versions'] = mvpa2.externals.versions
    # save to a file identified by version of PyMVPA and nibabel and hash of
    # all other versions
    out_filename = filename or get_testing_fmri_dataset_filename()
    h5save(out_filename, ds, compression=9)
    # ATM it produces >700kB .hdf5 which is this large because of
    # the ds.a.mapper with both Flatten and StaticFeatureSelection occupying
    # more than 190kB each, with ds.a.mapper as a whole generating 570kB file
    # Among those .ca seems to occupy notable size, e.g. 130KB for the FlattenMapper
```

```
    # even though no heavy storage is really needed for any available value --
    # primarily all is meta-information embedded into hdf5 to describe our things
    return ds, out_filename

generate_testing_fmri_dataset.__test__ = False

if __name__ == '__main__':
    generate_testing_fmri_dataset()
```

4.17 Annoy

Annoy（Approximate Nearest Neighbors Oh Yeah）是一个带有 Python 绑定的 C++ 库，用于在空间中找到和已知的查询点临近的点。它还可以创建大型的基于文件的只读数据结构，并映射至内存，以便多个进程能共同使用相同的数据。

1. 工作原理

使用随机投影和建立树（见图 4.24）。在树的每个中间节点上，选择一个随机超平面，将空间划分为两个子空间。这个超平面是通过从子集中抽取两个点，然后取距离相等的超平面来选择的。这样重复做 k 次，就得到了一片树林。k 必须根据你的需要进行调整，在精度和性能之间进行权衡。Hamming 距离（Martin Aumüller）将数据打包成 64 位整数，并使用内置的位计数原语，因此速度非常快。所有的劈叉都是轴向对齐的。点积距离（Peter Sobot）将提供的向量从点（或"内积"）空间减少到更便于查询的余弦空间。

图 4.24　Annoy 原理示意图

2. Annoy 安装

Annoy 软件下载地址为 https://pypi.org/project/annoy/，安装指令如下：

```
pip install --user annoy
```

3. 建立索引过程

Annoy 的目标是建立一个数据结构，使得查询一个点的最近邻点的时间复杂度为次线性。Annoy 通过建立一棵二叉树来使得每个点查找时间复杂度为 $O(\log n)$。在划分的子空间内进行不停地递归迭代继续划分，直到每个子空间最多只剩下 K 个数据节点。通过多次递归迭代划分，最终原始数据会形成类似图 4.25 所示的二叉树结构。二叉树底层是叶子节点，记录原始数据节点，其他中间节点记录的是分割超平面的信息。Annoy 建立这样的二叉树结构是希望满足一个假设：相似的数据节点应该在二叉树上位置更接近，一个分割超平面不应该把相似的数据节点分割到二叉树的不同分支上。

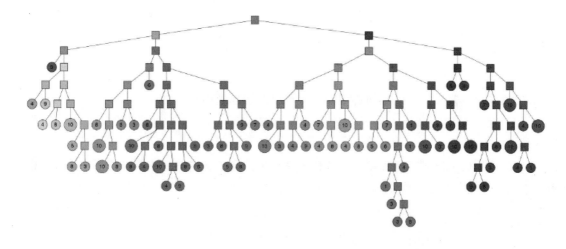

图 4.25　Annoy 构建的二叉树

4. 查询过程

上面已完成节点索引的建立过程。如何对一个数据点查找相似节点集合呢？查找的过程就是不断地看在分割超平面的哪一边。从二叉树索引结构来看，就是从根节点不停地往叶子节点遍历的过程。通过对二叉树每个中间节点（分割超平面相关信息）和查询数据节点进行相关计算来确定二叉树遍历过程是往这个中间节点的左孩子节点走还是往右孩子节点走。通过以上方式完成查询过程，如图 4.26 Annoy 所示。

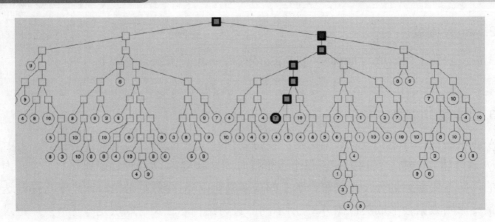

图 4.26　Annoy 查询

但上述描述存在两个问题：

（1）查询过程最终落到叶子节点的数据节点数小于需要的 Top N 相似邻居节点数目怎么办？

（2）两个相近的数据节点划分到二叉树不同分支上怎么办？

针对这个问题可以通过以下方法来解决：

（1）如果分割超平面的两边都很相似，可以两边都遍历。

（2）建立多棵二叉树，构成一片森林（见图 4.27），每棵树建立机制都如上面所述的那样。

（3）采用优先队列机制：采用一个优先队列来遍历二叉树，从根节点往下的路径，根据查询节点与当前分割超平面距离（Margin）进行排序。

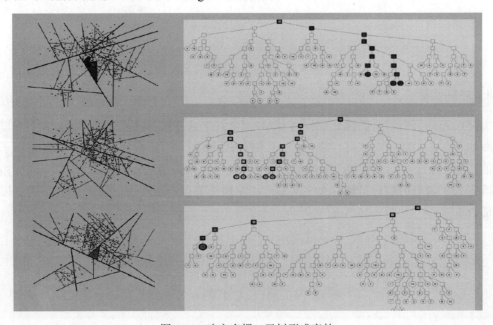

图 4.27　建立多棵二叉树形成森林

5. 返回最终近邻节点

每棵树都返回一堆近邻点后，如何得到最终的 Top N 相似集合呢？首先所有树返回近邻点都插入优先队列中，求并集以便去掉重复的近邻点，然后计算和查询点距离，最终根据距离值从近距离到远距离排序，返回 Top N 近邻节点集合。

【例 4.19】

```
#!usr/bin/env python
#_*_coding:utf-8_*_
from __future__ import print_function
import random, time
from annoy import AnnoyIndex

try:
    xrange
except NameError:
    # Python 3 compat
    xrange = range

n, f = 100000, 40

t = AnnoyIndex(f)
for i in xrange(n):
    v = []
    for z in xrange(f):
        v.append(random.gauss(0, 1))
    t.add_item(i, v)

t.build(2 * f)
t.save('test.tree')

limits = [10, 100, 1000, 10000]
k = 10
prec_sum = {}
prec_n = 1000
time_sum = {}

for i in xrange(prec_n):
    j = random.randrange(0, n)

    closest = set(t.get_nns_by_item(j, k, n))
    for limit in limits:
        t0 = time.time()
        toplist = t.get_nns_by_item(j, k, limit)
        T = time.time() - t0
```

```
        found = len(closest.intersection(toplist))
        hitrate = 1.0 * found / k
        prec_sum[limit] = prec_sum.get(limit, 0.0) + hitrate
        time_sum[limit] = time_sum.get(limit, 0.0) + T

for limit in limits:
    print('limit: %-9d precision: %6.2f%% avg time: %.6fs'
        % (limit, 100.0 * prec_sum[limit] / (i + 1),
            time_sum[limit] / (i + 1)))
```

结果输出:

```
limit: 10         precision:  13.66% avg time: 0.000052s
limit: 100        precision:  20.57% avg time: 0.000092s
limit: 1000       precision:  56.11% avg time: 0.000279s
limit: 10000      precision:  96.95% avg time: 0.001855s
```

4.18 Deap

Deap 是一个创新的、仍在发展中的计算框架，用于快速构建原型和测试方法。它旨在使算法和数据结构更加清晰透明。它与并行机制（如多进程和 SCOOP 模块）完美协调。

1. 学习 DEAP 框架资源

- GitHub 源码: https://github.com/deap/deap。
- Deap 文档: http://deap.gel.ulaval.ca/doc/dev/index.html。
- Deap 下载: https://pypi.python.org/pypi/deap/。

2. 安装

可以使用 easy_install 或者 pip 安装 Deap。其他的安装程序，如 apt-get、yum 等，通常安装的都是过时的版本。

```
pip install deap
```

最近的版本可以按照下面的命令安装:

```
pip install git+https://github.com/DEAP/deap@master
```

如果希望可以编译源代码，可以先下载或者克隆源码，然后执行:

```
python setup.py install
```

3. 举例

Deap 实例地址为 https://deap.readthedocs.io/en/master/examples/index.html。通过下面的代码可以快速学习如何使用 Deap 的遗传算法实现 Onemax 问题优化。

【例 4.20】使用 Deap 的遗传算法实现 Onemax 问题优化

```
import random
from deap import creator, base, tools, algorithms

creator.create("FitnessMax", base.Fitness, weights=(1.0,))
creator.create("Individual", list, fitness=creator.FitnessMax)

toolbox = base.Toolbox()

toolbox.register("attr_bool", random.randint, 0, 1)
toolbox.register("individual", tools.initRepeat, creator.Individual,
toolbox.attr_bool, n=100)
toolbox.register("population", tools.initRepeat, list, toolbox.individual)

def evalOneMax(individual):
    return sum(individual),

toolbox.register("evaluate", evalOneMax)
toolbox.register("mate", tools.cxTwoPoint)
toolbox.register("mutate", tools.mutFlipBit, indpb=0.05)
toolbox.register("select", tools.selTournament, tournsize=3)

population = toolbox.population(n=300)

NGEN=40
for gen in range(NGEN):
    offspring = algorithms.varAnd(population, toolbox, cxpb=0.5, mutpb=0.1)
    fits = toolbox.map(toolbox.evaluate, offspring)
    for fit, ind in zip(fits, offspring):
        ind.fitness.values = fit
    population = toolbox.select(offspring, k=len(population))
top10 = tools.selBest(population, k=10)
```

4.19 Pattern

Pattern 是一个 Python 的网络挖掘模块。它绑定了数据挖掘（Google + Twitter + Wikipedia API，网络爬虫，HTML DOM 解析器）、自然语言处理（词性标注，n gram 搜索，语义分析，WordNet）、机器学习（向量空间模型，K-Means 聚类，Naive Bayes + k-NN + SVM 分类器）和网络分析（图核心性 Graph Centrality 和可视化）等工具。

1. 安装

```
pip install pattern
```

2. pattern.web

pattern.web 模块是一个 Web 工具包，包含 API（Google、Gmail、Bing、Twitter、Facebook、Wikipedia、Wiktionary、DBPedia、Flickr，…）一个健壮的 HTML DOM 解析器和一个 Web 爬虫程序。

```
>>> from pattern.web import Twitter, plaintext
>>> twitter = Twitter(language='en')
>>> for tweet in twitter.search('"more important than"', cached=False):
>>>     print plaintext(tweet.text)
```

输出：

```
'The mobile web is more important than mobile apps.'
'Start slowly, direction is more important than speed.'
'Imagination is more important than knowledge. - Albert Einstein'
...
```

3. pattern.en

pattern.en 是一个面向英语的自然语言处理（NLP）工具包。因为语言是模糊的（例如，I can ↔ a can），可以使用统计方法+正则表达式。这意味着它速度快，相当准确，偶尔也会出错。它有一个词性标注，用于标识单词类型（例如，名词、动词、形容词）、单词屈折变化（共轭、单数化）和 WordNet API。

```
from pattern.en import parse
s = 'The mobile web is more important than mobile apps.'
s = parse(s, relations=True, lemmata=True)
print (s)
```

输出：

```
The/DT/B-NP/O/NP-SBJ-1/the mobile/JJ/I-NP/O/NP-SBJ-1/mobile
web/NN/I-NP/O/NP-SBJ-1/web is/VBZ/B-VP/O/VP-1/be more/RBR/B-ADJP/O/O/more
important/JJ/I-ADJP/O/O/important than/IN/B-PP/B-PNP/O/than
mobile/JJ/B-NP/I-PNP/O/mobile apps/NN/I-NP/I-PNP/O/apps ././O/O/O/.
```

word	tag	chunk	role	id	pnp	lemma
The	DT	NP	SBJ	1	-	*the*
mobile	JJ	NP^	SBJ	1	-	*mobile*
web	NN	NP^	SBJ	1	-	*web*
is	VBZ	VP	-	1	-	*be*

more	RBR	ADJP	-	-	-	*more*
important	JJ	ADJP^	-	-	-	*important*
than	IN	PP	-	-	PNP	*than*
mobile	JJ	NP	-	-	PNP	*mobile*
apps	NNS	NP^	-	-	PNP	*app*
.	.	-	-	-	-	.

文本中已经标注了单词类型，例如名词（NN）、动词（VB）、形容词（JJ）和限定词（DT）、单词类型（例如句子主语 SBJ）和介词名词短语（PNP）。要遍历标注文本中的部分，可以构造一个解析树。

4. pattern.search

pattern.search 模块包含一个搜索算法，用于从标注文本中检索单词序列（称为 n-gram）。Python2.7 下运行以下代码：

```
>>> from pattern.en import parsetree
>>> from pattern.search import search
>>>
>>> s = 'The mobile web is more important than mobile apps.'
>>> s = parsetree(s, relations=True, lemmata=True)
>>>
>>> for match in search('NP be RB?+ important than NP', s):
>>>     print match.constituents()[-1], '=>', \
>>>         match.constituents()[0]
输出：
Chunk('mobile apps/NP') => Chunk('The mobile web/NP-SBJ-1')
```

5. pattern.vector

pattern.vector 模块是一种机器学习工具，基于带加权特征（如 TF-IDF）和距离度量（如余弦相似度、信息增益）的袋装文档的向量空间模型。模型可用于聚类（K 均值、层次结构）、分类（朴素贝叶斯、感知器、k-NN、SVM）和潜在语义分析（LSA）。在 Python2.7 下运行以下代码：

```
>>> from pattern.web    import Twitter
>>> from pattern.en     import tag
>>> from pattern.vector import KNN, count
>>>
>>> twitter, knn = Twitter(), KNN()
>>>
>>> for i in range(1, 10):
>>>     for tweet in twitter.search('#win OR #fail', start=i, count=100):
>>>         s = tweet.text.lower()
```

```
>>>         p = '#win' in s and 'WIN' or 'FAIL'
>>>         v = tag(s)
>>>         v = [word for word, pos in v if pos == 'JJ'] # JJ = adjective
>>>         v = count(v)
>>>         if v:
>>>             knn.train(v, type=p)
>>>
>>> print knn.classify('sweet potato burger')
>>> print knn.classify('stupid autocorrect')
输出：
'WIN'
'FAIL'
```

6. pattern.graph

pattern.graph 模块提供了一个表示节点之间关系的图数据结构（例如，术语、概念）。图形可以导出为 HTML <canvas>动画（演示）。在图 4.28 的示例中，更多的中心节点（更多的输入信息）被涂成蓝色。

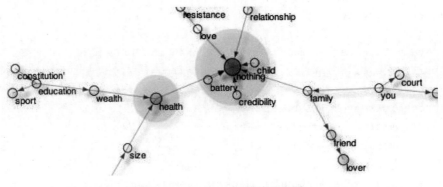

图 4.28　pattern.graph 模块节点图

在 Python2.7 下运行以下代码：

```
>>> from pattern.web    import Bing, plaintext
>>> from pattern.en     import parsetree
>>> from pattern.search import search
>>> from pattern.graph  import Graph
>>>
>>> g = Graph()
>>> for i in range(10):
>>>     for result in Bing().search('"more important than"', start=i+1, count=50):
>>>         s = r.text.lower()
>>>         s = plaintext(s)
>>>         s = parsetree(s)
>>>         p = '{NP} (VP) more important than {NP}'
>>>         for m in search(p, s):
```

```
>>>             x = m.group(1).string # NP left
>>>             y = m.group(2).string # NP right
>>>             if x not in g:
>>>                 g.add_node(x)
>>>             if y not in g:
>>>                 g.add_node(y)
>>>             g.add_edge(g[x], g[y], stroke=(0,0,0,0.75)) # R,G,B,A
>>>
>>> g = g.split()[0] # Largest subgraph.
>>>
>>> for n in g.sorted()[:40]: # Sort by Node.weight.
>>>     n.fill = (0, 0.5, 1, 0.75 * n.weight)
>>>
>>> g.export('test', directed=True, weighted=0.6)
```

有些关系（如边）可以用一些额外的后加工，例如"nothing is more important than life，nothing is not more important than life"。

4.20 Requests

Requests 库是用 Python 语言基于 urllib 编写的，采用的是 Apache2 Licensed 开源协议的 HTTP 库，可自动爬取 HTML 页面，自动网络请求提交，是网络数据爬取和网页解析的基本库，常用于网络爬虫与信息提取。Requests 支持 Python3。

1. 安装 Requests

通过 pip 安装：

```
pip install requests
```

或者，下载代码后安装：

```
$ git clone git://github.com/kennethreitz/requests.git
$ cd requests
$ python setup.py install
```

2. 发送请求与传递参数

（1）Response 对象

Response 对象包含服务器返回的所有信息，也包含请求的 Request 信息，其有 5 个属性：

- r.status_code，HTTP 请求的返回状态。
- r.text，HTTP 响应内容的字符串形式，即 URL 对应的页面内容。
- r.encoding，从 HTTP header 中猜测的响应内容编码方式。

- r.apparent_encoding,从内容中分析出的响应内容编码方式(备选编码方式)。
- r.content,HTTP 响应内容的二进制形式。

(2)连接异常

requests 库支持 6 种常见的连接异常:

- requests.ConnectionError,网络连接错误异常,如 DNS 查询失败、拒绝连接等。
- requests.HTTPError,HTTP 错误异常。
- requests.URLRequired,URL 缺失异常。
- requests.TooManyRedirects,超过最大重定向次数,产生重定向异常。
- requests.ConnectTimeout,连接远程服务器超时异常。
- requests.Timeout,请求 URL 超时,产生超时异常。

3. requests 库方法

requests 库一共有 7 个主要方法(见表 4.1),除最基础的 request() 方法外,其余 6 个与 HTTP 协议方法的功能一一对应。

表 4.1　requests 7 个库方法

方法	描述	与 HTTP 对应方法
requests.request()	构造一个请求,支撑以下各方法的基础方法	无
requests.get()	获取 HTML 网页的主要方法	GET
requests.head()	获取 HTML 网页头信息的方法	HEAD
requests.post()	向 HTML 网页提交 POST 请求的方法	POST
requests.put()	向 HTML 网页提交 PUT 请求的方法	PUT
requests.patch()	向 HTML 网页提交局部修改请求	PATCH
requests.delete()	向 HTML 网页提交删除请求	DELETE

(1)request()方法

request()方法是所有方法的基础方法,其余的所有方法都是基于 request() 方法来封装的。

格式:requests.request(method, url, **kwargs)。

- method:请求方式,不同的方式对应不同的 HTTP 请求功能。一共有 7 个请求方式,即 GET、HEAD、POST、PUT、PATCH、DELETE、OPTIONS,分别对应 HTTP 中的请求功能。如果想发送一个请求,除了改变 request() 方法中的请求方式外,还可以直接使用 request 库中对应的方法。例如:r=r.request(GET,"http://www.baidu.com")与 r=r.get("http://www.baidu.com")等价。
- url:拟获取页面的 url 链接。
- **kwargs:控制访问的参数,共 13 个。它是可选的,使用时,需要用命名方法来调用,例如 params=kv(kv 为事先定义好的一个字典)。
    ➢ Params,字典或字节序列,作为参数增加到 url 中。
    ➢ Data,字典、字节序列或文件对象,作为 Request 的内容。

- Json,JSON 格式的数据,作为 Request 的内容。
- Headers,字典,可以用来控制访问链接的 HTTP 头。
- Cookies,字典或 CookieJar,Request 中的 cookie。
- Auth,元组,支持 HTTP 认证功能。
- Files,字典类型,向服务器传输文件时使用的字段。
- Timeout,设定超时时间,以秒为单位,如果设定时间内未返回,将产生一个 TimeoutError 异常。
- Proxies,字典类型,设定访问代理服务器,还可增加登录认证,能有效地隐藏用户爬取网页的源 IP,从而防止对爬虫的逆追踪。
- allow_redirects,重定向开关,取值为 True/False,默认为 True。
- stream,获取内容立即下载开关,默认为 True。
- verify,认证 SSL 证书开关,默认为 True。
- cert,保存本地 SSL 证书路径的字段。

(2)get()方法

get() 方法通过给定 url 来构造一个向 HTTP 请求资源的 Request 对象,返回一个包含 HTTP 资源的 Response 对象,包含从 HTTP 返回的所有相关资源。

格式:r = requests.get(url, **kwargs)。

(3)head() 方法

head() 方法通过给定 url 来构造一个向 HTTP 请求资源的 Request 对象,返回一个包含 HTTP 资源的 Response 对象,包含从 HTTP 返回的所有头部资源。

格式:requests.head(url, **kwargs)。

(4)post() 方法

post() 方法向 url 提交新增数据,格式为字典,自动编码为 form(表单)。

格式:requests.post(url, **kwargs)。

(5)put() 方法

put() 方法向 url 提交新增数据,格式为字符串,自动编码为 data。

格式:requests.put(url,, **kwargs)。

(6)patch() 方法

patch() 方法向 url 提交局部修改请求。

格式:requests.patch(url,, **kwargs)。

(7)delete() 方法

delete() 方法向 url 提交删除请求。

格式:requests.delete(url, **kwargs)。

4. 编码

网络上的资源均有其特定的编码，如果没有编码，将无法通过有效的解析方式使其可读，因此需要通过编码来进行解析：

- r.encoding 通过对 header 进行分析，如果其中不存在 charset 字段，则认为编码为 ISO‐8859‐1。
- r.apparent_encoding 根据网页内容分析出编码方式，其解析方式要比 r.encoding 更为准确，可看作是 r.encoding 的备选。
- r.text 根据 r.encoding 来显示网页内容。

5. 通用爬取框架

通用爬取框架实质上就是一组代码，其最大的作用是使用户更有效、稳定、可靠地爬取网页上的内容。在使用 requests 库进行网页访问时，经常使用 get() 函数来获得 URL 的相关内容，但这样的内容不是一定成立的，因为网络连接有一定的异常，此时异常处理就极为重要。Response 对象提供了一个与处理异常的方法：

```
r.raise_for_status()
```

其功能是判断返回的 Response 类型是不是 200，如果不是 200，就将产生异常 requests.HTTPError，因此有了如下的通用代码框架：

```python
#!usr/bin/env python
#_*_coding:utf-8_*_
import requests

def getHTMLText(url):
    try:
        r=requests.get(url,timeout=30)
        r.raise_for_status()#如果状态不是200，就会引发HTTPError异常
        r.encoding=r.apparent_encoding
        return r.text
    except:
        return "产生异常"

if __name__=="__main__":
    url="http://www.baidu.com"
    print(getHTMLText(url))
```

【例 4.21】案例

```
#!usr/bin/env python
#_*_coding:utf-8_*_
import requests

URL = 'http://ip.taobao.com/service/getIpInfo.php'  # 淘宝 IP 地址库 API
```

```
try:
    r = requests.get(URL, params={'ip': '8.8.8.8'}, timeout=1)
    r.raise_for_status()     # 如果响应状态码不是 200，就主动抛出异常
except requests.RequestException as e:
    print(e)
else:
    result = r.json()
    print(type(result), result, sep='\n')
```

输出：

```
<class 'dict'>
{'code': 0, 'data': {'ip': '8.8.8.8', 'country': '美国', 'area': '', 'region': 'XX',
'city': 'XX', 'county': 'XX', 'isp': 'Level3', 'country_id': 'US', 'area_id': '',
'region_id': 'xx', 'city_id': 'xx', 'county_id': 'xx', 'isp_id': '200053'}}
```

4.21 Seaborn

Seaborn 同 Matplotlib 一样，也是 Python 进行数据可视化分析的重要第三方包。Seaborn 在 Matplotlib 的基础上进行了更高级的 API 封装，使得绘图更加容易，图形更加漂亮。应该把 Seaborn 视为 Matplotlib 的补充，而不是替代物。

1. 安装 Seaborn

```
pip install seaborn
```

2. 举例

定义一个简单的方程来绘制一些偏置的正弦波（见图 4.29），用来帮助查看不同的画图风格。

【例 4.22】

```
#!usr/bin/env python
#_*_coding:utf-8_*_
import numpy as np
import matplotlib as mpl
import matplotlib.pyplot as plt
import seaborn as sns
np.random.seed(sum(map(ord, "aesthetics")))
def sinplot(flip=1):
    x = np.linspace(0, 14, 100)
    for i in range(1, 7):
        plt.plot(x, np.sin(x + i * .5) * (7 - i) * flip)
sinplot()
```

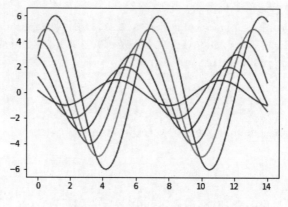

图 4.29　Seaborn 绘制正弦波

Seaborn 将 Matplotlib 的参数划分为两个独立的组合：第一组设置绘图的外观风格；第二组主要将绘图的各种元素按比例缩放的，以嵌入到不同的背景环境中。

操控这些参数的接口主要有两对方法：

- 控制风格：axes_style()和 set_style()。
- 缩放绘图：plotting_context()和 set_context()。

每对方法中的第一个方法（axes_style()和 plotting_context()）都会返回一组字典参数，而第二个方法（set_style()和 set_context()）会设置 Matplotlib 的默认参数。

3. Seaborn 的 5 种绘图风格

Seaborn 有 5 种绘图风格，分别是 darkgrid、whitegrid、dark、white、ticks，它们各自适合不同的应用。默认的主题是 darkgrid。

（1）whitegrid（见图 4.30）

```
sns.set_style("whitegrid")
data = np.random.normal(size=(20, 6)) + np.arange(6) / 2
sns.boxplot(data=data);
```

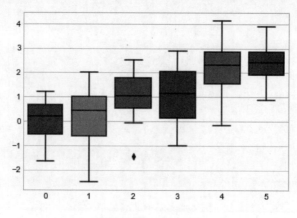

图 4.30　whitegrid

（2）dark（见图4.31）

```
import numpy as np
import matplotlib.pyplot as plt
import seaborn as sns
def sinplot(flip=1):
    x = np.linspace(0, 14, 100)
    for i in range(1, 7):
        plt.plot(x, np.sin(x + i * .5) * (7 - i) * flip)
sns.set_style("dark")
sinplot()
```

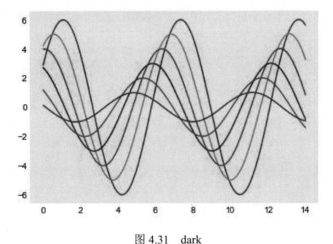

图4.31　dark

（3）white（见图4.32）

```
import numpy as np
import matplotlib.pyplot as plt
import seaborn as sns
def sinplot(flip=1):
    x = np.linspace(0, 14, 100)
    for i in range(1, 7):
        plt.plot(x, np.sin(x + i * .5) * (7 - i) * flip)
sns.set_style("white")
sinplot()
```

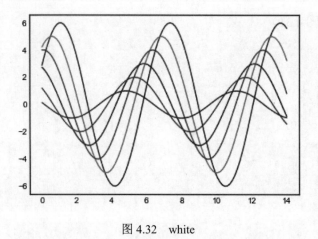

图 4.32 white

（4）ticks

```
import numpy as np
import matplotlib.pyplot as plt
import seaborn as sns
def sinplot(flip=1):
    x = np.linspace(0, 14, 100)
    for i in range(1, 7):
        plt.plot(x, np.sin(x + i * .5) * (7 - i) * flip)
sns.set_style("ticks")
sinplot()
```

4. 移除轴脊柱

white 和 ticks 两个风格都能够移除顶部和右侧不必要的轴脊柱。通过 Matplotlib 参数是做不到这一点的，但是可以使用 Seaborn 的 despine()方法来移除它们（见图 4.33）：

```
import numpy as np
import matplotlib.pyplot as plt
import seaborn as sns
def sinplot(flip=1):
    x = np.linspace(0, 14, 100)
    for i in range(1, 7):
        plt.plot(x, np.sin(x + i * .5) * (7 - i) * flip)
sinplot()
sns.despine()
```

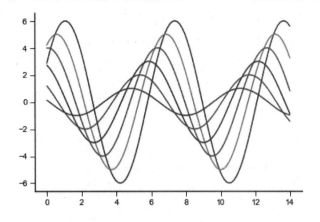

图 4.33　Seaborn 的 despine()方法来移除轴脊柱

5. 临时设置绘图风格（见图 4.34）

来回切换风格很容易，同时也可以在一个 with 语句中使用 axes_style()方法来临时设置绘图参数。这也允许用不同风格的轴来绘图：

```
import numpy as np
import matplotlib.pyplot as plt
import seaborn as sns
def sinplot(flip=1):
    x = np.linspace(0, 14, 100)
    for i in range(1, 7):
        plt.plot(x, np.sin(x + i * .5) * (7 - i) * flip)
with sns.axes_style("darkgrid"):
    plt.subplot(211)
    sinplot()
plt.subplot(212)
sinplot(-1)
```

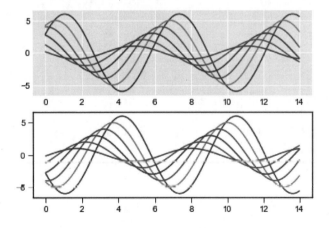

图 4.34　with 语句中使用 axes_style()方法临时设置绘图风格

6. 覆盖 Seaborn 风格元素（见图 4.35）

如果想定制化 Seaborn 风格，可以将一个字典参数传递给 axes_style()和 set_style()的参数 rc，而且只能通过这个方法来覆盖风格定义中的部分参数。

```
sns.axes_style()
{'axes.axisbelow': True,
 'axes.edgecolor': '.8',
 'axes.facecolor': 'white',
 'axes.grid': True,
 'axes.labelcolor': '.15',
 'axes.linewidth': 1.0,
 'figure.facecolor': 'white',
 'font.family': [u'sans-serif'],
 'font.sans-serif': [u'Arial',
  u'DejaVu Sans',
  u'Liberation Sans',
  u'Bitstream Vera Sans',
  u'sans-serif'],
 'grid.color': '.8',
 'grid.linestyle': u'-',
 'image.cmap': u'rocket',
 'legend.frameon': False,
 'legend.numpoints': 1,
 'legend.scatterpoints': 1,
 'lines.solid_capstyle': u'round',
 'text.color': '.15',
 'xtick.color': '.15',
 'xtick.direction': u'out',
 'xtick.major.size': 0.0,
 'xtick.minor.size': 0.0,
 'ytick.color': '.15',
 'ytick.direction': u'out',
 'ytick.major.size': 0.0,
 'ytick.minor.size': 0.0}
```

然后，可以设置这些参数的不同版本。

```
import numpy as np
import matplotlib.pyplot as plt
import seaborn as sns
def sinplot(flip=1):
    x = np.linspace(0, 14, 100)
    for i in range(1, 7):
        plt.plot(x, np.sin(x + i * .5) * (7 - i) * flip)
```

```
with sns.axes_style("darkgrid"):
    plt.subplot(211)
    sinplot()
sns.set_style("darkgrid", {"axes.facecolor": ".9"})
sinplot()
```

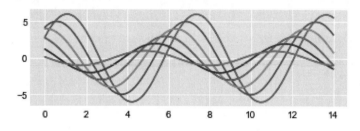

图 4.35　覆盖 Seaborn 风格元素

7. 绘图元素比例

通过 set()重置默认的参数可以控制绘图元素的比例。

```
sns.set()
```

有 4 个预置的环境,按大小从小到大排列,分别为 paper、notebook、talk、poster。其中,notebook 是默认的。

（1）paper（见图 4.36）

```
import numpy as np
import matplotlib.pyplot as plt
import seaborn as sns
def sinplot(flip=1):
    x = np.linspace(0, 14, 100)
    for i in range(1, 7):
        plt.plot(x, np.sin(x + i * .5) * (7 - i) * flip)
with sns.axes_style("darkgrid"):
    plt.subplot(211)
    sinplot()
sns.set_context("paper")
sinplot()
```

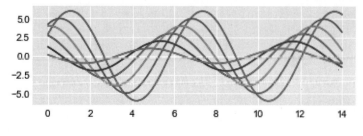

图 4.36　paper 预置的环境

(2) talk（见图4.37）

```
import numpy as np
import matplotlib.pyplot as plt
import seaborn as sns
def sinplot(flip=1):
    x = np.linspace(0, 14, 100)
    for i in range(1, 7):
        plt.plot(x, np.sin(x + i * .5) * (7 - i) * flip)
with sns.axes_style("darkgrid"):
    plt.subplot(211)
    sinplot()
sns.set_context("talk")
sinplot()
```

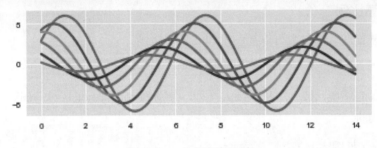

图4.37　talk预置的环境

本节主要介绍了Seaborn的5种绘图风格，以及移除轴脊柱、临时设置绘图风格、覆盖Seaborn风格元素和绘图元素比例缩放等方法。

4.22　本章小结

开源是技术创新和快速发展的核心。本章展示Python机器学习开源项目以及在分析过程中发现的见解和趋势。为了紧跟这种日新月异的发展速度，一个保持了解和学习机器学习前沿的方法是，参与到开源社区，为被很多专业人士使用的开源项目和工具做出贡献。在此举例分析发布的排名靠前、比较重要的常用Python机器学习开源项目。

第 5 章
Kaggle平台机器学习实战

目前机器学习悄无声息地进入部分数据挖掘领域。当然，国外数据挖掘已经很成熟了，机器算法应用的范围更加广泛，分别有网络搜索、邮件分类、机器人、生物和医药学研究等。机器学习主要包括两个任务：分类和回归。分类非常容易理解，就是在一个预测任务中把数据分类；回归主要是统计意义上的，用于预测数据。回归一个非常重要的任务——数据拟合曲线：通过给定的数据集合拟合出最优曲线，使得该曲线尽量能够反映数据的趋势，在不过度拟合的情况下让给定的数据集落在线附近。

5.1 Kaggle 信用卡欺诈检测

信用卡欺诈检测又叫异常检测。可以简单想一下，异常检测无非就是正常和异常，是一个二分类任务，显然正常的占绝大部分，异常的只占很少的比例，要检测的就是这些异常的。明确了任务后，就是进行二分类的处理。

5.1.1 Kaggle 信用卡欺诈检测准备

1. 数据准备

数据集是来自 Kaggle 上的信用卡进行交易的数据。此数据集显示两天内发生的交易，其中 284 807 笔交易中有 492 笔被盗刷。数据集非常不平衡，被盗刷占所有交易的 0.172%。其中，数据特征 v1,v2,...,v28 是某些特征，银行为了保密，并没有提供具体代表的内容；Class 是响应变量，如果发生被盗刷，那么取值 1，否则为 0；Amount 为消费金额。

Kaggle 上的数据：https://www.kaggle.com/mlg-ulb/creditcardfraud。

拿到的数据集是经过银行初步筛选拿到的数据集，因为基于银行数据会有相关隐私，这个是可以理解的，其实这并不耽误运行模型和进行预测。

数据显示代码.

```
In[1]
#!usr/bin/env python
#_*_coding:utf-8_*_
import pandas as pd
```

```
import matplotlib.pyplot as plt
import numpy as np
data = pd.read_csv("D:\Anaconda3\workspace\creditcard.csv")
data.head(6)
```

前 6 行的结果如图 5.1 所示。从中可以观察到前面有一列时间序列对于异常来说是没有意义的。Amount 序列数值浮动比较大，要做标准化或归一化，因为对于数值较大的值，计算机会误认为它的权重大，需要把数据的大小尽量均衡。Class 这一列中 0 占的百分比相当高，根据前面的分析，0 是正常的样本，1 为异常的。

Out[1]

	Time	V1	V2	V3	V4	V5	V6	V7	V8	V9	...	V21	V22	V23
0	0.0	-1.359807	-0.072781	2.536347	1.378155	-0.338321	0.462388	0.239599	0.098698	0.363787	...	-0.018307	0.277838	-0.110474
1	0.0	1.191857	0.266151	0.166480	0.448154	0.060018	-0.082361	-0.078803	0.085102	-0.255425	...	-0.225775	-0.638672	0.101288
2	1.0	-1.358354	-1.340163	1.773209	0.379780	-0.503198	1.800499	0.791461	0.247676	-1.514654	...	0.247998	0.771679	0.909412
3	1.0	-0.966272	-0.185226	1.792993	-0.863291	-0.010309	1.247203	0.237609	0.377436	-1.387024	...	-0.108300	0.005274	-0.190321
4	2.0	-1.158233	0.877737	1.548718	0.403034	-0.407193	0.095921	0.592941	-0.270533	0.817739	...	-0.009431	0.798278	-0.137458
5	2.0	-0.425966	0.960523	1.141109	-0.168252	0.420987	-0.029728	0.476201	0.260314	-0.568671	...	-0.208254	-0.559825	-0.026398

6 rows × 31 columns

V6	V7	V8	V9	...	V21	V22	V23	V24	V25	V26	V27	V28	Amount	Class
2388	0.239599	0.098698	0.363787	...	-0.018307	0.277838	-0.110474	0.066928	0.128539	-0.189115	0.133558	-0.021053	149.62	0
2361	-0.078803	0.085102	-0.255425	...	-0.225775	-0.638672	0.101288	-0.339846	0.167170	0.125895	-0.008983	0.014724	2.69	0
0499	0.791461	0.247676	-1.514654	...	0.247998	0.771679	0.909412	-0.689281	-0.327642	-0.139097	-0.055353	-0.059752	378.66	0
7203	0.237609	0.377436	-1.387024	...	-0.108300	0.005274	-0.190321	-1.175575	0.647376	-0.221929	0.062723	0.061458	123.50	0
5921	0.592941	-0.270533	0.817739	...	-0.009431	0.798278	-0.137458	0.141267	-0.206010	0.502292	0.219422	0.215153	69.99	0
9728	0.476201	0.260314	-0.568671	...	-0.208254	-0.559825	-0.026398	-0.371427	-0.232794	0.105915	0.253844	0.081080	3.67	0

图 5.1 creditcard.csv 数据样本

2. 绘制欺诈类直方图

```
In[2]
#统计这一列中有多少不同的值，并排列出来
count_classes = pd.value_counts(data['Class'], sort = True).sort_index()
count_classes.plot(kind = 'bar')
plt.title("Fraud class histogram")
plt.xlabel("Class")
plt.ylabel("Frequency")


Out[2]
Text(0,0.5,'Frequency')
```

在数据样本中，有明确的 label 列指定了 class 为 0 代表正常情况、class 为 1 代表发生了欺诈行为的样本。从图 5.2 看出 class 为 1 的并不是没有，而是太少了，少到基本看不出来，由样本极度不均衡所致。正负样本不均衡，可以通过上下采样调整样本分布均匀。在数据分析

中，常用的两种方法是下采样和过采样。

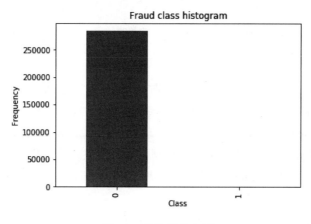

图 5.2 欺诈类直方图

3. 样本不均衡问题

对样本不均衡的处理通常有以下 3 种方式。

- 欠采样

抛弃数据集中样本数量较多的类别来缓解不平衡问题，缺点是会丢失多数类样本中的一些重要信息。

- 过采样

对训练集里面过少的样本进行新的数据合成来达到数据平衡的问题。这里比较经典的算法是 SMOTE 算法，它会从相近的几个样本中加入随机噪声，随机扰动一个特征来生成新的数据实例。

- 权重值的调整

也就是说调整权重值，将少数样本权重设置为一个较大权重、多数样本设置为一个较小权重。

对于信用卡欺诈这个问题，当不平衡超过 500 时，用欠采样，就会丢弃 20 多万条数据，很可能会丢失很多重要的信息；如果采用权重值的调整，也并不容易找到合适的权重值；可以采用过采样来合成新数据。

4. 数据预处理

在合成数据之前需要对数据进行常规的预处理，将可能的特征属性进行标准化处理，因为算法都假设所有数据集的所有特征集中在 0 附近，并且有相同的方差，如果某个特征方差远大于其他特征方差，那么该特征可能在目标函数中所占权重更大，而且差距太大的话会对收敛速度产生很大的影响，甚至不收敛，这里采用 SKlearn 自带的 StandardScaler 来进行处理。

```
from sklearn.preprocessing import StandardScaler
```

```
data['Amount'] = StandardScaler().fit_transform(data['Amount'].reshape(-1, 1))
data.drop(['Time'],axis=1)
```

处理后该列数据会变成均值为 0、方差为 1 的数据。

5. 模型的生成与调参

对于监督学习算法，过度拟合比欠拟合有时更难处理，尤其是过多的特征与过少的数据，最容易导致过度拟合问题。解决过度拟合问题，通常可以增加数据集和减少模型复杂度。正则化是减少模型复杂度的一种方法，可以保留所有的特征变量，但是会减小特征变量的数量级。SKlearn 逻辑回归算法提供了 c 值，也就是正则化系数倒数，c 值越小，对应越强的正则化，越能得到一个简单的假设曲线，也就越能减少过度拟合的风险。这里衡量精度采用 recall_score，也就是召回率，而不是准确率和正确率。

接下来，把预测结果的结果精度显示在一个混淆矩阵里面。

5.1.2　Kaggle 信用卡欺诈检测实例

【例 5.1】信用卡欺诈检测实例代码在 Spyder（Python3.6）环境中实现

```
#!usr/bin/env python
#_*_coding:utf-8_*_
import numpy as np
import pandas as pd
import matplotlib.pyplot as plt
from sklearn.preprocessing import StandardScaler

'''
from sklearn.cross_validation import train_test_split
from sklearn.cross_validation import KFold
因为库更新后会抛出 error
'''
from sklearn.model_selection import train_test_split
from sklearn.model_selection import KFold, cross_val_score
from sklearn.linear_model import LogisticRegression
from sklearn.metrics import confusion_matrix,recall_score,classification_report

 data = pd.read_csv("D:\python 语言与大数据\数据集-训练
\creditcard.csv\creditcard.csv", engine='python')
# print(data.head())#打印查看数据形式
def test1()::#画条形图查看 0 和 1 的数据分布情况
    count_classes = pd.value_counts(data['Class'],sort=True).sort_index()
    count_classes.plot(kind='bar')#画一张条形图
    plt.title('Fraud class histogram')
    plt.xlabel("Class")
```

```python
    plt.ylabel("Frequency")
    plt.show()
# test1()
#reshape 函数可以重新调整矩阵的行数、列数、维数-1, 1 意思：指定为1列，行数根据数据自动处理，
生成矩阵
#fit_transform 对数据进行变换
#将 Amount 这列数据作为一个列向量保存起来
'''
data['Amount'].reshape(-1, 1)会抛出 error
Series 数据类型没有 reshape 函数
解决办法：
用 values 方法将 Series 对象转化成 numpy 的 ndarray,再用 ndarray 的 reshape 方法。
data['Amount'].values.reshape(-1, 1)
另外，pandas 有两种对象：Series 和 DataFrame。
可以这么理解，DataFrame 像一张表，Series 是指表里的某一行数据或某一列数据
'''
data['normAount'] =
StandardScaler().fit_transform(data['Amount'].values.reshape(-1, 1))
#将原始数据中去除掉 time 和 amount 没用的整列数据，axis 的1为列、0为行
data = data.drop(['Time','Amount'],axis=1)

X = data.iloc[:, data.columns!='Class']
y = data.iloc[:, data.columns=='Class']
'''
下取样方式效果使得01样本同样少
下采样是有一种误杀率大的问题，即 FP 大问题，即查准率 P 低
'''
#Number of data points in the minority class
number_records_fraud = len(data[data.Class == 1])#取 class=1 的数据个数
#将 class=1 的索引取出来作为一个列向量保存
fraud_indices = np.array(data[data.Class==1].index)

#Pincking the indices of the normal classes
normal_indices = data[data.Class == 0].index

#Out of the indices we picked, randomly select 'x' number (number_records_fraud)
#在 normal_indices 中随机选择 number_records_fraud 个数据
random_normal_indices =
np.random.choice(normal_indices,number_records_fraud,replace = False)
random_normal_indices = np.array(random_normal_indices)#转化为列向量的格式

#Appending the 2 indices
under_sample_indices = np.concatenate([fraud_indices,random_normal_indices])
```

```python
#Under sample dataset
#根据组合后的样本索引到原始数据中取出整行数据保存
under_sample_data = data.iloc[under_sample_indices,:]

X_undersample = under_sample_data.iloc[:,under_sample_data.columns != 'Class']
y_undersample = under_sample_data.iloc[:,under_sample_data.columns == 'Class']

#showing ratio
print("Percentage of normal transactions: ",
#打印正样本数目
len(under_sample_data[under_sample_data.Class == 0])/len(under_sample_data))
print("Percentage of fraud transactions: ",
#打印负样本数目
len(under_sample_data[under_sample_data.Class == 1])/len(under_sample_data))
print("Total number of transactions in resampledata: ",len(under_sample_data))

'''交叉验证'''
#Whole dadtaset
#random_state=0 给定一个伪随机参数0来打乱数据，在原始数据中抽取0.7部分作为训练集
X_train,X_test,y_train,y_test =
train_test_split(X,y,test_size=0.3,random_state=0)

print("Number transactions train dataset: ",len(X_train))
print("Number transactions test dataset: ",len(X_test))
print("Total number of transactions: ",len(X_train)+len(X_test))

#Undersampled dataset
#下采样中抽取0.3作为测试集
X_train_undersample,X_test_undersample,y_train_undersample,y_test_undersample =
train_test_split(X_undersample,y_undersample,test_size=0.3,random_state=0)
print("")
print("Number transactions train dataset: ",len(X_train_undersample))
print("Number transactions test dataset: ",len(X_test_undersample))
print("Total number of transactions:
",len(X_train_undersample)+len(X_test_undersample))

'''正则化惩罚，用于判断相同效果下哪个模型参数更加稳定'''

'''
失败的版本1
def plot_confusion_matrix(cm, title='Confusion Matrix', cmap=plt.cm.binary):
    plt.imshow(cm, interpolation='nearest', cmap=cmap)
```

```python
    plt.title(title)
    plt.colorbar()
    xlocations = np.array(range(len(labels)))
    plt.xticks(xlocations, labels, rotation=90)
    plt.yticks(xlocations, labels)
    plt.ylabel('True label')
    plt.xlabel('Predicted label')
'''
import itertools
def plot_confusion_matrix(cm, classes,
                          normalize=False,
                          title='Confusion matrix',
                          cmap=plt.cm.Blues):
    """
    This function prints and plots the confusion matrix.
    Normalization can be applied by setting `normalize=True`.
    """
    if normalize:
        cm = cm.astype('float') / cm.sum(axis=1)[:, np.newaxis]
        print("Normalized confusion matrix")
    else:
        print('Confusion matrix, without normalization')

    print(cm)

    plt.imshow(cm, interpolation='nearest', cmap=cmap)
    plt.title(title)
    plt.colorbar()
    tick_marks = np.arange(len(classes))
    plt.xticks(tick_marks, classes, rotation=45)
    plt.yticks(tick_marks, classes)

    fmt = '.2f' if normalize else 'd'
    thresh = cm.max() / 2.
    for i, j in itertools.product(range(cm.shape[0]), range(cm.shape[1])):
        plt.text(j, i, format(cm[i, j], fmt),
                 horizontalalignment="center",
                 color="white" if cm[i, j] > thresh else "black")

    plt.ylabel('True label')
    plt.xlabel('Predicted label')
    plt.tight_layout()
```

```python
#传入原始训练集再进行切分
def printing_Kfold_scores(X_train_data,Y_train_data):
    fold = KFold(5, shuffle=False)
    #正则化中不同的惩罚项参数
    c_param_range =[0.01,0.1,1,10,100]

results_table=pd.DataFrame(index=range(len(c_param_range)),columns=['C_parameter','Mean recall score'])
    results_table['C_parameter'] = c_param_range
    #the k-fold will gives 2 lists:
train_indexes=indexes[0],test_indexes=indexes[1]
    j=0
    for c_param in c_param_range:
        print('-------------------------------------------------')
        print('C parameter:',c_param)
        print('-------------------------------------------------')
        print('')
        recall_accs=[]
        for iteration, indices in enumerate(fold.split(X_train_data)):
            # Call the logistic regression model with a certain C parameter
            '''
            缺少指定solver='liblinear'会抛出如下警告：
            FutureWarning: Default solver will be changed to 'lbfgs' in 0.22
            然后参考了LogisticRegression对solver说明：
            solver str , {'newton-cg','lbfgs','liblinear','sag','saga'}默认
            'liblinear'求解优化问题使用的算法。
            小数据集使用liblinear是不错的选择，但是sag和saga对于较大的数据集速度更快。
            多分类问题只能使用newton-cg、sag、saga和lbfgs处理多分类损失，liblinear限于
            one-versus-rest方案。
            newton-cg、lbfgs和seg只支持 L2 正则，liblinear和saga只支持 L1 正则。
            sag和saga快速卷积只有在特征规模大致相同时才能保证快速收敛。可以使用
            sklearn.preprocessing的缩放器对数据进行预处理。
            '''
            # Call the logistic regression model with a certain C parameter
            # L1 惩罚，L2 惩罚
            lr = LogisticRegression(C=c_param, penalty='l1',solver='liblinear')
            # Use the training data to fit the model. In this case, we use the portion
of the fold to train the model
            # with indices[0]. We then predict on the portion assigned as the 'test
cross validation' with indices[1]
            lr.fit(X_train_data.iloc[indices[0], :],
Y_train_data.iloc[indices[0], :].values.ravel())
            # Predict values using the test indices in the training data
```

```python
        Y_pred_undersample = lr.predict(X_train_data.iloc[indices[1], :].values)

        #绘制confusion matrix
        cnf_matrix = confusion_matrix( Y_train_data.iloc[indices[1], :].values,Y_pred_undersample)
        np.set_printoptions(precision=2)

        #显示出confusion matrix矩阵图
        class_names = [0,1]
        plt.figure()
        MYtitle='Confusion matrix C_param_range='+str(c_param)
        plot_confusion_matrix(cnf_matrix,classes=class_names,title=MYtitle)
        plt.show()

        # Calculate the recall score and append it to a list for recall scores representing the current c_parameter
        recall_acc = recall_score(Y_train_data.iloc[indices[1], :].values,Y_pred_undersample)
        recall_accs.append(recall_acc)

        print('Iteration',iteration,':recall score =',recall_acc)
    results_table.loc[j,'Mean recall score']=np.mean(recall_accs)
    j+=1
    print('')
    print('Mean recall score',np.mean(recall_accs))
    print('')
  best_c=results_table.loc[results_table['Mean recall score'].astype('float64').idxmax()]['C_parameter']
  #Finally,we can check which C parameter is the best amongst the chosen
  print('*********************************************************')
  print('Best model to choose from cross validation is with C parameter =',best_c)
  print('*********************************************************')
  return best_c

best_c = printing_Kfold_scores(X_train_undersample,y_train_undersample)

# print('')
# print("********原始数据直接测试*************")
# print('')
# #证明效果很不好
# best_c = printing_Kfold_scores(X_train,y_train)
def testfazhi():#设置阈值
    lr = LogisticRegression(C=0.01, penalty='l1',solver='liblinear')
```

```python
    lr.fit(X_train_undersample, y_train_undersample.values.ravel())
    # 原来是预测类别值，而此处是预测概率，方便后续比较
    y_pred_undersample_proba = lr.predict_proba(X_test_undersample.values)

    thresholds = [0.1, 0.2, 0.3, 0.4, 0.5, 0.6, 0.7, 0.8, 0.9]

    plt.figure(figsize=(10, 10))

    j = 1
    for i in thresholds:
        y_test_predictions_high_recall = y_pred_undersample_proba[:, 1] > i

        plt.subplot(3, 3, j)
        j += 1

        # Compute confusion matrix
        cnf_matrix = confusion_matrix(y_test_undersample, y_test_predictions_high_recall)
        np.set_printoptions(precision=2)

        print("Recall metric in the testing dataset: ", cnf_matrix[1, 1] / (cnf_matrix[1, 0] + cnf_matrix[1, 1]))

        # Plot non-normalized confusion matrix
        class_names = [0, 1]
        plot_confusion_matrix(cnf_matrix
                              , classes=class_names
                              , title='Threshold >= %s' % i)

    plt.show()

testfazhi()
```

输出：

```
Percentage of normal transactions: 0.5
Percentage of fraud transactions: 0.5
Total number of transactions in resampledata: 984
Number transactions train dataset: 199364
Number transactions test dataset: 85443
Total number of transactions: 284807

Number transactions train dataset: 688
Number transactions test dataset: 296
```

```
Total number of transactions: 984
--------------------------------------------------
C parameter: 0.01
--------------------------------------------------

Confusion matrix, without normalization
[[45 20]
 [ 3 70]]
```

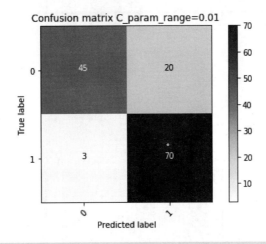

```
Iteration 0 :recall score = 0.958904109589041
Confusion matrix, without normalization
[[37 28]
 [ 5 68]]
```

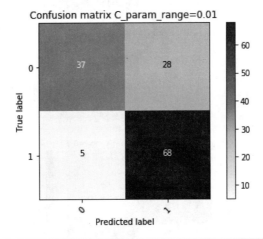

```
Iteration 1 :recall score = 0.9315068493150684
Confusion matrix, without normalization
[[41 38]
 [ 0 59]]
```

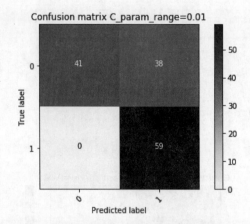

```
Iteration 2 :recall score = 1.0
Confusion matrix, without normalization
[[37 26]
 [ 2 72]]
```

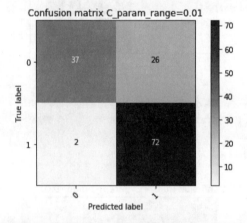

```
Iteration 3 :recall score = 0.972972972972973
Confusion matrix, without normalization
[[49 22]
 [ 3 63]]
```

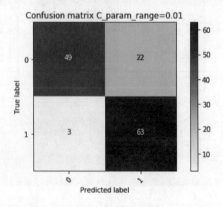

```
Iteration 4 :recall score = 0.9545454545454546

Mean recall score 0.9635858772845076

-------------------------------------------------
C parameter: 0.1
-------------------------------------------------

Confusion matrix, without normalization
[[65  0]
 [11 62]]
```

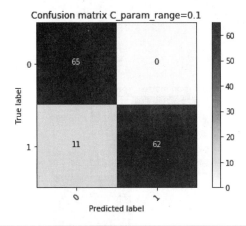

```
Iteration 0 :recall score = 0.8493150684931506
Confusion matrix, without normalization
[[65  0]
 [10 63]]
```

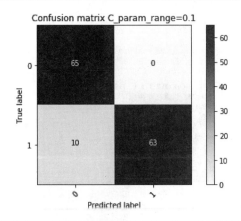

```
Iteration 1 :recall score = 0.9630136986301370
Confusion matrix, without normalization
[[78  1]
 [ 3 56]]
```

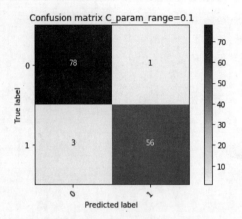

```
Iteration 2 :recall score = 0.9491525423728814
Confusion matrix, without normalization
[[60  3]
 [ 4 70]]
```

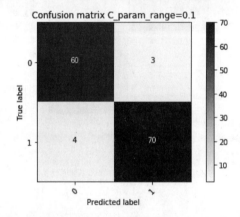

```
Iteration 3 :recall score = 0.9459459459459459
Confusion matrix, without normalization
[[71  0]
 [ 6 60]]
```

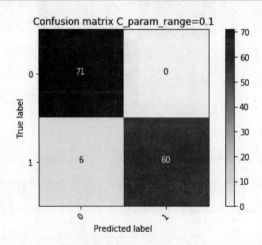

```
Iteration 4 :recall score = 0.9090909090909091

Mean recall score 0.9033036329066049

-------------------------------------------------
C parameter: 1
-------------------------------------------------

Confusion matrix, without normalization
[[64  1]
 [10 63]]
```

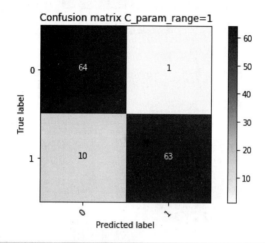

```
Iteration 0 :recall score = 0.863013698630137
Confusion matrix, without normalization
[[62  3]
 [ 9 64]]
```

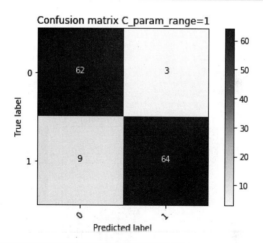

```
Iteration 1 :recall score = 0.8767123287671232
Confusion matrix, without normalization
[[77  2]
 [ 3 56]]
```

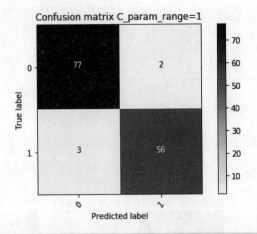

```
Iteration 2 :recall score = 0.9491525423728814
Confusion matrix, without normalization
[[59  4]
 [ 4 70]]
```

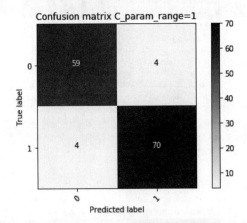

```
Iteration 3 :recall score = 0.9459459459459459
Confusion matrix, without normalization
[[70  1]
 [ 4 62]]
```

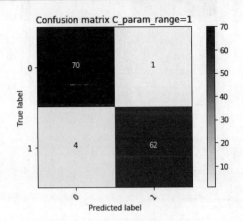

```
Iteration 4 :recall score = 0.9393939393939394

Mean recall score 0.9148436910220055

-------------------------------------------------
C parameter: 10
-------------------------------------------------

Confusion matrix, without normalization
[[63  2]
 [ 9 64]]
```

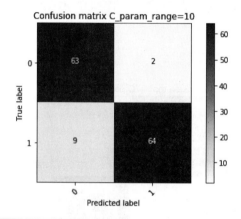

```
Iteration 0 :recall score = 0.8767123287671232
Confusion matrix, without normalization
[[59  6]
 [ 9 64]]
```

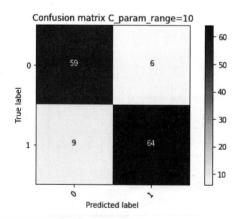

```
Iteration 1 :recall score = 0.8767123287671232
Confusion matrix, without normalization
[[76  3]
 [ 2 57]]
```

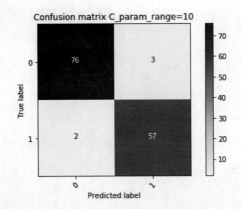

```
Iteration 2 :recall score = 0.9661016949152542
Confusion matrix, without normalization
[[60  3]
 [ 5 69]]
```

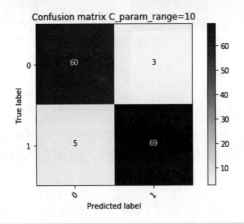

```
Iteration 3 :recall score = 0.9324324324324325
Confusion matrix, without normalization
[[70  1]
 [ 4 62]]
```

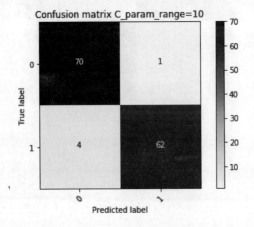

```
Iteration 4 :recall score = 0.9393939393939394

Mean recall score 0.9182705448551746

-------------------------------------------------
C parameter: 100
-------------------------------------------------

Confusion matrix, without normalization
[[63  2]
 [ 9 64]]
```

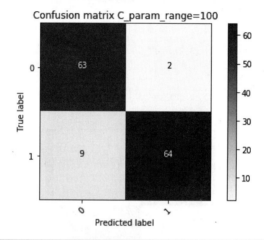

```
Iteration 0 :recall score = 0.8767123287671232
Confusion matrix, without normalization
[[59  6]
 [ 9 64]]
```

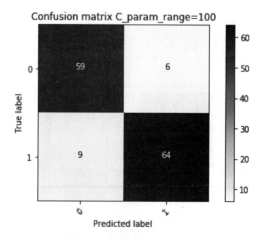

```
Iteration 1 :recall score = 0.8767123287671232
Confusion matrix, without normalization
```

```
[[76  3]
 [ 2 57]]
```

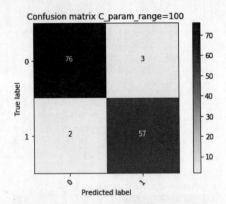

```
Iteration 2 :recall score = 0.9661016949152542
Confusion matrix, without normalization
[[60  3]
 [ 4 70]]
```

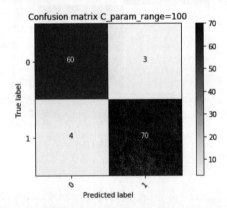

```
Iteration 3 :recall score = 0.9459459459459459
Confusion matrix, without normalization
[[70  1]
 [ 4 62]]
```

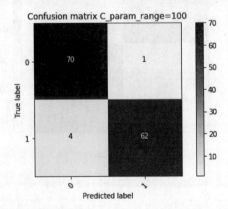

```
Iteration 4 :recall score = 0.9393939393939394

Mean recall score 0.9209732475578771

*****************************************************
Best model to choose from cross validation is with C parameter = 0.01
*****************************************************
Recall metric in the testing dataset: 1.0
Confusion matrix, without normalization
[[  0 149]
 [  0 147]]
Recall metric in the testing dataset: 1.0
Confusion matrix, without normalization
[[  0 149]
 [  0 147]]
Recall metric in the testing dataset: 1.0
Confusion matrix, without normalization
[[  1 148]
 [  0 147]]
Recall metric in the testing dataset: 0.9931972789115646
Confusion matrix, without normalization
[[ 41 108]
 [  1 146]]
Recall metric in the testing dataset: 0.9387755102040817
Confusion matrix, without normalization
[[127  22]
 [  9 138]]
Recall metric in the testing dataset: 0.8979591836734694
Confusion matrix, without normalization
[[145   4]
 [ 15 132]]
Recall metric in the testing dataset: 0.8367346938775511
Confusion matrix, without normalization
[[149   0]
 [ 24 123]]
Recall metric in the testing dataset: 0.7482993197278912
Confusion matrix, without normalization
[[149   0]
 [ 37 110]]
Recall metric in the testing dataset: 0.5714285714285714
Confusion matrix, without normalization
[[149   0]
 [ 63  84]]
```

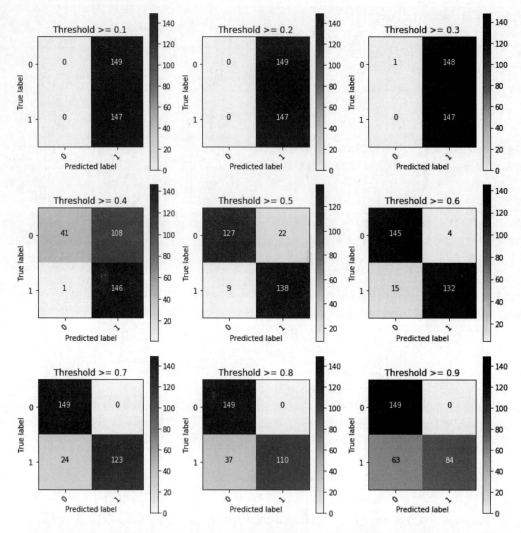

从结果图中可以看到不同的阈值产生的影响还是很大的。阈值较小，意味着模型非常严格，会使得绝大多数样本都被当成异常的样本，recall 很高，精度稍低；当阈值较大的时候模型稍微宽松些，这时会导致 recall 很低，精度稍高。综上所述，当使用逻辑回归算法的时候，还需要根据实际的应用场景来选择一个恰当的阈值。很明显，随着判断阈值增加，召回率是呈下降趋势的，但是误杀的概率会明显下降，综合来看 0.8 附近是相对比较合理的。

5.2 Kaggle 机器学习案例

在数据科学领域，可用的资源太多了：从 Datacamp 到 Udacity 中，再到 KDnuggets，有数千个在线的地方可以学习。Kaggle 可能是通过实践数据科学项目扩展技能的最佳选择。

5.2.1 Kaggle 机器学习概况

虽然 Kaggle 最初被称为机器学习竞赛的地方,但是它自称为"你的数据科学之家"——现在提供了一系列数据科学资源。

1. Kaggle 优势

虽然 Kaggle 提供的一系列数据科学资源专注于竞赛,但是 Kaggle 有很多优势:

- 数据集

可以免费下载和使用的数万种不同类型和大小的数据集。如果正在寻找有趣的数据来探索或测试你的建模技能,这是一个很好的去处。

- 机器学习竞赛

曾经是 Kaggle 的核心,这些建模技能测试是学习尖端机器学习技术和使用真实数据磨练能力的好方法。

- 学习

Jupyter Notebook 中教授的一系列数据科学学习课程,涵盖 SQL 到深度学习。

- 讨论

提供问题并从 Kaggle 社区的数千名数据科学家那里获得建议。

- 内核

在 Kaggle 服务器上运行的在线编程环境,可以在其中编写 Python / R 脚本或 Jupyter Notebook。这些内核完全免费运行(甚至可以添加 GPU)并且是一个很好的资源,因为不必担心在自己的计算机上设置数据科学环境。内核可用于分析任何数据集,参与机器学习竞赛或完成学习轨道。可以复制和构建来自其他用户的现有内核,并与社区共享你的内核以获得反馈。

2. Kaggle 的 3 项重要内容

Kaggle 主持了 3 项非常重要的内容:

- Datasets(https://www.kaggle.com/datasets)

它包含 9500 多个数据集,因此可以通过选择任何感兴趣的数据集来提高你的技能。

- Kernels(https://www.kaggle.com/kernels)

它是 Kaggle 的 Jupyter Notebook 版本,只是一种非常有效和酷炫的共享代码方式,以及大量的可视化、输出和解释。"内核"选项卡将参与者带到一个公共内核列表,在其中可以展示一些新工具或分享对某些特定数据集的专业知识或见解。

- Learn(https://www.kaggle.com/learn/overview)

此选项卡包含免费且实用的实践课程,这些课程涵盖了快速入门所需的最低前置课程。一切都是使用 Kaggle 的内核完成的。

> Kaggle Learn 的机器学习课程:
  https://www.kaggle.com/learn/machine-learning
> Python 课程:
  https://www.kaggle.com/learn/python

3. Kaggle 的 5 大板块

Kaggle 可分为 5 个大的板块:

- Competitions(竞赛)

分为商业竞赛、学术类竞赛、入门级竞赛和一些由大公司(如 Google、Facebook)不定时举办的邀请赛。

- Datasets(数据集)

公司或个人贡献的各类型的数据集。机器学习最怕找不到数据,Kaggle 提供了获取数据练习的一些捷径。

- Kernels(数据分析及建模)

有点类似 GitHub 的代码管理,说直白点就是给用户提供了云上的数据分析和建模的环境,不过涉及代码上传。

- Discussion(讨论区)

里面有全世界各地的数据科学、机器学习的专家和爱好者针对题目、算法、建模等热烈的讨论。

- Jobs(工作)

一些公司会直接在 Kaggle 上发出数据挖掘、机器学习类的岗位,基本都是欧美的中小型公司。

5.2.2 自行车租赁数据分析与可视化案例

【例 5.2】自行车租赁数据分析与可视化案例(实验环境 Jupyter Notebook)

(1)查看数据

```
In[1]
#!usr/bin/env python
#_*_coding:utf-8_*_
import pandas as pd          # 读取数据到 DataFrame
import urllib.request        # 获取网络数据
```

```python
import shutil              # 文件操作
import zipfile             # 压缩解压
import os

# 建立临时目录
try:
    os.system('mkdir bike_data')
except:
    os.system('rm -rf bike_data; mkdir bike_data')

data_source = 'http://archive.ics.uci.edu/ml/machine-learning-databases/00275/Bike-Sharing-Dataset.zip'       # 网络数据地址
zipname = 'bike_data/Bike-Sharing-Dataset.zip'      # 拼接文件和路径
urllib.request.urlretrieve(data_source, zipname)    # 获得数据

zip_ref = zipfile.ZipFile(zipname, 'r')     # 创建一个 ZipFile 对象处理压缩文件
#zip_ref.extractall(temp_dir)               # 解压
zip_ref.extractall('bike_data')
zip_ref.close()

daily_path = 'bike_data/day.csv'
daily_data = pd.read_csv(daily_path)        # 读取 csv 文件
# 把字符串数据转换成日期数据
daily_data['dteday'] = pd.to_datetime(daily_data['dteday'])
drop_list = ['instant', 'season', 'yr', 'mnth', 'holiday', 'workingday',
'weathersit', 'atemp', 'hum']                # 不关注的列
# inplace=true 在对象上直接操作
daily_data.drop(drop_list, inplace = True, axis = 1)

daily_data.head() # 看一看数据~!
```

Out[1]

	dteday	weekday	temp	windspeed	casual	registered	cnt
0	2011-01-01	6	0.344167	0.160446	331	654	985
1	2011-01-02	0	0.363478	0.248539	131	670	801
2	2011-01-03	1	0.196364	0.248309	120	1229	1349
3	2011-01-04	2	0.200000	0.160296	108	1454	1562
4	2011-01-05	3	0.226957	0.186900	82	1518	1600

（2）绘制散点图（见图 5.3）

```python
In[2]
#from matplotlib import font_manager
#fontP = font_manager.FontProperties()
```

```
#fontP.set_family('SimHei')
#fontP.set_size(14)

# 包装一个散点图的函数, 便于复用
def scatterplot(x_data, y_data, x_label, y_label, title):

    # 创建一个绘图对象
    fig, ax = plt.subplots()

    # 设置数据、点的大小、点的颜色和透明度
    ax.scatter(x_data, y_data, s = 10, color = '#539caf', alpha = 0.75) #
http://www.114la.com/other/rgb.htm

    # 添加标题和坐标说明
    ax.set_title(title)
    ax.set_xlabel(x_label)
    ax.set_ylabel(y_label)

# 绘制散点图
scatterplot(x_data = daily_data['temp'].values
           , y_data = daily_data['cnt'].values
           , x_label = 'Normalized temperature (C)'
           , y_label = 'Check outs'
           , title = 'Number of Check Outs vs Temperature')
```

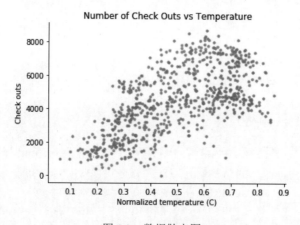

图 5.3　数据散点图

（3）线性回归

```
In[3]
# 线性回归
import statsmodels.api as sm # 最小二乘
from statsmodels.stats.outliers_influence import summary_table # 获得汇总信息
x = sm.add_constant(daily_data['temp']) # 线性回归增加常数项 y=kx+b
```

```python
y = daily_data['cnt']
regr = sm.OLS(y, x)  # 普通最小二乘模型
res = regr.fit()
# 从模型获得拟合数据
# 置信水平 alpha=5%, st 数据汇总, data 数据详情, ss2 数据列名
st, data, ss2 = summary_table(res, alpha=0.05)
fitted_values = data[:,2]

# 包装曲线绘制函数
def lineplot(x_data, y_data, x_label, y_label, title):
    # 创建绘图对象
    _, ax = plt.subplots()

    # 绘制拟合曲线, lw=linewidth, alpha=transparency
    ax.plot(x_data, y_data, lw = 2, color = '#539caf', alpha = 1)

    # 添加标题和坐标说明
    ax.set_title(title)
    ax.set_xlabel(x_label)
    ax.set_ylabel(y_label)

# 调用绘图函数（见图 5.4）
lineplot(x_data = daily_data['temp']
        , y_data = fitted_values
        , x_label = 'Normalized temperature (C)'
        , y_label = 'Check outs'
        , title = 'Line of Best Fit for Number of Check Outs vs Temperature')
```

```
Out[3]
d:\Anaconda3\lib\site-packages\numpy\core\fromnumeric.py:52: FutureWarning:
Method .ptp is deprecated and will be removed in a future version. Use numpy.ptp
instead.
  return getattr(obj, method)(*args, **kwds)
```

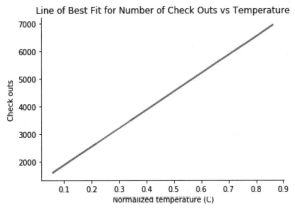

图 5.4　数据模型拟合

(4)创建置信区间 DataFrame

```
In[4]
# 获得5%置信区间的上下界
predict_mean_ci_low, predict_mean_ci_upp = data[:,4:6].T

# 创建置信区间 DataFrame 的上下界
CI_df = pd.DataFrame(columns = ['x_data', 'low_CI', 'upper_CI'])
CI_df['x_data'] = daily_data['temp']
CI_df['low_CI'] = predict_mean_ci_low
CI_df['upper_CI'] = predict_mean_ci_upp
CI_df.sort_values('x_data', inplace = True) # 根据x_data进行排序

# 绘制置信区间
def lineplotCI(x_data, y_data, sorted_x, low_CI, upper_CI, x_label, y_label, title):
    # 创建绘图对象
    _, ax = plt.subplots()

    # 绘制预测曲线
    ax.plot(x_data, y_data, lw = 1, color = '#539caf', alpha = 1, label = 'Fit')
    # 绘制置信区间,顺序填充
    ax.fill_between(sorted_x, low_CI, upper_CI, color = '#539caf', alpha = 0.4, label = '95% CI')
    # 添加标题和坐标说明
    ax.set_title(title)
    ax.set_xlabel(x_label)
    ax.set_ylabel(y_label)

    # 显示图例,配合label参数,loc='best'自适应方式
    ax.legend(loc = 'best')

# 调用函数创建曲线(见图5.5)
lineplotCI(x_data = daily_data['temp']
        , y_data = fitted_values
        , sorted_x = CI_df['x_data']
        , low_CI = CI_df['low_CI']
        , upper_CI = CI_df['upper_CI']
        , x_label = 'Normalized temperature (C)'
        , y_label = 'Check outs'
        , title = 'Line of Best Fit for Number of Check Outs vs Temperature')
```

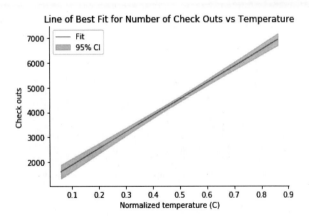

图 5.5 绘制预测曲线

（5）双纵坐标绘图（见图 5.6）

```
In[5]
# 双纵坐标绘图函数
def lineplot2y(x_data, x_label, y1_data, y1_color, y1_label, y2_data, y2_color,
y2_label, title):
    _, ax1 = plt.subplots()
    ax1.plot(x_data, y1_data, color = y1_color)
    # 添加标题和坐标说明
    ax1.set_ylabel(y1_label, color = y1_color)
    ax1.set_xlabel(x_label)
    ax1.set_title(title)

    ax2 = ax1.twinx() # 两个绘图对象共享横坐标轴
    ax2.plot(x_data, y2_data, color = y2_color)
    ax2.set_ylabel(y2_label, color = y2_color)
    # 右侧坐标轴可见
    ax2.spines['right'].set_visible(True)

# 调用绘图函数
lineplot2y(x_data = daily_data['dteday']
        , x_label = 'Day'
        , y1_data = daily_data['cnt']
        , y1_color = '#539caf'
        , y1_label = 'Check outs'
        , y2_data = daily_data['windspeed']
        , y2_color = '#7663b0'
        , y2_label = 'Normalized windspeed'
        , title = 'Check Outs and Windspeed Over Time')
```

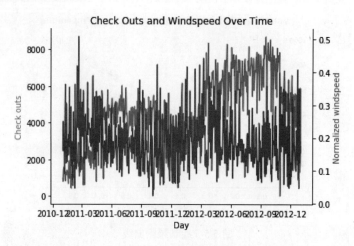

图 5.6 双纵坐标绘图

（6）绘制堆叠的直方图（见图 5.7）

```
In[6]
# 绘制堆叠的直方图
def overlaid_histogram(data1, data1_name, data1_color, data2, data2_name,
data2_color, x_label, y_label, title):
    # 归一化数据区间，对齐两个直方图的 bins
    max_nbins = 10
    #计算边界
    data_range = [min(min(data1), min(data2)), max(max(data1), max(data2))]
    binwidth = (data_range[1] - data_range[0]) / max_nbins
    # 生成直方图 bins 区间
    bins = np.arange(data_range[0], data_range[1] + binwidth, binwidth)

    # 创建曲线
    _, ax = plt.subplots()
    ax.hist(data1, bins = bins, color = data1_color, alpha = 1, label = data1_name)
    ax.hist(data2, bins = bins, color = data2_color, alpha = 0.75, label = data2_name)
    ax.set_ylabel(y_label)
    ax.set_xlabel(x_label)
    ax.set_title(title)
    ax.legend(loc = 'best')

# 调用函数创建曲线
overlaid_histogram(data1 = daily_data['registered']
            , data1_name = 'Registered'
            , data1_color = '#539caf'
            , data2 = daily_data['casual']
            , data2_name = 'Casual'
            , data2_color = '#7663b0'
```

```
          , x_label = 'Check outs'
          , y_label = 'Frequency'
          , title = 'Distribution of Check Outs By Type')
```

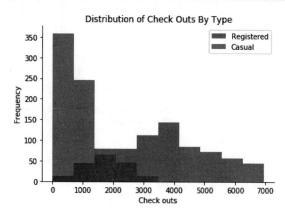

图 5.7　按类型结账堆叠的直方图

（7）计算概率密度（见图 5.8）

```
In[7]
# 计算概率密度
from scipy.stats import gaussian_kde
data = daily_data['registered']
density_est = gaussian_kde(data) # kernal density estimate:
https://en.wikipedia.org/wiki/Kernel_density_estimation
# 控制平滑程度，数值越大，越平滑
density_est.covariance_factor = lambda : .3
density_est._compute_covariance()
x_data = np.arange(min(data), max(data), 200)

# 绘制密度估计曲线
def densityplot(x_data, density_est, x_label, y_label, title):
    _, ax = plt.subplots()
    ax.plot(x_data, density_est(x_data), color = '#539caf', lw = 2)
    ax.set_ylabel(y_label)
    ax.set_xlabel(x_label)
    ax.set_title(title)

# 调用绘图函数
densityplot(x_data = x_data
          , density_est = density_est
          , x_label = 'Check outs'
          , y_label = 'Frequency'
          , title = 'Distribution of Registered Check Outs')
```

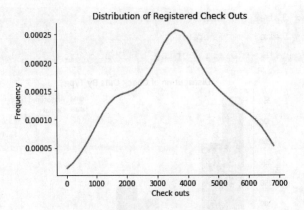

图 5.8 注册用户结账概率密度

（8）分天分析统计特征（见图 5.9）

```
In[8]
# 分天分析统计特征
mean_total_co_day = daily_data[['weekday',
'cnt']].groupby('weekday').agg([np.mean, np.std])
mean_total_co_day.columns = mean_total_co_day.columns.droplevel()

# 定义绘制柱状图的函数
def barplot(x_data, y_data, error_data, x_label, y_label, title):
    _, ax = plt.subplots()
    # 柱状图
    ax.bar(x_data, y_data, color = '#539caf', align = 'center')
    # 绘制方差
    # ls='none'去掉 bar 之间的连线
    ax.errorbar(x_data, y_data, yerr = error_data, color = '#297083', ls = 'none',
lw = 5)
    ax.set_ylabel(y_label)
    ax.set_xlabel(x_label)
    ax.set_title(title)

# 绘图函数调用
barplot(x_data = mean_total_co_day.index.values
        , y_data = mean_total_co_day['mean']
        , error_data = mean_total_co_day['std']
        , x_label = 'Day of week'
        , y_label = 'Check outs'
        , title = 'Total Check Outs By Day of Week (0 = Sunday)')
```

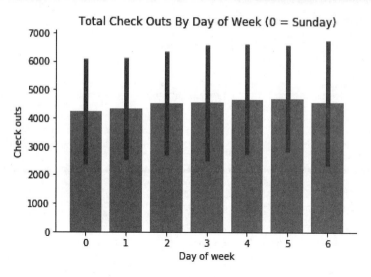

图 5.9　每周每天总结账

（9）分天统计注册和偶然使用的情况（见图 5.10）

```
In[9]
# 分天统计注册和偶然使用的情况
mean_by_reg_co_day = daily_data[['weekday', 'registered',
'casual']].groupby('weekday').mean()
# 分天统计注册和偶然使用的占比
mean_by_reg_co_day['total'] = mean_by_reg_co_day['registered'] +
mean_by_reg_co_day['casual']
mean_by_reg_co_day['reg_prop'] = mean_by_reg_co_day['registered'] /
mean_by_reg_co_day['total']
mean_by_reg_co_day['casual_prop'] = mean_by_reg_co_day['casual'] /
mean_by_reg_co_day['total']


# 绘制堆积柱状图
def stackedbarplot(x_data, y_data_list, y_data_names, colors, x_label, y_label,
title):
    _, ax = plt.subplots()
    # 循环绘制堆积柱状图
    for i in range(0, len(y_data_list)):
        if i == 0:
            ax.bar(x_data, y_data_list[i], color = colors[i], align = 'center', label
= y_data_names[i])
        else:
            # 采用堆积的方式，除了第一个分类，后面的分类都从前一个分类的柱状图接着画
            # 用归一化保证最终累积结果为 1
            ax.bar(x_data, y_data_list[i], color = colors[i], bottom = y_data_list[i
```

```
   - 1], align = 'center', label = y_data_names[i])
    ax.set_ylabel(y_label)
    ax.set_xlabel(x_label)
    ax.set_title(title)
    ax.legend(loc = 'upper right') # 设定图例位置

# 调用绘图函数
stackedbarplot(x_data = mean_by_reg_co_day.index.values
            , y_data_list = [mean_by_reg_co_day['reg_prop'],
mean_by_reg_co_day['casual_prop']]
            , y_data_names = ['Registered', 'Casual']
            , colors = ['#539caf', '#7663b0']
            , x_label = 'Day of week'
            , y_label = 'Proportion of check outs'
            , title = 'Check Outs By Registration Status and Day of Week (0 =
Sunday)')
```

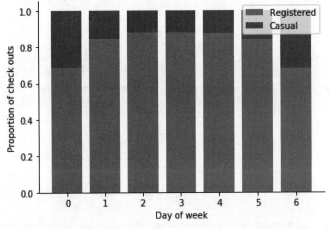

图 5.10　分天统计注册和偶然使用的情况

（10）绘制分组柱状图（见图 5.11）

```
In[10]
# 绘制分组柱状图的函数
def groupedbarplot(x_data, y_data_list, y_data_names, colors, x_label, y_label,
title):
    _, ax = plt.subplots()
    # 设置每一组柱状图的宽度
    total_width = 0.8
    # 设置每一个柱状图的宽度
    ind_width = total_width / len(y_data_list)
    # 计算每一个柱状图的中心偏移
    alteration = np.arange(-total_width/2+ind_width/2, total_width/2+ind_width/2,
ind_width)
```

```
    # 分别绘制每一个柱状图
    for i in range(0, len(y_data_list)):
        # 横向散开绘制
        ax.bar(x_data + alteration[i], y_data_list[i], color = colors[i], label = y_data_names[i], width = ind_width)
    ax.set_ylabel(y_label)
    ax.set_xlabel(x_label)
    ax.set_title(title)
    ax.legend(loc = 'upper right')
# 调用绘图函数
groupedbarplot(x_data = mean_by_reg_co_day.index.values
            , y_data_list = [mean_by_reg_co_day['registered'], mean_by_reg_co_day['casual']]
            , y_data_names = ['Registered', 'Casual']
            , colors = ['#539caf', '#7663b0']
            , x_label = 'Day of week'
            , y_label = 'Check outs'
            , title = 'Check Outs By Registration Status and Day of Week (0 = Sunday)')
```

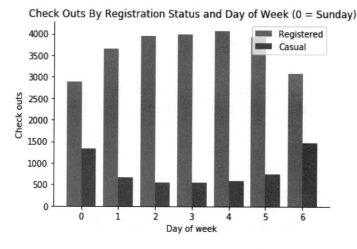

图 5.11 按注册状态和星期几结账

5.3 本章小结

本章主要介绍 Kaggle 机器学习平台上的两个实战案例：信用卡欺诈检测和 Kaggle 机器学习案例。

第 6 章
PaddlePaddle 平台机器学习实战

6.1 PaddlePaddle 平台安装

PaddlePaddle 是百度推出的开源深度学习平台，致力于让更多中国企业和开发者方便地完成深度学习应用。

1. PaddlePaddle 快速安装

PaddlePaddle 目前支持的 Python 版本包括 Python 2.7~3.7。PaddlePaddle 目前支持以下环境：

- Ubuntu 14.04 /16.04 /18.04。
- CentOS 7 / 6。
- MacOS 10.11 / 10.12 / 10.13 / 10.14。
- Windows 7 / 8/ 10（专业版/企业版）。

硬件环境要求是计算机拥有 64 位操作系统，处理器支持 AVX 指令集和 MKL。

如果希望使用 pip 安装 PaddlePaddle，可以直接使用以下命令：

- pip install paddlepaddle （CPU 版本最新）。
- pip install paddlepaddle-gpu （GPU 版本最新）。

注：pip install paddlepaddle-gpu 命令将安装支持 CUDA 9.0 cuDNN v7 的 PaddlePaddle。

如果希望通过 pip 方式安装老版本的 PaddlePaddle，可以使用如下命令：

- pip install paddlepaddle==[PaddlePaddle 版本号]（CPU 版）。
- pip install paddlepaddle-gpu==[PaddlePaddle 版本号]（GPU 版）。

如果希望使用 docker 安装 PaddlePaddle，可以直接使用以下命令：

```
docker run --name [Name of container] -it -v $PWD:/paddle hub.baidubce.com/paddlepaddle/
```

paddle:[docker 版本号] /bin/bash。

2. Fluid 编程

Fluid 和其他主流框架一样，使用 Tensor 数据结构来承载数据。在神经网络中传递的数据都是 Tensor，Tensor 可以简单理解成一个多维数组，一般而言可以有任意多的维度。不同的

Tensor 可以具有自己的数据类型和形状，同一 Tensor 中每个元素的数据类型都是一样的，Tensor 的形状就是 Tensor 的维度。

6.2 PaddlePaddle 平台手写体数字识别

1. 数字识别背景介绍

PaddlePaddle 机器学习（或深度学习）的入门教程一般都是 MNIST 数据库上的手写识别问题。原因是手写识别属于典型的图像分类问题，比较简单，同时 MNIST 数据集也很完备。MNIST 数据集作为一个简单的计算机视觉数据集，包含一系列如图 6.1 所示的手写数字图片和对应的标注。图片是 28×28 的像素矩阵，标注则对应着 0~9 的 10 个数字。每张图片都经过了大小归一化和居中处理。

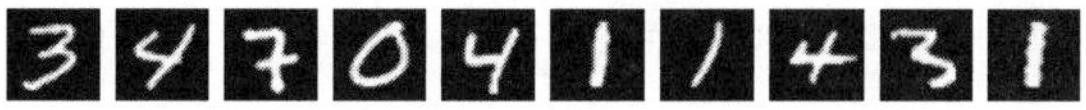

图 6.1 MNIST 手写数字图片示例

MNIST 数据集是从 NIST 的 Special Database 3（SD-3）和 Special Database 1（SD-1）构建而来的。由于 SD-3 是由美国人口调查局的员工进行标注的，SD-1 是由美国高中生进行标注的，因此 SD-3 比 SD-1 更干净，也更容易识别。Yann LeCun 等人从 SD-1 和 SD-3 中各取一半作为 MNIST 的训练集（60 000 条数据）和测试集（10 000 条数据），其中训练集来自 250 个不同的标注员，此外还保证了训练集和测试集的标注员是不完全相同的。

MNIST 吸引了大量的科学家，基于此数据集训练模型。1998 年，LeCun 分别用单层线性分类器、多层感知器（Multilayer Perceptron，MLP）和多层卷积神经网络 LeNet 进行实验，使测试集上的误差不断下降（从 12%下降到 0.7%）。在研究过程中，LeCun 提出了卷积神经网络（Convolutional Neural Network，CNN），大幅度地提高了手写字符的识别能力，因此成为深度学习领域的奠基人之一。此后，科学家们又基于 K 近邻（K-Nearest Neighbors）算法、支持向量机（SVM）、神经网络和 Boosting 方法等做了大量实验，并采用多种预处理方法（如去除歪曲、去噪、模糊等）来提高识别的准确率。

如今的深度学习领域，卷积神经网络占据了至关重要的地位，从最早 Yann LeCun 提出的简单 LeNet，到如今 ImageNet 大赛上的优胜模型 VGGNet、GoogLeNet、ResNet 等，人们在图像分类领域利用卷积神经网络取得了一系列惊人的结果。

本节中，从简单的 Softmax 回归模型开始，带大家了解手写字符识别，并介绍如何改进模型，利用多层感知机（MLP）和卷积神经网络（CNN）优化识别效果。

2. 数字识别模型

基于 MNIST 数据集训练一个分类器，在介绍本例使用的 3 个基本图像分类网络前，先给出一些定义：

- X 是输入：MNIST 图片是 28×28 的二维图像，为了进行计算，将其转化为 784 维向量，即 $X=(x_0,x_1,\cdots,x_{783})$。
- Y 是输出：分类器的输出是 10 类数字（0~9），即 $Y=(y_0,y_1,\cdots,y_9)$，每一维 y_i 代表图片分类为第 i 类数字的概率。
- Label 是图片的真实标注：Label$=(l_0,l_1,\cdots,l_9)$ 也是 10 维，但只有一维为 1，其他维都为 0。例如，某张图片上的数字为 2，则它的标注为 (0,0,1,0,0)。

（1）Softmax 回归（Softmax Regression）

最简单的 Softmax 回归模型是先将输入层经过一个全连接层得到特征，然后直接通过 softmax 函数计算多个类别的概率并输出。输入层的数据 X 传到输出层，在激活操作之前，会乘以相应的权重 W，并加上偏置变量 b，具体如下：

$$y_i = \text{softmax}(\sum_j W_{i,j} x_j + b_i) \tag{6.1}$$

$$\text{softmax}(x_i) = \frac{e^{x_i}}{\sum_j e^{x_j}} \tag{6.2}$$

图 6.2 为 Softmax 回归的网络图，图中权重用蓝线表示、偏置用红线表示、+1 代表偏置参数的系数为 1。

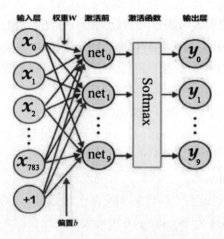

图 6.2 Softmax 回归的网络结构图

对于有 N 个类别的多分类问题，指定 N 个输出节点，N 维结果向量经过 Softmax 将归一化为 N 个[0,1]范围内的实数值，分别表示该样本属于这 N 个类别的概率。此处的 y_i 对应该图片为数字 i 的预测概率。

在分类问题中，一般采用交叉熵代价损失（Cross Entropy Loss）函数，公式如下：

$$L_{\text{cross-entropy}}(\text{label},y) = -\sum_i \text{label}_i \log(y_i) \tag{6.3}$$

（2）多层感知机（MultiLayer Perceptron，MLP）

Softmax 回归模型采用最简单的两层神经网络，即只有输入层和输出层，因此其拟合能力有限。为了达到更好的识别效果，考虑在输入层和输出层中间加上若干个隐藏层。

- 经过第一个隐藏层，可以得到 $H_1=\phi(W_1X+b_1)$，其中 ϕ 代表激活函数，常见的有 sigmoid、tanh 或 ReLU 等函数。
- 经过第二个隐藏层，可以得到 $H_2=\phi(W_2H_1+b_2)$。
- 经过输出层，得到 $Y=\text{softmax}(W_3H_2+b_3)$，即最后的分类结果向量。

图 6.3 为多层感知器的网络结构图，图中权重用蓝线表示、偏置用红线表示、+1 代表偏置参数的系数为 1。

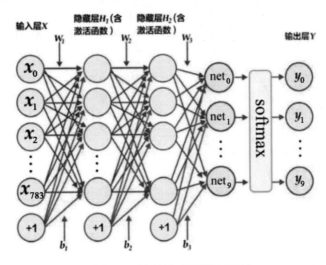

图 6.3 多层感知器网络结构图

（3）卷积神经网络（Convolutional Neural Network，CNN）

在多层感知器模型中，将图像展开成一维向量输入到网络中，忽略了图像的位置和结构信息，而卷积神经网络能够更好地利用图像的结构信息。图 6.4 是一个较简单的卷积神经网络，显示了其结构：输入的二维图像，先经过两次卷积层到池化层，再经过全连接层，最后使用 Softmax 分类作为输出层。下面我们主要介绍卷积层和池化层。

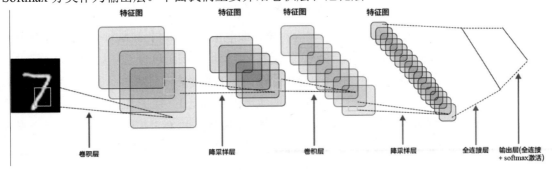

图 6.4 LeNet-5 卷积神经网络结构

（4）卷积层

卷积层是卷积神经网络的核心基石。在图像识别里提到的卷积是二维卷积，即离散二维滤波器（也称作卷积核）与二维图像做卷积操作，简单地讲是二维滤波器滑动到二维图像上所有位置，并在每个位置上与该像素点及其领域像素点做内积。卷积操作被广泛应用于图像处理领域，不同卷积核可以提取不同的特征，例如边沿、线性、角等。在深层卷积神经网络中，通过卷积操作可以提取出图像低级到复杂的特征。

图 6.5 给出一个卷积计算过程的示例图，输入图像大小为 $H=5$、$W=5$、$D=3$，即 5×5 大小的 3 通道（RGB，也称作深度）彩色图像。

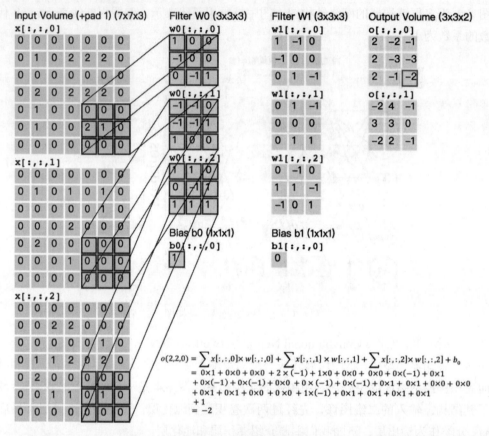

图 6.5 卷积层图片

这个示例图中包含两组（用 K 表示）卷积核，即 Filter W_0 和 Filter W_1。在卷积计算中，通常对不同的输入通道采用不同的卷积核，如图 6.5 中每组卷积核包含（$D=3$）个 3×3（用 $F×F$ 表示）大小的卷积核。另外，这个示例中卷积核在图像的水平方向（W 方向）和垂直方向（H 方向）的滑动步长为 2（用 S 表示）；对输入图像周围各填充 1（用 P 表示）个 0，即图中输入层原始数据为蓝色部分，灰色部分进行了大小为 1 的扩展，用 0 来扩展。经过卷积操作得到输出为 3×3×2（用 $H_o×W_o×K$ 表示）大小的特征图，即 3×3 大小的 2 通道特征图，其中 H_o 的计算公式为 $H_o=(H-F+2×P)/S+1$，W_o 同理。输出特征图中的每个像素是每组滤波器与输入图像每个特征图的内积再求和，然后加上偏置 b_o，偏置通常对于每个输出特征图是共享的。输出特征图 $o[:,:,0]$ 中的最后一个 –2 计算如图 6.5 右下角公式所示。

在卷积操作中卷积核是可学习的参数，经过上面示例的介绍，每层卷积的参数大小为 D

$\times F \times F \times K$。在多层感知器模型中,神经元通常是全部连接,参数较多。而卷积层的参数较少,这也是由卷积层的主要特性即局部连接和共享权重所决定的。

- 局部连接

每个神经元仅与输入神经元的一块区域连接,这块局部区域称作感受野(Receptive Field)。在图像卷积操作中,即神经元在空间维度(Spatial Dimension,图 6.5 的示例中 H 和 W 所在的平面)是局部连接,但在深度上是全部连接。对于二维图像本身而言,也是局部像素关联较强。这种局部连接保证了学习后的过滤器能够对于局部的输入特征有最强的响应。局部连接的思想也是受启发于生物学里面的视觉系统结构,视觉皮层的神经元就是局部接受信息的。

- 权重共享

计算同一个深度切片的神经元时采用的滤波器是共享的。例如,图 6.5 中计算 $o[:,:,0]$ 的每个神经元的滤波器均相同,都为 W_0,这样可以在很大程度上减少参数。共享权重在一定程度上讲是有意义的,例如图片的底层边缘特征与特征在图中的具体位置无关。但是在一些场景中是无意的,比如输入的图片是人脸,眼睛和头发位于不同的位置,希望在不同的位置学到不同的特征(参考斯坦福大学公开课)。注意,权重只是对于同一深度切片的神经元是共享的,在卷积层,通常采用多组卷积核提取不同特征,即对应不同深度切片的特征,不同深度切片的神经元权重是不共享的。另外,偏重对同一深度切片的所有神经元都是共享的。

通过卷积计算过程及其特性,可以看出卷积是线性操作,并具有平移不变性(Shift-Invariant)。平移不变性即在图像每个位置执行相同的操作。卷积层的局部连接和权重共享使得需要学习的参数大大减小,这样也有利于训练较大卷积神经网络。

(5)池化层

池化是非线性下采样的一种形式,主要作用是通过减少网络的参数来减小计算量,并且能够在一定程度上控制过度拟合。通常在卷积层的后面会加上一个池化层。池化包括最大池化、平均池化等。其中,最大池化是用不重叠的矩形框将输入层分成不同的区域,对于每个矩形框的数取最大值作为输出层,如图 6.6 所示。

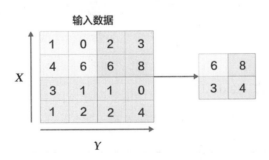

图 6.6　池化层图片

(6)常见激活函数介绍

- sigmoid 激活函数

$$f(x) = \text{sigmoid}(x) = \frac{1}{1+e^{-x}} \quad (6.4)$$

- tanh 激活函数

$$f(x) = \tanh(x) = \frac{e^x - e^{-x}}{e^x + e^{-x}} \quad (6.5)$$

- ReLU 激活函数

$$f(x) = \max(0, x) \quad (6.6)$$

实际上，tanh 函数只是规模变化的 sigmoid 函数，将 sigmoid 函数值放大 2 倍之后再向下平移 1 个单位：$\tanh(x) = 2\text{sigmoid}(2x) - 1$ 。

3. 数据介绍

PaddlePaddle 在 API 中提供了自动加载 MNIST 数据的模块 paddle.dataset.mnist。加载后的数据位于 /home/username/.cache/paddle/dataset/mnist 下：

- train-images-idx3-ubyte 训练数据图片，60000 条数据。
- train-labels-idx1-ubyte 训练数据标注，60000 条数据。
- t10k-images-idx3-ubyte 测试数据图片，10000 条数据。
- t10k-labels-idx1-ubyte 测试数据标注，10000 条数据。

4. Fluid API 概述

演示将使用最新的 Fluid API。Fluid API 是最新的 PaddlePaddle API。它在不牺牲性能的情况下简化了模型配置。

下面是 Fluid API 中几个重要概念的概述：

- inference_program: 指定如何从数据输入中获得预测的函数，这是指定网络流的地方。
- train_program: 指定如何从 inference_program 和标注值中获取 loss 的函数，这是指定损失计算的地方。
- optimizer_func: 指定优化器配置的函数，优化器负责减少损失并驱动训练，Paddle 支持多种不同的优化器。

5. 配置说明

加载 PaddlePaddle 的 Fluid API 包。

```
import os
from PIL import Image # 导入图像处理模块
import matplotlib.pyplot as plt
import numpy
import paddle # 导入paddle模块
import paddle.fluid as fluid
from __future__ import print_function # 将Python3中的print特性导入当前版本
```

（1）Program Functions 配置

项目需要设置 inference_program 函数，用这个程序来演示 3 个不同的分类器，每个分类器都定义为 Python 函数。需要将图像数据输入到分类器中。Paddle 为读取数据提供了一个特殊的层 layer.data 层来创建一个数据层，以读取图像并将其连接到分类网络。

- Softmax 回归

只通过一层简单的以 softmax 为激活函数的全连接层就可以得到分类的结果。

```
def softmax_regression():
    """
    定义softmax分类器:
        一个以softmax为激活函数的全连接层
    Return:
        predict_image -- 分类的结果
    """
    # 输入的原始图像数据，大小为28*28*1
    img = fluid.layers.data(name='img', shape=[1, 28, 28], dtype='float32')
    # 以softmax为激活函数的全连接层，输出层的大小必须为数字的个数10
    predict = fluid.layers.fc(
        input=img, size=10, act='softmax')
    return predict
```

- 多层感知器

下面的代码实现了一个含有两个隐藏层（全连接层）的多层感知器。其中，两个隐藏层的激活函数均采用 ReLU，输出层的激活函数用 softmax。

```
def multilayer_perceptron():
    """
    定义多层感知机分类器:
        含有两个隐藏层（全连接层）的多层感知器
        其中前两个隐藏层的激活函数采用relu，输出层的激活函数用softmax

    Return:
        predict_image -- 分类的结果
    """
    # 输入的原始图像数据，大小为28*28*1
    img = fluid.layers.data(name='img', shape=[1, 28, 28], dtype='float32')
    # 第一个全连接层，激活函数为ReLU
    hidden = fluid.layers.fc(input=img, size=200, act='relu')
    # 第二个全连接层，激活函数为ReLU
    hidden = fluid.layers.fc(input=hidden, size=200, act='relu')
    # 以softmax为激活函数的全连接输出层，输出层的大小必须为数字的个数10
    prediction = fluid.layers.fc(input=hidden, size=10, act='softmax')
```

```
    return prediction
```

- 卷积神经网络 LeNet-5

输入的二维图像，首先经过两次卷积层到池化层，再经过全连接层，最后使用以 softmax 为激活函数的全连接层作为输出层。

```
def convolutional_neural_network():
    """
    定义卷积神经网络分类器：
        输入的二维图像，经过两个卷积-池化层，使用以softmax为激活函数的全连接层作为输出层

    Return:
        predict -- 分类的结果
    """
    # 输入的原始图像数据，大小为28*28*1
    img = fluid.layers.data(name='img', shape=[1, 28, 28], dtype='float32')
    # 第一个卷积-池化层
    # 使用20个5*5的滤波器，池化大小为2，池化步长为2，激活函数为relu
    conv_pool_1 = fluid.nets.simple_img_conv_pool(
        input=img,
        filter_size=5,
        num_filters=20,
        pool_size=2,
        pool_stride=2,
        act="relu")
    conv_pool_1 = fluid.layers.batch_norm(conv_pool_1)
    # 第二个卷积-池化层
    # 使用50个5*5的滤波器，池化大小为2，池化步长为2，激活函数为relu
    conv_pool_2 = fluid.nets.simple_img_conv_pool(
        input=conv_pool_1,
        filter_size=5,
        num_filters=50,
        pool_size=2,
        pool_stride=2,
        act="relu")
    # 以softmax为激活函数的全连接输出层，输出层的大小必须为数字的个数10
    prediction = fluid.layers.fc(input=conv_pool_2, size=10, act='softmax')
    return prediction
```

（2）Train Program 配置

然后设置训练程序 train_program。它首先从分类器中进行预测。在训练期间，它将从预测中计算 avg_cost。训练程序应该返回一个数组，第一个返回参数必须是 avg_cost。训练器使用它来计算梯度。

下面的代码测试 Softmax 回归 softmax_regression、MLP 和卷积神经网络（Convolutional Neural Network）分类器之间的不同结果。

```
def train_program():
    """
    配置train_program

    Return:
        predict -- 分类的结果
        avg_cost -- 平均损失
        acc -- 分类的准确率

    """
    # 标注层，名称为label，对应输入图片的类别标注
    label = fluid.layers.data(name='label', shape=[1], dtype='int64')

    # predict = softmax_regression()           # 取消注释将使用softmax回归
    # predict = multilayer_perceptron()        # 取消注释将使用多层感知器
    predict = convolutional_neural_network()   # 取消注释将使用LeNet5卷积神经网络

    # 使用类交叉熵函数计算predict和label之间的损失函数
    cost = fluid.layers.cross_entropy(input=predict, label=label)
    # 计算平均损失
    avg_cost = fluid.layers.mean(cost)
    # 计算分类准确率
    acc = fluid.layers.accuracy(input=predict, label=label)
    return predict, [avg_cost, acc]
```

（3）Optimizer Function 配置

下面的 Adam optimizer、learning_rate 是学习率，它的大小与网络的训练收敛速度相关。

```
def optimizer_program():
    return fluid.optimizer.Adam(learning_rate=0.001)
```

（4）数据集 Feeders 配置

下一步，开始训练过程。paddle.dataset.mnist.train()和 paddle.dataset.mnist.test()分别进行训练和测试数据集。这两个函数各自返回一个 reader——PaddlePaddle 中的 reader 是一个 Python 函数，每次调用的时候返回一个 Python yield generator。

下面的 shuffle 是一个 reader decorator，接受一个 reader A，返回另一个 reader B。reader B 每次读入 buffer_size 条训练数据到一个 buffer 里，然后随机打乱其顺序，并且逐条输出。

batch 是一个特殊的 decorator，输入是一个 reader，输出是一个 batched reader。在 PaddlePaddle 里，一个 reader 每次产生（yield）一条训练数据，而一个 batched reader 每次产生一个 minibatch。

```
# 一个 minibatch 中有 64 个数据
BATCH_SIZE = 64

# 每次读取训练集中的 500 个数据并随机打乱，传入 batched reader 中，batched reader 每次产生
64 个数据
train_reader = paddle.batch(
        paddle.reader.shuffle(
            paddle.dataset.mnist.train(), buf_size=500),
        batch_size=BATCH_SIZE)
# 读取测试集的数据，每次产生 64 个数据
test_reader = paddle.batch(
            paddle.dataset.mnist.test(), batch_size=BATCH_SIZE)
```

6. 构建训练过程

现在，需要构建一个训练过程。将使用到前面定义的训练程序 train_program、place 和优化器 optimizer，包含训练迭代、检查训练期间测试误差以及保存所需要用来预测的模型参数。

（1）Event Handler 配置

可以在训练期间通过调用一个 handler 函数来监控训练进度。将在这里演示两个 event_handler 程序。event_handler 用来在训练过程中输出训练结果。

```
def event_handler(pass_id, batch_id, cost):
    # 打印训练的中间结果、训练轮次、batch 数、损失函数
    print("Pass %d, Batch %d, Cost %f" % (pass_id,batch_id, cost))
from paddle.utils.plot import Ploter

train_prompt = "Train cost"
test_prompt = "Test cost"
cost_ploter = Ploter(train_prompt, test_prompt)

# 将训练过程绘图表示
def event_handler_plot(ploter_title, step, cost):
    cost_ploter.append(ploter_title, step, cost)
    cost_ploter.plot()
```

（2）开始训练

加入设置的 event_handler 和 data reader，然后就可以开始训练模型了。设置一些运行需要的参数，配置数据描述 feed_order 用于将数据目录映射到 train_program 以创建一个反馈训练过程中误差的 train_test。

① 定义网络结构

```
# 该模型运行在单个 CPU 上
```

```python
use_cuda = False # 如果想使用GPU，就设置为 True
place = fluid.CUDAPlace(0) if use_cuda else fluid.CPUPlace()

# 调用 train_program 获取预测值、损失值
prediction, [avg_loss, acc] = train_program()

# 输入的原始图像数据，大小为 28*28*1
img = fluid.layers.data(name='img', shape=[1, 28, 28], dtype='float32')
# 标注层，名称为 label，对应输入图片的类别标注
label = fluid.layers.data(name='label', shape=[1], dtype='int64')
# 告知网络传入的数据分为两部分，第一部分是 img 值，第二部分是 label 值
feeder = fluid.DataFeeder(feed_list=[img, label], place=place)

# 选择 Adam 优化器
optimizer = fluid.optimizer.Adam(learning_rate=0.001)
optimizer.minimize(avg_loss)
```

② 设置训练过程的超参

```python
PASS_NUM = 5 #训练5轮
epochs = [epoch_id for epoch_id in range(PASS_NUM)]

# 将模型参数存储在名为 save_dirname 的文件中
save_dirname = "recognize_digits.inference.model"
def train_test(train_test_program,
               train_test_feed, train_test_reader):

    # 将分类准确率存储在 acc_set 中
    acc_set = []
    # 将平均损失存储在 avg_loss_set 中
    avg_loss_set = []
    # 将测试 reader yield 出的每一个数据传入网络中进行训练
    for test_data in train_test_reader():
        acc_np, avg_loss_np = exe.run(
            program=train_test_program,
            feed=train_test_feed.feed(test_data),
            fetch_list=[acc, avg_loss])
        acc_set.append(float(acc_np))
        avg_loss_set.append(float(avg_loss_np))
    # 获得测试数据上的准确率和损失值
    acc_val_mean = numpy.array(acc_set).mean()
    avg_loss_val_mean = numpy.array(avg_loss_set).mean()
    # 返回平均损失值、平均准确率
    return avg_loss_val_mean, acc_val_mean
```

③ 创建执行器

```
exe = fluid.Executor(place)
exe.run(fluid.default_startup_program())
```

④ 设置 main_program 和 test_program

```
main_program = fluid.default_main_program()
test_program = fluid.default_main_program().clone(for_test=True)
```

⑤ 开始训练

```
lists = []
step = 0
for epoch_id in epochs:
    for step_id, data in enumerate(train_reader()):
        metrics = exe.run(main_program,
                          feed=feeder.feed(data),
                          fetch_list=[avg_loss, acc])
        if step % 100 == 0: #每训练100次 打印一次log
            print("Pass %d, Batch %d, Cost %f" % (step, epoch_id, metrics[0]))
            event_handler_plot(train_prompt, step, metrics[0])
        step += 1

    # 测试每个 epoch 的分类效果
    avg_loss_val, acc_val = train_test(train_test_program=test_program,
                                       train_test_reader=test_reader,
                                       train_test_feed=feeder)

    print("Test with Epoch %d, avg_cost: %s, acc: %s" %(epoch_id, avg_loss_val, acc_val))
    event_handler_plot(test_prompt, step, metrics[0])

    lists.append((epoch_id, avg_loss_val, acc_val))

    # 保存训练好的模型参数,用于预测
    if save_dirname is not None:
        fluid.io.save_inference_model(save_dirname,
                                      ["img"], [prediction], exe,
                                      model_filename=None,
                                      params_filename=None)

# 选择效果最好的 pass
best = sorted(lists, key=lambda list: float(list[1]))[0]
print('Best pass is %s, testing Avgcost is %s' % (best[0], best[1]))
print('The classification accuracy is %.2f%%' % (float(best[2]) * 100))
```

训练过程是完全自动的，event_handler 里打印的日志如下所示。Pass 表示训练轮次，Batch 表示训练全量数据的次数，cost 表示当前 pass 的损失值。每训练完一个 Epoch 后，计算一次平均损失和分类准确率。

```
Pass 0, Batch 0, Cost 0.125650
Pass 100, Batch 0, Cost 0.161387
Pass 200, Batch 0, Cost 0.040036
Pass 300, Batch 0, Cost 0.023391
Pass 400, Batch 0, Cost 0.005856
Pass 500, Batch 0, Cost 0.003315
Pass 600, Batch 0, Cost 0.009977
Pass 700, Batch 0, Cost 0.020959
Pass 800, Batch 0, Cost 0.105560
Pass 900, Batch 0, Cost 0.239809
Test with Epoch 0, avg_cost: 0.053097883707459624, acc: 0.9822850318471338
```

训练之后，检查模型的预测准确度。用 MNIST 训练的时候，一般 Softmax 回归模型的分类准确率约为 92.34%，多层感知器为 97.66%，卷积神经网络可以达到 99.20%。

7. 应用模型

可以使用训练好的模型对手写体数字图片进行分类。下面的程序展示了如何使用训练好的模型进行推断。

（1）生成预测输入数据

infer_3.png 是数字 3 的一个示例图像。把它变成一个 NumPy 数组以匹配数据 feed 格式。

```python
def load_image(file):
    im = Image.open(file).convert('L')
    im = im.resize((28, 28), Image.ANTIALIAS)
    im = numpy.array(im).reshape(1, 1, 28, 28).astype(numpy.float32)
    im = im / 255.0 * 2.0 - 1.0
    return im

cur_dir = os.getcwd()
tensor_img = load_image(cur_dir + '/image/infer_3.png')
```

（2）Inference 创建及预测

```python
inference_scope = fluid.core.Scope()
with fluid.scope_guard(inference_scope):
    # 使用 fluid.io.load_inference_model 获取 inference program desc
    # feed_target_names 用于指定需要传入网络的变量名
    # fetch_targets 指定希望从网络中 fetch 出的变量名
    [inference_program, feed_target_names,
     fetch_targets] = fluid.io.load_inference_model(
```

```
                save_dirname, exe, None, None)

    # 将 feed 构建成字典 {feed_target_name: feed_target_data}
    # 结果将包含一个与 fetch_targets 对应的数据列表
    results = exe.run(inference_program,
                      feed={feed_target_names[0]: tensor_img},
                      fetch_list=fetch_targets)
    lab = numpy.argsort(results)

    # 打印 infer_3.png 这张图片的预测结果
    img=Image.open('image/infer_3.png')
    plt.imshow(img)
    print("Inference result of image/infer_3.png is: %d" % lab[0][0][-1])
```

（3）预测结果

如果顺利，预测结果输入如下：Inference result of image/infer_3.png is: 3，说明预测网络成功地识别出了这张图片。

本节的 Softmax 回归、多层感知机和卷积神经网络是最基础的深度学习模型，复杂的神经网络都是从它们衍生出来的，因此这几个模型对之后的学习大有裨益。同时，也观察到从最简单的 Softmax 回归变换到稍复杂的卷积神经网络的时候，MNIST 数据集上的识别准确率有了大幅度的提升，原因是卷积层具有局部连接和共享权重的特性。在之后学习新模型时，希望也要深入到新模型相比原模型带来效果提升的关键之处。此外，本节还介绍了 PaddlePaddle 模型搭建的基本流程，从 dataprovider 的编写、网络层的构建，到最后的训练和预测。对这个流程熟悉以后，就可以用自己的数据定义自己的网络模型，并完成自己的训练和预测任务了。

完整的训练代码如下：

```
In[1]
#   Copyright (c) 2018 PaddlePaddle Authors. All Rights Reserved.
#
# Licensed under the Apache License, Version 2.0 (the "License");
# you may not use this file except in compliance with the License.
# You may obtain a copy of the License at
#
#     http://www.apache.org/licenses/LICENSE-2.0
#
# Unless required by applicable law or agreed to in writing, software
# distributed under the License is distributed on an "AS IS" BASIS,
# WITHOUT WARRANTIES OR CONDITIONS OF ANY KIND, either express or implied.
# See the License for the specific language governing permissions and
# limitations under the License.

from __future__ import print_function

import os
from PIL import Image
```

```python
import numpy
import paddle
import paddle.fluid as fluid

BATCH_SIZE = 64
PASS_NUM = 5


def loss_net(hidden, label):
    prediction = fluid.layers.fc(input=hidden, size=10, act='softmax')
    loss = fluid.layers.cross_entropy(input=prediction, label=label)
    avg_loss = fluid.layers.mean(loss)
    acc = fluid.layers.accuracy(input=prediction, label=label)
    return prediction, avg_loss, acc


def multilayer_perceptron(img, label):
    img = fluid.layers.fc(input=img, size=200, act='tanh')
    hidden = fluid.layers.fc(input=img, size=200, act='tanh')
    return loss_net(hidden, label)


def softmax_regression(img, label):
    return loss_net(img, label)


def convolutional_neural_network(img, label):
    conv_pool_1 = fluid.nets.simple_img_conv_pool(
        input=img,
        filter_size=5,
        num_filters=20,
        pool_size=2,
        pool_stride=2,
        act="relu")
    conv_pool_1 = fluid.layers.batch_norm(conv_pool_1)
    conv_pool_2 = fluid.nets.simple_img_conv_pool(
        input=conv_pool_1,
        filter_size=5,
        num_filters=50,
        pool_size=2,
        pool_stride=2,
        act="relu")
    return loss_net(conv_pool_2, label)


def train(nn_type,
          use_cuda,
          save_dirname=None,
          model_filename=None,
          params_filename=None):
    if use_cuda and not fluid.core.is_compiled_with_cuda():
```

```python
    return

img = fluid.layers.data(name='img', shape=[1, 28, 28], dtype='float32')
label = fluid.layers.data(name='label', shape=[1], dtype='int64')

if nn_type == 'softmax_regression':
    net_conf = softmax_regression
elif nn_type == 'multilayer_perceptron':
    net_conf = multilayer_perceptron
else:
    net_conf = convolutional_neural_network

prediction, avg_loss, acc = net_conf(img, label)

test_program = fluid.default_main_program().clone(for_test=True)

optimizer = fluid.optimizer.Adam(learning_rate=0.001)
optimizer.minimize(avg_loss)

def train_test(train_test_program, train_test_feed, train_test_reader):
    acc_set = []
    avg_loss_set = []
    for test_data in train_test_reader():
        acc_np, avg_loss_np = exe.run(
            program=train_test_program,
            feed=train_test_feed.feed(test_data),
            fetch_list=[acc, avg_loss])
        acc_set.append(float(acc_np))
        avg_loss_set.append(float(avg_loss_np))
    # get test acc and loss
    acc_val_mean = numpy.array(acc_set).mean()
    avg_loss_val_mean = numpy.array(avg_loss_set).mean()
    return avg_loss_val_mean, acc_val_mean

place = fluid.CUDAPlace(0) if use_cuda else fluid.CPUPlace()

exe = fluid.Executor(place)

train_reader = paddle.batch(
    paddle.reader.shuffle(paddle.dataset.mnist.train(), buf_size=500),
    batch_size=BATCH_SIZE)
test_reader = paddle.batch(
    paddle.dataset.mnist.test(), batch_size=BATCH_SIZE)
feeder = fluid.DataFeeder(feed_list=[img, label], place=place)

exe.run(fluid.default_startup_program())
main_program = fluid.default_main_program()
epochs = [epoch_id for epoch_id in range(PASS_NUM)]

lists = []
step = 0
```

```python
    for epoch_id in epochs:
        for step_id, data in enumerate(train_reader()):
            metrics = exe.run(
                main_program,
                feed=feeder.feed(data),
                fetch_list=[avg_loss, acc])
            if step % 100 == 0:
                print("Pass %d, Batch %d, Cost %f" % (step, epoch_id,
                                                     metrics[0]))
            step += 1
        # test for epoch
        avg_loss_val, acc_val = train_test(
            train_test_program=test_program,
            train_test_reader=test_reader,
            train_test_feed=feeder)

        print("Test with Epoch %d, avg_cost: %s, acc: %s" %
            (epoch_id, avg_loss_val, acc_val))
        lists.append((epoch_id, avg_loss_val, acc_val))
        if save_dirname is not None:
            fluid.io.save_inference_model(
                save_dirname, ["img"], [prediction],
                exe,
                model_filename=model_filename,
                params_filename=params_filename)

    # find the best pass
    best = sorted(lists, key=lambda list: float(list[1]))[0]
    print('Best pass is %s, testing Avgcost is %s' % (best[0], best[1]))
    print('The classification accuracy is %.2f%%' % (float(best[2]) * 100))

def infer(use_cuda,
          save_dirname=None,
          model_filename=None,
          params_filename=None):
    if save_dirname is None:
        return

    place = fluid.CUDAPlace(0) if use_cuda else fluid.CPUPlace()
    exe = fluid.Executor(place)

    def load_image(file):
        im = Image.open(file).convert('L')
        im = im.resize((28, 28), Image.ANTIALIAS)
        im = numpy.array(im).reshape(1, 1, 28, 28).astype(numpy.float32)
        im = im / 255.0 * 2.0 - 1.0
        return im

    cur_dir = os.path.dirname(os.path.realpath(__file__))
    tensor_img = load_image(cur_dir + '/image/infer_3.png')
```

```python
    inference_scope = fluid.core.Scope()
    with fluid.scope_guard(inference_scope):
        # Use fluid.io.load_inference_model to obtain the inference program desc,
        # the feed_target_names (the names of variables that will be feeded
        # data using feed operators), and the fetch_targets (variables that
        # we want to obtain data from using fetch operators).
        [inference_program, feed_target_names,
         fetch_targets] = fluid.io.load_inference_model(
            save_dirname, exe, model_filename, params_filename)

        # Construct feed as a dictionary of {feed_target_name: feed_target_data}
        # and results will contain a list of data corresponding to fetch_targets.
        results = exe.run(
            inference_program,
            feed={feed_target_names[0]: tensor_img},
            fetch_list=fetch_targets)
        lab = numpy.argsort(results)
        print("Inference result of image/infer_3.png is: %d" % lab[0][0][-1])


def main(use_cuda, nn_type):
    model_filename = None
    params_filename = None
    save_dirname = "recognize_digits_" + nn_type + ".inference.model"

    # call train() with is_local argument to run distributed train
    train(
        nn_type=nn_type,
        use_cuda=use_cuda,
        save_dirname=save_dirname,
        model_filename=model_filename,
        params_filename=params_filename)
    infer(
        use_cuda=use_cuda,
        save_dirname=save_dirname,
        model_filename=model_filename,
        params_filename=params_filename)


if __name__ == '__main__':
    use_cuda = False
    # predict = 'softmax_regression' # uncomment for Softmax
    # predict = 'multilayer_perceptron' # uncomment for MLP
    predict = 'convolutional_neural_network' # uncomment for LeNet5
    main(use_cuda=use_cuda, nn_type=predict)
```

6.3 PaddlePaddle 平台图像分类

图像相比文字能够提供更加生动、容易理解及更具艺术感的信息,是人们传递与交换信息的重要来源。本节专注于图像识别领域的一个重要问题,即图像分类。

图像分类根据图像的语义信息将不同类别图像区分开来,是计算机视觉中重要的基本问题,也是图像检测、图像分割、物体跟踪、行为分析等其他高层视觉任务的基础。图像分类在很多领域都有广泛应用,包括安防领域的人脸识别和智能视频分析、交通领域的交通场景识别、互联网领域基于内容的图像检索和相册自动归类、医学领域的图像识别等。

一般来说,图像分类通过手工提取特征或特征学习方法对整个图像进行全部描述,然后使用分类器判断物体类别,因此如何提取图像的特征至关重要。在深度学习算法之前使用较多的是基于词袋(Bag of Words)模型的物体分类方法。词袋方法从自然语言处理中引入,即一句话可以用一个装了词的袋子表示其特征,袋子中的词为句子中的单词、短语或字。对于图像而言,词袋方法需要构建字典。最简单的词袋模型框架可以设计为底层特征抽取、特征编码、分类器设计 3 个过程。

基于深度学习的图像分类方法可以通过有监督或无监督的方式学习层次化的特征描述,从而取代手工设计或选择图像特征的工作。深度学习模型中的卷积神经网络(Convolution Neural Network,CNN)近年来在图像领域取得了惊人的成绩,直接利用图像像素信息作为输入,最大程度上保留了输入图像的所有信息,通过卷积操作进行特征的提取和高层抽象,模型输出直接是图像识别的结果。这种基于"输入-输出"直接端到端的学习方法取得了非常好的效果,得到了广泛的应用。

本节主要介绍图像分类的深度学习模型,以及如何使用 PaddlePaddle 训练 CNN 模型。

图像分类包括通用图像分类、细粒度图像分类等。

一个好的模型既要对不同类别识别正确,同时也应该能够对不同视角、光照、背景、变形或部分遮挡的图像正确识别(这里统一称作图像扰动)。图 6.7 展示了一些图像的扰动,较好的模型会像聪明的人类一样能够正确识别。

图 6.7 一些图像的扰动

1. 图像分类模型

图像识别领域大量的研究成果都是建立在 PASCAL VOC、ImageNet 等公开的数据集上，很多图像识别算法通常在这些数据集上进行测试和比较。PASCAL VOC 是 2005 年发起的一个视觉挑战赛，ImageNet 是 2010 年发起的大规模视觉识别竞赛（ILSVRC）的数据集，本节基于这些竞赛的一些论文介绍图像分类模型。

在 2012 年之前的传统图像分类方法可以用背景描述中提到的三步完成，但通常完整建立图像识别模型一般包括底层特征提取、特征编码、空间约束、分类器设计等几个阶段。

- 底层特征提取

通常从图像中按照固定步长、尺度提取大量局部特征描述。常用的局部特征包括 SIFT（Scale-Invariant Feature Transform，尺度不变特征转换）、HOG（Histogram of Oriented Gradient，方向梯度直方图）、LBP（Local Binary Pattern，局部二值模式）等，一般也采用多种特征描述，防止丢失过多的有用信息。

- 特征编码

底层特征中包含了大量冗余与噪声，为了提高特征表达的鲁棒性，需要使用一种特征变换算法对底层特征进行编码，称作特征编码。常用的特征编码方法包括向量量化编码、稀疏编码、局部线性约束编码、Fisher 向量编码等。

- 空间约束

特征编码之后一般会经过空间特征约束，也称作特征汇聚。特征汇聚是指在一个空间范围内，对每一维特征取最大值或者平均值，可以获得一定特征不变形的特征表达。金字塔特征匹配是一种常用的特征汇聚方法，这种方法提出将图像均匀分块，在分块内做特征汇聚。

- 分类器设计

经过前面的步骤之后，一张图像可以用一个固定维度的向量进行描述，接下来就是用分类器对图像进行分类。通常使用的分类器包括 SVM（Support Vector Machine，支持向量机）、随机森林等。而使用核方法的 SVM 是最为广泛的分类器，在传统图像分类任务上性能很好。

这种传统的图像分类方法在 PASCAL VOC 竞赛中的图像分类算法中被广泛使用。NEC 实验室在 ILSVRC2010 中采用 SIFT 和 LBP 特征，两个非线性编码器以及 SVM 分类器获得图像分类的冠军。

Alex Krizhevsky 在 2012 年 ILSVRC 提出的 CNN 模型取得了历史性的突破，效果大幅度超越传统方法，获得了 ILSVRC2012 冠军，该模型被称作 AlexNet。这也是首次将深度学习用于大规模图像分类中。从 AlexNet 之后，涌现了一系列 CNN 模型，不断地在 ImageNet 上刷新成绩，如图 6.8 所示。

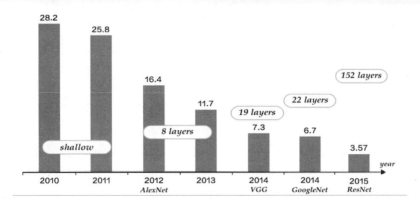

图 6.8　2012 年后 ImageNet 上不断刷新的成绩

随着模型变得越来越深（以及精妙的结构设计），Top-5 的错误率越来越低，降到了 3.5% 附近。在同样的 ImageNet 数据集上，人眼的辨识错误率大概在 5.1%，也就是目前的深度学习模型的识别能力已经超过了人眼。

2. 卷积神经网络

传统的卷积神经网络（CNN）包含卷积层、全连接层等组件，并采用 Softmax 多类别分类器和多类交叉熵损失函数。一个典型的卷积神经网络如图 6.9 所示。

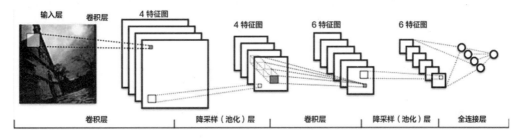

图 6.9　典型的卷积神经网络

构造卷积神经网络的常见组件如下：

- 卷积层（Convolution Layer）

执行卷积操作，提取底层到高层的特征，发掘出图片局部关联性质和空间不变性质。

- 池化层（Pooling Layer）

执行降采样操作。通过取卷积输出特征图中局部区块的最大值（Max-Pooling）或者均值（Avg-Pooling）。降采样也是图像处理中常见的一种操作，可以过滤掉一些不重要的高频信息。

- 全连接层（Fully-Connected Layer，或者 FC Layer）

输入层到隐藏层的神经元是全部连接的。

- 非线性变化

卷积层、全连接层后面一般都会接非线性变化函数，例如 Sigmoid、Tanh、ReLu 等，来增强网络的表达能力，在 CNN 里最常使用的为 ReLu 激活函数。

- Dropout

在模型训练阶段随机让一些隐藏层节点权重不工作，提高网络的泛化能力，在一定程度上防止过度拟合。

另外，在训练过程中每层参数不断更新，会导致下一次输入分布发生变化，因此需要精心设计超参数。例如，在 2015 年 Sergey Ioffe 和 Christian Szegedy 提出的 Batch Normalization（BN）算法中，每个 batch 对网络中的每一层特征都做了归一化，使得每层分布相对稳定。BN 算法不仅起到一定的正则作用，而且弱化了一些超参数的设计。经过实验证明，BN 算法加速了模型收敛过程，在后来较深的模型中被广泛使用。

接下来主要介绍 VGG、GoogleNet 和 ResNet 网络结构。

3. VGG

牛津大学 VGG（Visual Geometry Group）组在 2014 年 ILSVRC 提出的模型被称作 VGG 模型。该模型相比以往模型进一步加宽和加深了网络结构，核心是五组卷积操作，每两组之间做 Max-Pooling 空间降维。同一组内采用多次连续的 3×3 卷积，卷积核的数目由较浅组的 64 增多到最深组的 512，同一组内的卷积核数目是一样的。卷积之后接两层全连接层，之后是分类层。由于每组的内卷积层不同，因此有 11、13、16、19 层几种模型，图 6.10 展示了一个 16 层的网络结构。VGG 模型结构相对简洁，提出之后也有很多文章基于此模型进行研究，如在 ImageNet 上首次公开超过人眼识别的模型就是借鉴 VGG 模型的结构。

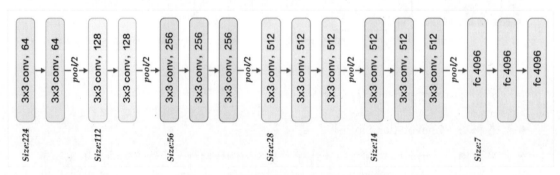

图 6.10 基于 ImageNet 的 VGG16 模型

4. GoogleNet

GoogleNet 在 2014 年的 ILSVRC 获得了冠军，在介绍该模型之前先来了解 NIN（Network in Network）模型和 Inception 模块，因为 GoogleNet 模型由多组 Inception 模块组成，模型设计借鉴了 NIN 的一些思想。

NIN 模型主要有两个特点：

（1）引入多层感知卷积网络（Multi-Layer Perceptron Convolution，MLPconv）代替一层

线性卷积网络。MLPconv 是一个微小的多层卷积网络，即在线性卷积后面增加若干层 1×1 的卷积，以提取出高度非线性特征。

（2）传统的卷积神经网络（CNN）最后几层一般都是全连接层，参数较多。NIN 模型设计最后一层卷积层包含类别维度大小的特征图，然后采用全局均值池化（Avg-Pooling）替代全连接层，得到类别维度大小的向量，再进行分类。这种替代全连接层的方式有利于减少参数。

Inception 模块如图 6.11 所示。图 6.11（a）是最简单的设计，输出是 3 个卷积层和一个池化层的特征拼接。这种设计的缺点是池化层不会改变特征通道数，拼接后会导致特征的通道数较大，经过几层这样的模块堆积后，通道数会越来越大，导致参数和计算量也随之增大。为了改善这个缺点，图 6.11（b）引入 3 个 1×1 卷积层进行降维（所谓的降维，就是减少通道数），同时如 NIN 模型中提到的 1×1 卷积也可以修正线性特征。

GoogleNet 由多组 Inception 模块堆积而成。另外，在网络最后也没有采用传统的多层全连接层，而是像 NIN 网络一样采用了均值池化层；与 NIN 不同的是，GoogleNet 在池化层后加了一个全连接层来映射类别数。除了这两个特点之外，网络中间层特征也很有判别性，GoogleNet 在中间层添加了两个辅助分类器，在后向传播中增强梯度并且增强正则化，而整个网络的损失函数是这 3 个分类器的损失加权求和。

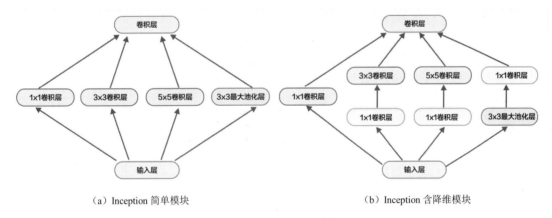

(a) Inception 简单模块　　　　　　　(b) Inception 含降维模块

图 6.11　Inception 模块

GoogleNet 整体网络结构如图 6.12 所示，总共 22 层网络：开始由 3 层普通的卷积组成；接下来由 3 组子网络组成，第一组子网络包含 2 个 Inception 模块，第二组包含 5 个 Inception 模块，第三组包含 2 个 Inception 模块；然后接均值池化层、全连接层。

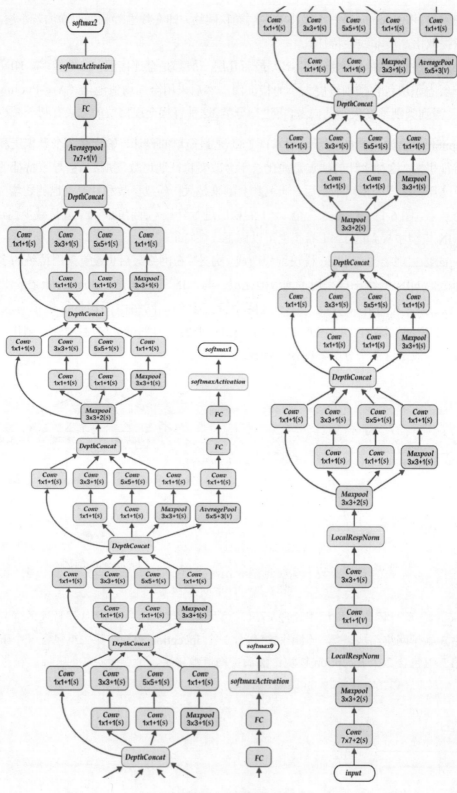

图 6.12 GoogleNet

上面介绍的是 GoogleNet 第一版模型（称作 GoogleNet-v1）。GoogleNet-v2 引入 BN 层；GoogleNet-v3 对一些卷积层做了分解，进一步提高网络非线性能力和加深网络；GoogleNet-v4 引入下面要讲的 ResNet（Residual Network）设计思路。从 v1 到 v4，每一版的改进都带来了准确度的提升，限于篇幅，这里不详细介绍 v2 到 v4 的结构。

5. ResNet

ResNet 是 2015 年 ImageNet 图像分类、图像物体定位和图像物体检测比赛的冠军。针对随着网络训练加深导致准确度下降的问题，ResNet 提出了残差学习方法来减轻训练深层网络的困难。在已有设计思路（BN，小卷积核，全卷积网络）的基础上，引入了残差模块。每个残差模块包含两条路径，其中一条路径是输入特征的直连通路，另一条路径对该特征做两到三次卷积操作，得到该特征的残差，最后将两条路径上的特征相加。

残差模块如图 6.13 所示。左边是基本模块连接方式，由两个输出通道数相同的 3×3 卷积组成。右边是瓶颈模块（Bottleneck）连接方式，之所以称为瓶颈，是因为上面的 1×1 卷积用来降维（256→64），下面的 1×1 卷积用来升维（64→256），中间 3×3 卷积的输入和输出通道数都较小（64→64）。

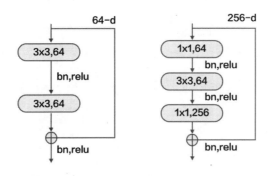

图 6.13 残差模块

图 6.14 展示了 50、101、152 层网络连接示意图，使用的是瓶颈模块。这 3 个模型的区别在于每组中残差模块的重复次数不同（见图 6.14 右上角）。ResNet 训练收敛较快，成功地训练了上百乃至近千层的卷积神经网络。

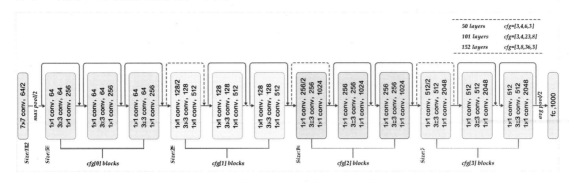

图 6.14 基于 ImageNet 的 ResNet 模型

6. 数据准备

通用图像分类公开的标准数据集常用的有 CIFAR、ImageNet、COCO 等，常用的细粒度图像分类数据集包括 CUB-200-2011、Stanford Dog、Oxford-flowers 等。其中，ImageNet 数据集规模相对较大，大量研究成果基于 ImageNet。ImageNet 数据从 2010 年来稍有变化，常用的是 ImageNet-2012 数据集，该数据集包含 1 000 个类别：训练集包含 1 281 167 张图片，每个类别数据为 732 至 1300 张不等，验证集包含 50 000 张图片，平均每个类别 50 张图片。ImageNet 数据集较大、下载和训练较慢，为了方便大家学习，本节使用 CIFAR10 数据集。CIFAR10 数据集包含 60 000 张 32×32 的彩色图片，10 个类别，每个类包含 6 000 张。其中，50 000 张图片作为训练集，10 000 张作为测试集。图 6.15 从每个类别中随机抽取 10 张图片，展示了所有的类别。

图 6.15　CIFAR10 数据集

Paddle API 提供了自动加载 cifar 数据集模块 paddle.dataset.cifar。通过输入 python train.py 就可以开始训练模型了，下面将详细介绍 train.py 的相关内容。

7. 模型结构

（1）Paddle 初始化

从导入 Paddle Fluid API 和辅助模块开始。

```
import paddle
import paddle.fluid as fluid
import numpy
import sys
from __future__ import print_function
```

本节中提供了 VGG 和 ResNet 两个模型的配置。

(2) VGG

首先介绍 VGG 模型结构。由于 CIFAR10 图片大小和数量相比 ImageNet 数据小很多，因此这里的模型针对 CIFAR10 数据做了一定的适配。卷积部分引入了 BN 和 Dropout 操作。VGG 核心模块的输入是数据层，vgg_bn_drop 定义了 16 层 VGG 结构，每层卷积后面引入 BN 层和 Dropout 层，详细的定义如下：

```python
def vgg_bn_drop(input):
    def conv_block(ipt, num_filter, groups, dropouts):
        return fluid.nets.img_conv_group(
            input=ipt,
            pool_size=2,
            pool_stride=2,
            conv_num_filter=[num_filter] * groups,
            conv_filter_size=3,
            conv_act='relu',
            conv_with_batchnorm=True,
            conv_batchnorm_drop_rate=dropouts,
            pool_type='max')

    conv1 = conv_block(input, 64, 2, [0.3, 0])
    conv2 = conv_block(conv1, 128, 2, [0.4, 0])
    conv3 = conv_block(conv2, 256, 3, [0.4, 0.4, 0])
    conv4 = conv_block(conv3, 512, 3, [0.4, 0.4, 0])
    conv5 = conv_block(conv4, 512, 3, [0.4, 0.4, 0])

    drop = fluid.layers.dropout(x=conv5, dropout_prob=0.5)
    fc1 = fluid.layers.fc(input=drop, size=512, act=None)
    bn = fluid.layers.batch_norm(input=fc1, act='relu')
    drop2 = fluid.layers.dropout(x=bn, dropout_prob=0.5)
    fc2 = fluid.layers.fc(input=drop2, size=512, act=None)
    predict = fluid.layers.fc(input=fc2, size=10, act='softmax')
    return predict
```

首先定义了一组卷积网络，即 conv_block。卷积核大小为 3×3，池化窗口大小为 2×2，窗口滑动大小为 2，groups 决定每组 VGG 模块是几次连续的卷积操作，dropouts 指定 Dropout 操作的概率。所使用的 img_conv_group 是在 paddle.networks 中预定义的模块，由若干组 Conv→BN→ReLu→Dropout 和一组 Pooling 组成。

然后定义了 5 组卷积操作，即 5 个 conv_block：第一、二组采用两次连续的卷积操作；第三、四、五组采用 3 次连续的卷积操作。每组最后一个卷积后面 dropout 的概率为 0，即不使用 Dropout 操作。最后接两层 512 维的全连接。

在这里，VGG 网络首先提取高层特征，随后在全连接层中将其映射到和类别维度大小一致的向量上，最后通过 Softmax 方法计算图片划为每个类别的概率。

(3) ResNet

ResNet 模型的第 1、3、4 步和 VGG 模型相同,这里不再介绍,主要介绍第 2 步,即 CIFAR10 数据集上的 ResNet 核心模块。

这里先介绍 resnet_cifar10 中的一些基本函数,再介绍网络连接过程。

- conv_bn_layer

带 BN 的卷积层。

- shortcut

残差模块的"直连"路径。"直连"实际分两种形式:残差模块输入和输出特征通道数不等时,采用 1×1 卷积的升维操作;残差模块输入和输出通道相等时,采用直连操作。

- basicblock

一个基础残差模块,如图 6.13 左边所示,由两组 3×3 卷积组成的路径和一条"直连"路径组成。

- layer_warp

一组残差模块,由若干个残差模块堆积而成。每组中第一个残差模块滑动窗口大小可以与其他不同,以用来减少特征图在垂直和水平方向的大小。

```
def conv_bn_layer(input,
          ch_out,
          filter_size,
          stride,
          padding,
          act='relu',
          bias_attr=False):
    tmp = fluid.layers.conv2d(
        input=input,
        filter_size=filter_size,
        num_filters=ch_out,
        stride=stride,
        padding=padding,
        act=None,
        bias_attr=bias_attr)
    return fluid.layers.batch_norm(input=tmp, act=act)

def shortcut(input, ch_in, ch_out, stride):
    if ch_in != ch_out:
        return conv_bn_layer(input, ch_out, 1, stride, 0, None)
    else:
        return input
```

```
def basicblock(input, ch_in, ch_out, stride):
    tmp = conv_bn_layer(input, ch_out, 3, stride, 1)
    tmp = conv_bn_layer(tmp, ch_out, 3, 1, 1, act=None, bias_attr=True)
    short = shortcut(input, ch_in, ch_out, stride)
    return fluid.layers.elementwise_add(x=tmp, y=short, act='relu')
def layer_warp(block_func, input, ch_in, ch_out, count, stride):
    tmp = block_func(input, ch_in, ch_out, stride)
    for i in range(1, count):
        tmp = block_func(tmp, ch_out, ch_out, 1)
    return tmp
```

resnet_cifar10 的连接结构主要有以下几个过程：

- 底层输入连接一层 conv_bn_layer，即带 BN 的卷积层。
- 然后连接 3 组残差模块，即下面配置 3 组 layer_warp，每组由图 6.13 左边所示的残差模块组成。
- 最后对网络做均值池化并返回该层。

> **注 意**
>
> 除第一层卷积层和最后一层全连接层之外，要求 3 组 layer_warp 总的含参层数能够被 6 整除，即 resnet_cifar10 的 depth 要满足 (depth−2)(depth−2)。

```
def resnet_cifar10(ipt, depth=32):
    # depth should be one of 20, 32, 44, 56, 110, 1202
    assert (depth - 2) % 6 == 0
    n = (depth - 2) // 6
    nStages = {16, 64, 128}
    conv1 = conv_bn_layer(ipt, ch_out=16, filter_size=3, stride=1, padding=1)
    res1 = layer_warp(basicblock, conv1, 16, 16, n, 1)
    res2 = layer_warp(basicblock, res1, 16, 32, n, 2)
    res3 = layer_warp(basicblock, res2, 32, 64, n, 2)
    pool = fluid.layers.pool2d(
        input=res3, pool_size=8, pool_type='avg', pool_stride=1)
    predict = fluid.layers.fc(input=pool, size=10, act='softmax')
    return predict
```

（4）Inference Program 配置

网络输入定义为 data_layer（数据层），在图像分类中即为图像像素信息。CIFRAR10 是 RGB 3 通道 32×32 大小的彩色图，因此输入数据大小为 3072（3×32×32）。

```
def inference_program():
    # The image is 32 * 32 with RGB representation.
    data_shape = [3, 32, 32]
    images = fluid.layers.data(name='pixel', shape=data_shape, dtype='float32')

    predict = resnet_cifar10(images, 32)
    # predict = vgg_bn_drop(images) # un-comment to use vgg net
    return predict
```

(5) Train Program 配置

然后需要设置训练程序 train_program。它首先从推理程序中进行预测。在训练期间，它将从预测中计算 avg_cost。在有监督训练中需要输入图像对应的类别信息，同样通过 fluid.layers.data 来定义。训练中采用多类交叉熵作为损失函数，并作为网络的输出，预测阶段定义网络的输出为分类器得到的概率信息。训练程序应该返回一个数组，第一个返回参数必须是 avg_cost。训练器使用它来计算梯度。

```
def train_program():
    predict = inference_program()

    label = fluid.layers.data(name='label', shape=[1], dtype='int64')
    cost = fluid.layers.cross_entropy(input=predict, label=label)
    avg_cost = fluid.layers.mean(cost)
    accuracy = fluid.layers.accuracy(input=predict, label=label)
    return [avg_cost, accuracy]
```

(6) Optimizer Function 配置

在下面的 Adam optimizer，learning_rate 是学习率，与网络的训练收敛速度有关。

```
def optimizer_program():
    return fluid.optimizer.Adam(learning_rate=0.001)
```

8. 训练模型

(1) Data Feeders 配置

cifar.train10() 每次产生一条样本，在完成 shuffle 和 batch 之后作为训练的输入。

```
# Each batch will yield 128 images
BATCH_SIZE = 128

# Reader for training
train_reader = paddle.batch(
    paddle.reader.shuffle(paddle.dataset.cifar.train10(), buf_size=50000),
    batch_size=BATCH_SIZE)

# Reader for testing. A separated data set for testing.
```

```
test_reader = paddle.batch(
    paddle.dataset.cifar.test10(), batch_size=BATCH_SIZE)
```

（2）Trainer 程序的实现

需要为训练过程制定一个 main_program。同样的，还需要为测试程序配置一个 test_program。定义训练的 place，并使用先前定义的优化器 optimizer_func。

```
use_cuda = False
place = fluid.CUDAPlace(0) if use_cuda else fluid.CPUPlace()

feed_order = ['pixel', 'label']

main_program = fluid.default_main_program()
star_program = fluid.default_startup_program()

avg_cost, acc = train_program()

# Test program
test_program = main_program.clone(for_test=True)

optimizer = optimizer_program()
optimizer.minimize(avg_cost)

exe = fluid.Executor(place)

EPOCH_NUM = 2

# For training test cost
def train_test(program, reader):
    count = 0
    feed_var_list = [
        program.global_block().var(var_name) for var_name in feed_order
    ]
    feeder_test = fluid.DataFeeder(
        feed_list=feed_var_list, place=place)
    test_exe = fluid.Executor(place)
    accumulated = len([avg_cost, acc]) * [0]
    for tid, test_data in enumerate(reader()):
        avg_cost_np = test_exe.run(program=program,
                              feed=feeder_test.feed(test_data),
                              fetch_list=[avg_cost, acc])
        accumulated = [x[0] + x[1][0] for x in zip(accumulated, avg_cost_np)]
        count += 1
    return [x / count for x in accumulated]
```

（3）训练主循环以及过程输出

在接下来的主训练循环中，将通过输出来观察训练过程或进行测试等，也可以使用 plot，利用回调数据来打点画图：

```
params_dirname = "image_classification_resnet.inference.model"

from paddle.utils.plot import Ploter

train_prompt = "Train cost"
test_prompt = "Test cost"
plot_cost = Ploter(test_prompt,train_prompt)

# main train loop.
def train_loop():
    feed_var_list_loop = [
        main_program.global_block().var(var_name) for var_name in feed_order
    ]
    feeder = fluid.DataFeeder(
        feed_list=feed_var_list_loop, place=place)
    exe.run(star_program)

    step = 0
    for pass_id in range(EPOCH_NUM):
        for step_id, data_train in enumerate(train_reader()):
            avg_loss_value = exe.run(main_program,
                            feed=feeder.feed(data_train),
                            fetch_list=[avg_cost, acc])
            if step % 1 == 0:
                plot_cost.append(train_prompt, step, avg_loss_value[0])
                plot_cost.plot()
            step += 1

        avg_cost_test, accuracy_test = train_test(test_program,
                                        reader=test_reader)
        plot_cost.append(test_prompt, step, avg_cost_test)

        # save parameters
        if params_dirname is not None:
            fluid.io.save_inference_model(params_dirname, ["pixel"],
                                [predict], exe)
```

（4）训练

通过 trainer_loop 函数训练，这里只进行了 2 个 Epoch，一般在实际应用上会执行上百个

Epoch。每个 Epoch 在 CPU 中将花费 15～20 分钟。这部分可能需要一段时间。在 GPU 上运行测试，可随意修改代码，以提高训练速度。

`train_loop()`

一轮训练日志（log）的示例如下所示，经过 1 个 pass，训练集上平均 Accuracy 为 0.59，测试集上平均 Accuracy 为 0.6。

```
Pass 0, Batch 0, Cost 3.869598, Acc 0.164062
..............................................................................
...........................
Pass 100, Batch 0, Cost 1.481038, Acc 0.460938
..............................................................................
...........................
Pass 200, Batch 0, Cost 1.340323, Acc 0.523438
..............................................................................
...........................
Pass 300, Batch 0, Cost 1.223424, Acc 0.593750
..............................................................................
............
Test with Pass 0, Loss 1.1, Acc 0.6
```

图 6.16 是训练的分类错误率曲线图，运行到第 200 个 pass 后基本收敛，最终得到测试集上的分类错误率为 8.54%。

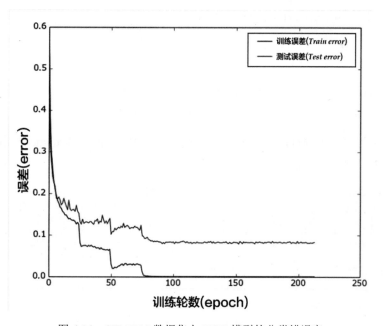

图 6.16　CIFAR10 数据集上 VGG 模型的分类错误率

9. 应用模型

可以使用训练好的模型对图片进行分类。下面的程序展示了如何加载已经训练好的网络和参数进行推断。

（1）生成预测输入数据

dog.png 是一张小狗的图片，可以将它转换成 NumPy 数组，以满足 feeder 的格式。

```
# Prepare testing data.
from PIL import Image
import os

def load_image(file):
    im = Image.open(file)
    im = im.resize((32, 32), Image.ANTIALIAS)

    im = numpy.array(im).astype(numpy.float32)
    # The storage order of the loaded image is W(width),
    # H(height), C(channel). PaddlePaddle requires
    # the CHW order, so transpose them.
    im = im.transpose((2, 0, 1))  # CHW
    im = im / 255.0

    # Add one dimension to mimic the list format.
    im = numpy.expand_dims(im, axis=0)
    return im

cur_dir = os.getcwd()
img = load_image(cur_dir + '/image/dog.png')
```

（2）推理程序（Inferencer）的配置和预测

与训练过程类似，推理程序需要构建相应的过程。从 params_dirname 加载网络和经过训练的参数。可以简单地插入前面定义的推理程序。现在准备做预测。

```
place = fluid.CUDAPlace(0) if use_cuda else fluid.CPUPlace()
exe = fluid.Executor(place)
inference_scope = fluid.core.Scope()

with fluid.scope_guard(inference_scope):

    [inference_program, feed_target_names,
     fetch_targets] = fluid.io.load_inference_model(params_dirname, exe)

        # The input's dimension of conv should be 4-D or 5-D.
        # Use inference_transpiler to speedup
```

```python
inference_transpiler_program = inference_program.clone()
t = fluid.transpiler.InferenceTranspiler()
t.transpile(inference_transpiler_program, place)

    # Construct feed as a dictionary of {feed_target_name: feed_target_data}
    # and results will contain a list of data corresponding to fetch_targets.
results = exe.run(inference_program,
            feed={feed_target_names[0]: img},
            fetch_list=fetch_targets)

transpiler_results = exe.run(inference_transpiler_program,
                    feed={feed_target_names[0]: img},
                    fetch_list=fetch_targets)

assert len(results[0]) == len(transpiler_results[0])
for i in range(len(results[0])):
    numpy.testing.assert_almost_equal(
        results[0][i], transpiler_results[0][i], decimal=5)

# infer label
label_list = [
    "airplane", "automobile", "bird", "cat", "deer", "dog", "frog", "horse",
    "ship", "truck"
]

print("infer results: %s" % label_list[numpy.argmax(results[0])])
```

传统图像分类方法由多个阶段构成，框架较为复杂，而端到端的 CNN 模型结构可一步到位，而且大幅度提升了分类准确率。本节首先介绍 VGG、GoogleNet、ResNet 三个经典的模型；然后基于 CIFAR10 数据集介绍如何使用 PaddlePaddle 配置和训练 CNN 模型，尤其是 VGG 和 ResNet 模型；最后介绍如何使用 PaddlePaddle 的 API 接口对图片进行预测和特征提取。对于其他数据集（比如 ImageNet），配置和训练流程是同样的，大家可以自行进行实验。

6.4 PaddlePaddle 平台词向量

词向量是自然语言处理中常见的一个操作，是搜索引擎、广告系统、推荐系统等互联网服务背后常见的基础技术。本节介绍词的向量表征，也称为 Word Embedding。

项目源码来源为 https://github.com/PaddlePaddle/book/tree/develop/04.word2vec。

在这些互联网服务里，经常要比较两个词或者两段文本之间的相关性。为了做这样的比较，往往先要把词表示成计算机适合处理的方式。最自然的方式恐怕莫过于向量空间模型（Vector

Space Model）。在这种方式里，每个词被表示成一个实数向量（One-hot Vector），其长度为字典大小，每个维度对应一个字典里的每个词，除了这个词对应维度上的值是1，其他元素都是0。

One-hot Vector虽然自然，但是用处有限。比如，在互联网广告系统里，如果用户输入的query是"母亲节"，而有一个广告的关键词是"康乃馨"。虽然按照常理，知道这两个词之间是有联系的——母亲节通常应该送给母亲一束康乃馨；但是这两个词对应的One-Hot Vector之间的距离度量无论是欧氏距离还是余弦相似度（Cosine Similarity），由于其向量正交，都认为这两个词毫无相关性。得出这种相悖结论的根本原因是：每个词本身的信息量都太小。所以，仅仅给定两个词，不足以准确判别它们是否相关。要想精确计算相关性，还需要更多的信息——从大量数据里通过机器学习方法归纳出来的知识。

在机器学习领域里，各种"知识"被各种模型表示，词向量模型（Word Embedding Model）就是其中的一类。通过词向量模型可将一个One-Hot Vector映射到一个维度更低的实数向量（Embedding Vector），如embedding(母亲节)=[0.3,4.2,−1.5,...]，embedding(康乃馨)=[0.2,5.6,−2.3,...]，embedding(母亲节)=[0.3,4.2,−1.5,...]，embedding(康乃馨)=[0.2,5.6,−2.3,...]。在这个映射到的实数向量表示中，希望两个语义（或用法）上相似的词对应的词向量"更像"，这样如"母亲节"和"康乃馨"的对应词向量的余弦相似度就不再为零了。

词向量模型可以是概率模型、共生矩阵（Co-Occurrence Matrix）模型或神经元网络模型。在用神经网络求词向量之前，传统做法是统计一个词语的共生矩阵 X。X是一个 $|V|\times|V|$ 大小的矩阵，X_{ij}表示在所有语料中词汇表 V(Vocabulary)中第 i 个词和第 j 个词同时出现的词数，$|V|$为词汇表的大小。对 X 做矩阵分解（如奇异值分解，Singular Value Decomposition），得到的 U 即视为所有词的词向量：

$$X=USV^{T} \tag{6.7}$$

这样的传统做法有很多问题：

- 由于很多词没有出现，导致矩阵极其稀疏，因此需要对词频做额外处理来达到好的矩阵分解效果。
- 矩阵非常大，维度太高（通常达到 $10^6 \times 10^6$ 的数量级）。
- 需要手动去掉停用词（如although, a,...），不然这些频繁出现的词也会影响矩阵分解的效果。

基于神经网络的模型不需要计算和存储一个在全语料上统计产生的大表，而是通过学习语义信息得到词向量，因此能够很好地解决以上问题。本节将展示基于神经网络训练词向量的细节，以及如何用PaddlePaddle训练一个词向量模型。

1. 模型概述

本节介绍3个训练词向量的模型：N-gram模型、CBOW模型和Skip-gram模型。它们的中心思想都是通过上下文得到一个词出现的概率。对于N-gram模型，先介绍语言模型的概念，并在之后的训练模型中用PaddlePaddle实现它。后两个模型是近年来最有名的神经元词向量模型，由Tomas Mikolov在Google研发，虽然它们很浅很简单，但是训练效果很好。

(1) 语言模型

在介绍词向量模型之前，先来引入一个概念：语言模型。语言模型旨在为语句的联合概率函数 $P(w_1,...,w_T)$ 建模，其中 w_i 表示句子中的第 i 个词。语言模型的目标是，希望模型对有意义的句子赋予大概率，对没有意义的句子赋予小概率。这样的模型可以应用于很多领域，如机器翻译、语音识别、信息检索、词性标注、手写识别等，它们都希望能得到一个连续序列的概率。以信息检索为例，当你在搜索"how long is a football bame"时（bame 是一个医学名词），搜索引擎会提示你是否希望搜索"how long is a football game"，这是因为根据语言模型计算出"how long is a football bame"的概率很低，而与 bame 近似的，可能引起错误的词中，game 会使该句生成的概率最大。

对语言模型的目标概率 $P(w_1,...,w_T)$，如果假设文本中每个词都是相互独立的，那么整句话的联合概率可以表示为其中所有词语条件概率的乘积，即：

$$P(w_1,...,w_T) = \prod_{t=1}^{T} P(w_t) \tag{6.8}$$

然而语句中的每个词出现的概率都与其前面的词紧密相关，所以实际上通常用条件概率表示语言模型：

$$P(w_1,...,w_T) = \prod_{t=1}^{T} P(w_t \mid w_1,...,w_{t-1}) \tag{6.9}$$

(2) n-gram 神经网络模型

在计算语言学中，n-gram 是一种重要的文本表示方法，表示一个文本中连续的 n 个项。基于具体的应用场景，每一项可以是一个字母、单词或者音节。n-gram 模型也是统计语言模型中的一种重要方法，用 n-gram 训练语言模型时，一般用每个 n-gram 的历史 $n-1$ 个词语组成的内容来预测第 n 个词。

Yoshua Bengio 等科学家于 2003 年在著名论文"Neural Probabilistic Language Models"中介绍如何学习一个神经元网络表示的词向量模型。文中的神经概率语言模型（Neural Network Language Model，NNLM）通过一个线性映射和一个非线性隐藏层连接，同时学习了语言模型和词向量，即通过学习大量语料得到词语的向量表达，通过这些向量得到整个句子的概率。因为所有的词语都用一个低维向量来表示，所以用这种方法学习语言模型可以克服维度灾难（Curse of Dimensionality）。由于"神经概率语言模型"说法较为泛泛，因此在这里不用 NNLM 的本名，考虑到其具体做法，本文中称该模型为 n-gram 神经网络模型。

在上文中已经讲到用条件概率建模语言模型，即一句话中第 t 个词的概率和该句话的前 $t-1$ 个词相关。实际上，越远的词语对该词的影响越小，如果考虑一个 n-gram，那么每个词都只受其前面 $n-1$ 个词的影响，则有：

$$P(w_1,...,w_T) = \prod_{t=n}^{T} P(w_t \mid w_{t-1},w_{t-2},...,w_{t-n+1}) \tag{6.10}$$

给定一些真实语料，这些语料中都是有意义的句子，n-gram 模型的优化目标则是最大化目标函数：

$$\frac{1}{T}\sum_t f(w_t, w_{t-1}, ..., w_{t-n+1}; \theta) + R(\theta) \tag{6.11}$$

其中，$f(w_t, w_{t-1}, ..., w_{t-n+1})$ 表示根据历史 n-1 个词得到当前词 w_t 的条件概率，$R(\theta)$ 表示参数正则项。

图 6.17 展示了 n-gram 神经网络模型，从下往上看，该模型分为以下几个部分：对于每个样本，模型输入 $w_{t-n+1}, ..., w_{t-1}$，输出句子第 t 个词在字典中 $|V|$ 个词上的概率分布。

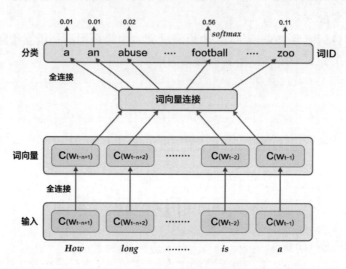

图 6.17　n-gram 神经网络模型

每个输入词 $w_{t-n+1}, ..., w_{t-1}$ 首先通过映射矩阵映射到词向量 $C(w_{t-n+1}), ..., C(w_{t-1})$。

然后所有词语的词向量拼接成一个大向量，并经过一个非线性映射得到历史词语的隐藏层表示：

$$g = U\tanh(\theta Tx + b_1) + Wx + b_2 \tag{6.12}$$

其中，x 为所有词语的词向量拼接成的大向量，表示文本历史特征；θ、U、b_1、b_2 和 W 分别为词向量层到隐藏层连接的参数；g 表示未经归一化的所有输出单词概率，g_i 表示未经归一化的字典中第 i 个单词的输出概率。

根据 Softmax 回归的定义，通过归一化 g_i，生成目标词 w_t 的概率为：

$$P(w_t | w_1, ..., w_{t-n+1}) = \frac{e^{g_{w_t}}}{\sum_i^{|V|} e^{g_i}} \tag{6.13}$$

整个网络的损失值（Cost）为多类分类交叉熵，用公式表示为：

$$J(\theta) = -\sum_{i=1}^{N}\sum_{k=1}^{|V|} y_k^i \log(\text{softmax}(g_k^i)) \tag{6.14}$$

其中，y_k^i 表示第 i 个样本第 k 类的真实标注（0 或 1），$\text{softmax}(g_k^i)$ 表示第 i 个样本第 k

类 softmax 输出的概率。

（3）Continuous Bag-of-Words model（CBOW，连续词袋模型）

CBOW 模型通过一个词的上下文（各 N 个词）预测当前词。当 N=2 时，模型如图 6.18 所示。

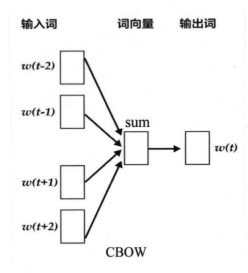

图 6.18　N=2 时 CBOW 模型

具体来说，不考虑上下文的词语输入顺序，CBOW 是用上下文词语的词向量的均值来预测当前词，即：

$$\text{Context} = \frac{x_{t-1} + x_{t-2} + x_{t+1} + x_{t+2}}{4} \tag{6.15}$$

其中，x_t 为第 t 个词的词向量，分类分数（Score）向量 $z = U * \text{context}$，最终的分类 y 采用 softmax，损失函数采用多类分类交叉熵。

（4）Skip-gram 模型

CBOW 的好处是对上下文词语的分布在词向量上进行了平滑，去掉了噪声，因此在小数据集上很有效。而 Skip-gram 的方法中，用一个词预测其上下文，得到了当前词上下文的很多样本，因此可用于更大的数据集。

如图 6.19 所示，Skip-gram 模型的具体做法是将一个词的词向量映射到 $2n$ 个词的词向量（$2n$ 表示当前输入词的前后各 n 个词），然后分别通过 softmax 得到这 $2n$ 个词的分类损失值之和。

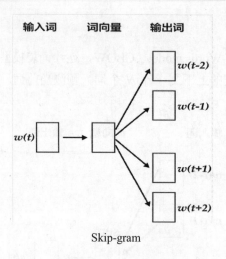

图 6.19　Skip-gram 模型

2. 数据准备

（1）数据介绍

本文使用 Penn Treebank（PTB，经 Tomas Mikolov 预处理过的版本）数据集。PTB 数据集较小，训练速度快，应用于 Mikolov 的公开语言模型训练工具中。其统计情况如下：

- 训练数据，ptb.train.txt，42068 句。
- 验证数据，ptb.valid.txt，3370 句。
- 测试数据，ptb.test.txt，3761 句。

（2）数据预处理

本节训练的是 5-gram 模型，表示在 PaddlePaddle 训练时，每条数据的前 4 个词用来预测第 5 个词。PaddlePaddle 提供了对应 PTB 数据集的 Python 包 paddle.dataset.imikolov，自动做数据的下载与预处理，方便大家使用。

预处理会把数据集中的每一句话前后加上开始符号<s>以及结束符号<e>，然后依据窗口大小（5），从头到尾每次向右滑动窗口并生成一条数据。

例如，"I have a dream that one day"一句提供了 5 条数据：

```
<s> I have a dream
I have a dream that
have a dream that one
a dream that one day
dream that one day <e>
```

最后，每个输入会按其单词在字典里的位置转化成整数的索引序列，作为 PaddlePaddle 的输入。

3. 编程实现

本配置的模型结构如图 6.20 所示。

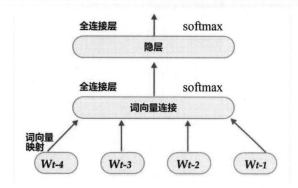

图 6.20 模型配置中的 n-gram 神经网络模型

(1) 加载所需要的包：

```
import paddle as paddle
import paddle.fluid as fluid
import six
import numpy
import math

from __future__ import print_function
```

(2) 定义参数：

```
EMBED_SIZE = 32          # embedding 维度
HIDDEN_SIZE = 256        # 隐藏层大小
N = 5                    # ngram 大小，这里固定取 5
BATCH_SIZE = 100         # batch 大小
PASS_NUM = 100           # 训练轮数

use_cuda = False         # 如果用 GPU 训练，就设置为 True

word_dict = paddle.dataset.imikolov.build_dict()
dict_size = len(word_dict)
```

更大的 BATCH_SIZE 将使得训练更快收敛，但也会消耗更多内存。由于词向量计算规模较大，如果环境允许，就开启使用 GPU 进行训练，能更快得到结果。不同于之前的 PaddlePaddle v2 版本，在新的 Fluid 版本里不必再手动计算词向量。PaddlePaddle 提供了一个内置的方法 fluid.layers.embedding，可以直接用来构造 n-gram 神经网络。

(3) 定义 n-gram 神经网络结构。这个结构在训练和预测中都会使用到。因为词向量比较稀疏，所以传入参数 is_sparse==True，可以加速稀疏矩阵的更新。

```
def inference_program(words, is_sparse):

    embed_first = fluid.layers.embedding(
```

```
        input=words[0],
        size=[dict_size, EMBED_SIZE],
        dtype='float32',
        is_sparse=is_sparse,
        param_attr='shared_w')
    embed_second = fluid.layers.embedding(
        input=words[1],
        size=[dict_size, EMBED_SIZE],
        dtype='float32',
        is_sparse=is_sparse,
        param_attr='shared_w')
    embed_third = fluid.layers.embedding(
        input=words[2],
        size=[dict_size, EMBED_SIZE],
        dtype='float32',
        is_sparse=is_sparse,
        param_attr='shared_w')
    embed_fourth = fluid.layers.embedding(
        input=words[3],
        size=[dict_size, EMBED_SIZE],
        dtype='float32',
        is_sparse=is_sparse,
        param_attr='shared_w')

    concat_embed = fluid.layers.concat(
        input=[embed_first, embed_second, embed_third, embed_fourth], axis=1)
    hidden1 = fluid.layers.fc(input=concat_embed,
                              size=HIDDEN_SIZE,
                              act='sigmoid')
    predict_word = fluid.layers.fc(input=hidden1, size=dict_size, act='softmax')
    return predict_word
```

(4) 基于以上的神经网络结构,可以定义如下的训练方法:

```
def train_program(predict_word):
    # 'next_word'的定义必须要在inference_program的声明之后,
    # 否则train program输入数据的顺序就变成了[next_word, firstw, secondw,
    # thirdw, fourthw], 这是不正确的
    next_word = fluid.layers.data(name='nextw', shape=[1], dtype='int64')
    cost = fluid.layers.cross_entropy(input=predict_word, label=next_word)
    avg_cost = fluid.layers.mean(cost)
    return avg_cost

def optimizer_func():
```

```
    return fluid.optimizer.AdagradOptimizer(
        learning_rate=3e-3,
        regularization=fluid.regularizer.L2DecayRegularizer(8e-4))
```

（5）现在可以开始训练了。PaddlePaddle 有现成的训练和测试集：paddle.dataset.imikolov.train()和 paddle.dataset.imikolov.test()。两者都会返回一个读取器。读取器是一个 Python 的函数，每次调用都会读取下一条数据，是一个 Python 的 generator。

paddle.batch 会读入一个读取器，然后输出一个批次化了的读取器。还可以在训练过程中输出每个步骤、批次的训练情况。

```
def train(if_use_cuda, params_dirname, is_sparse=True):
    place = fluid.CUDAPlace(0) if if_use_cuda else fluid.CPUPlace()

    train_reader = paddle.batch(
        paddle.dataset.imikolov.train(word_dict, N), BATCH_SIZE)
    test_reader = paddle.batch(
        paddle.dataset.imikolov.test(word_dict, N), BATCH_SIZE)

    first_word = fluid.layers.data(name='firstw', shape=[1], dtype='int64')
    second_word = fluid.layers.data(name='secondw', shape=[1], dtype='int64')
    third_word = fluid.layers.data(name='thirdw', shape=[1], dtype='int64')
    forth_word = fluid.layers.data(name='fourthw', shape=[1], dtype='int64')
    next_word = fluid.layers.data(name='nextw', shape=[1], dtype='int64')

    word_list = [first_word, second_word, third_word, forth_word, next_word]
    feed_order = ['firstw', 'secondw', 'thirdw', 'fourthw', 'nextw']

    main_program = fluid.default_main_program()
    star_program = fluid.default_startup_program()

    predict_word = inference_program(word_list, is_sparse)
    avg_cost = train_program(predict_word)
    test_program = main_program.clone(for_test=True)

    sgd_optimizer = optimizer_func()
    sgd_optimizer.minimize(avg_cost)

    exe = fluid.Executor(place)

    def train_test(program, reader):
        count = 0
        feed_var_list = [
            program.global_block().var(var_name) for var_name in feed_order
```

```python
    ]
    feeder_test = fluid.DataFeeder(feed_list=feed_var_list, place=place)
    test_exe = fluid.Executor(place)
    accumulated = len([avg_cost]) * [0]
    for test_data in reader():
        avg_cost_np = test_exe.run(
            program=program,
            feed=feeder_test.feed(test_data),
            fetch_list=[avg_cost])
        accumulated = [
            x[0] + x[1][0] for x in zip(accumulated, avg_cost_np)
        ]
        count += 1
    return [x / count for x in accumulated]

def train_loop():
    step = 0
    feed_var_list_loop = [
        main_program.global_block().var(var_name) for var_name in feed_order
    ]
    feeder = fluid.DataFeeder(feed_list=feed_var_list_loop, place=place)
    exe.run(star_program)
    for pass_id in range(PASS_NUM):
        for data in train_reader():
            avg_cost_np = exe.run(
                main_program, feed=feeder.feed(data), fetch_list=[avg_cost])

            if step % 10 == 0:
                outs = train_test(test_program, test_reader)

                print("Step %d: Average Cost %f" % (step, outs[0]))

                # 整个训练过程要花费几个小时，如果平均损失低于5.8，
                # 就认为模型已经达到很好的效果，可以停止训练了。
                # 注意，5.8是一个相对较高的值，为了获取更好的模型，
                # 可以将这里的阈值设为3.5，但训练时间也会更长。
                if outs[0] < 5.8:
                    if params_dirname is not None:
                        fluid.io.save_inference_model(params_dirname, [
                            'firstw', 'secondw', 'thirdw', 'fourthw'
                        ], [predict_word], exe)
                    return
            step += 1
```

```
            if math.isnan(float(avg_cost_np[0])):
                sys.exit("got NaN loss, training failed.")

    raise AssertionError("Cost is too large {0:2.2}".format(avg_cost_np[0]))

train_loop()
```

train_loop 将会开始训练，打印训练过程的日志如下：

```
Step 0: Average Cost 7.337213
Step 10: Average Cost 6.136128
Step 20: Average Cost 5.766995
...
```

4. 模型应用

在模型训练后，可以用来做一些预测。

可以用训练过的模型在得知之前的 n-gram 后预测下一个词。

```
def infer(use_cuda, params_dirname=None):
    place = fluid.CUDAPlace(0) if use_cuda else fluid.CPUPlace()

    exe = fluid.Executor(place)

    inference_scope = fluid.core.Scope()
    with fluid.scope_guard(inference_scope):
        # 使用 fluid.io.load_inference_model 获取 inference program,
        # feed 变量的名称 feed_target_names 和从 scope 中 fetch 的对象 fetch_targets
        [inferencer, feed_target_names,
         fetch_targets] = fluid.io.load_inference_model(params_dirname, exe)

        # 设置输入，用 4 个 LoDTensor 来表示 4 个词语。这里每个词都是一个 id,
        # 用来查询 embedding 表获取对应的词向量，因此其形状大小是[1]。
        # recursive_sequence_lengths 设置的是基于长度的 LoD，因此都应该设为[[1]]
        # 注意，recursive_sequence_lengths 是列表的列表
        data1 = [[211]]  # 'among'
        data2 = [[6]]    # 'a'
        data3 = [[96]]   # 'group'
        data4 = [[4]]    # 'of'
        lod = [[1]]

        first_word = fluid.create_lod_tensor(data1, lod, place)
        second_word = fluid.create_lod_tensor(data2, lod, place)
        third_word = fluid.create_lod_tensor(data3, lod, place)
        fourth_word = fluid.create_lod_tensor(data4, lod, place)
```

```python
    assert feed_target_names[0] == 'firstw'
    assert feed_target_names[1] == 'secondw'
    assert feed_target_names[2] == 'thirdw'
    assert feed_target_names[3] == 'fourthw'

    # 构造 feed 词典 {feed_target_name: feed_target_data}
    # 预测结果包含在 results 之中
    results = exe.run(
        inferencer,
        feed={
            feed_target_names[0]: first_word,
            feed_target_names[1]: second_word,
            feed_target_names[2]: third_word,
            feed_target_names[3]: fourth_word
        },
        fetch_list=fetch_targets,
        return_numpy=False)

    print(numpy.array(results[0]))
    most_possible_word_index = numpy.argmax(results[0])
    print(most_possible_word_index)
    print([
        key for key, value in six.iteritems(word_dict)
        if value == most_possible_word_index
    ][0])
```

由于词向量矩阵本身比较稀疏，因此训练的过程如果要达到一定的精度耗时会比较长。为了能简单看到效果，教程只设置了经过很少的训练就结束并得到如下的预测。模型预测 among a group of 的下一个词是 the。这比较符合文法规律。如果训练时间更长，比如几个小时，那么会得到的下一个预测是 workers。预测输出的格式如下：

```
[[0.03768077 0.03463154 0.00018074 ... 0.00022283 0.00029888 0.02967956]]
0
the
```

其中，第一行表示预测词在词典上的概率分布，第二行表示概率最大的词对应的 id，第三行表示概率最大的词。

整个程序的入口：

```python
def main(use_cuda, is_sparse):
    if use_cuda and not fluid.core.is_compiled_with_cuda():
        return

    params_dirname = "word2vec.inference.model"
```

```
train(
    if_use_cuda=use_cuda,
    params_dirname=params_dirname,
    is_sparse=is_sparse)

infer(use_cuda=use_cuda, params_dirname=params_dirname)

main(use_cuda=use_cuda, is_sparse=True)
```

本节中，我们介绍了词向量、语言模型和词向量的关系以及如何通过训练神经网络模型获得词向量。在信息检索中，可以根据向量间的余弦夹角来判断 query 和文档关键词这两者间的相关性。在句法分析和语义分析中，训练好的词向量可以用来初始化模型，以得到更好的效果。在文档分类中，有了词向量之后，可以用聚类的方法将文档中同义词进行分组，也可以用 n-gram 来预测下一个词。

6.5 PaddlePaddle 平台个性化推荐

项目源代码的地址为 https://github.com/PaddlePaddle/book/tree/develop/05.recommender_system。

在网络技术不断发展和电子商务规模不断扩大的背景下，商品数量和种类快速增长，用户需要花费大量时间才能找到自己想买的商品，这就是信息超载问题。为了解决这个难题，个性化推荐系统（Recommender System）应运而生。

个性化推荐系统是信息过滤系统（Information Filtering System）的子集，可以用在很多领域，如电影、音乐、电商和 Feed 流推荐等。个性化推荐系统通过分析、挖掘用户行为，发现用户的个性化需求与兴趣特点，将用户可能感兴趣的信息或商品推荐给用户。与搜索引擎不同，个性化推荐系统不需要用户准确地描述出自己的需求，而是根据用户的历史行为进行建模，主动提供满足用户兴趣和需求的信息。

1994 年明尼苏达大学推出的 GroupLens 系统一般被认为是个性化推荐系统成为一个相对独立的研究方向的标志。该系统首次提出了基于协同过滤来完成推荐任务的思想。此后，基于该模型的协同过滤推荐引领了个性化推荐系统十几年的发展方向。

传统的个性化推荐系统方法主要有：

- 协同过滤推荐（Collaborative Filtering Recommendation）：该方法是应用最广泛的技术之一，需要收集和分析用户的历史行为、活动和偏好。它通常可以分为两个子类：基于用户（User-Based）的推荐和基于物品（Item-Based）的推荐。该方法的一个关键优势是它不依赖于机器去分析物品的内容特征，因此无须理解物品本身也能够准确地推荐诸如电影之类的复杂物品；缺点是对于没有任何行为的新用户存在冷启动的问题，同时也存在用户与商品之间的交互数据不够多造成的稀疏问题。值得一提的是，社交网络或地理位置等上下文信息都可以结合到协同过滤中去。

- 基于内容过滤推荐（Content-based Filtering Recommendation）：该方法利用商品的内容描述，抽象出有意义的特征，通过计算用户的兴趣和商品描述之间的相似度来给用户做推荐。优点是简单直接，不需要依据其他用户对商品的评价，而是通过商品属性进行商品相似度度量，从而推荐给用户所感兴趣商品的相似商品；缺点是对于没有任何行为的新用户同样存在冷启动的问题。
- 组合推荐（Hybrid Recommendation）：运用不同的输入和技术共同进行推荐，以弥补各自推荐技术的缺点。

近些年来，深度学习在很多领域都取得了巨大的成功。学术界和工业界都在尝试将深度学习应用于个性化推荐系统领域中。深度学习具有优秀的自动提取特征的能力，能够学习多层次的抽象特征表示，并对异质或跨域的内容信息进行学习，可以在一定程度上处理个性化推荐系统冷启动问题。本节主要介绍个性化推荐的深度学习模型，以及如何使用 PaddlePaddle 实现模型。

使用包含用户信息、电影信息与电影评分的数据集作为个性化推荐的应用场景。当训练好模型后，只需要输入对应的用户 ID 和电影 ID，就可以得出一个匹配的分数（范围[0,5]，分数越高视为兴趣越大），然后根据所有电影的推荐得分排序，推荐给用户可能感兴趣的电影。

1. 模型概述

首先介绍 YouTube 的视频个性化推荐系统，然后介绍实现的融合推荐模型。

（1）YouTube 的深度神经网络个性化推荐系统

YouTube 是世界上最大的视频上传、分享和发现网站。YouTube 个性化推荐系统为超过 10 亿用户从不断增长的视频库中推荐个性化的内容。整个系统由两个神经网络组成：候选生成网络和排序网络。候选生成网络从百万量级的视频库中生成上百个候选，排序网络对候选进行打分排序，输出排名最高的数十个结果。

① 候选生成网络（Candidate Generation Network）

候选生成网络将推荐问题建模为一个类别数极大的多类分类问题：对于一个 YouTube 用户，使用其观看历史（视频 ID）、搜索词记录（Search Tokens）、人口学信息（如地理位置、用户登录设备）、二值特征（如性别，是否登录）和连续特征（如用户年龄）等，对视频库中所有视频进行多分类，得到每一类别的分类结果（每一个视频的推荐概率），最终输出概率较高的几百个视频。

首先，将观看历史及搜索词记录这类历史信息映射为向量后取平均值得到定长表示；同时，输入人口学特征以优化新用户的推荐效果，并将二值特征和连续特征归一化处理到[0, 1]范围。接下来，将所有特征表示拼接为一个向量，并输入给非线性多层感知器（MLP，详见识别数字教程）处理。最后，训练时将 MLP 的输出给 softmax 进行分类，预测时计算用户的综合特征（MLP 的输出）与所有视频的相似度，取得分最高的 k 个作为候选生成网络的筛选结果。

图 6.21 显示了候选生成网络结构。

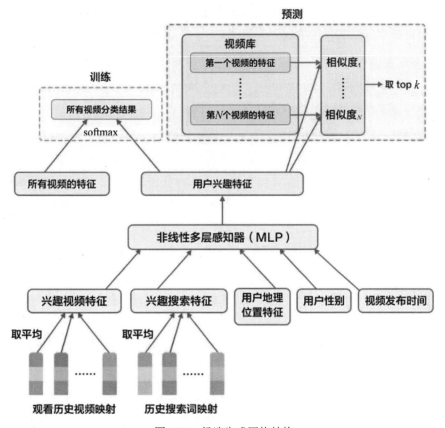

图 6.21 候选生成网络结构

对于一个用户 U，预测此刻用户要观看的视频 ω 为视频 i 的概率公式为：

$$P(\omega=i|u)=\frac{e^{v_i u}}{\sum j\in V^{e^{v_j u}}} \quad (6.16)$$

其中，u 为用户 U 的特征表示，V 为视频库集合，v_i 为视频库中第 i 个视频的特征表示。u 和 v_i 为长度相等的向量，两者点积可以通过全连接层实现。考虑到 softmax 分类的类别数非常多，为了保证一定的计算效率，要做到以下两点：

- 训练阶段，使用负样本类别采样将实际计算的类别数缩小至数千。
- 推荐（预测）阶段，忽略 softmax 的归一化计算（不影响结果），将类别打分问题简化为点积（Dot Product）空间中的最近邻（Nearest Neighbor）搜索问题，取与 u 最近的 k 个视频作为生成的候选。

② 排序网络（Ranking Network）

排序网络的结构类似于候选生成网络，但是它的目标是对候选进行更细致的打分排序。和传统广告排序中的特征抽取方法类似，这里也构造了大量的用于视频排序的相关特征（如视频

ID、上次观看时间等）。这些特征的处理方式和候选生成网络类似，不同之处是排序网络的顶部是一个加权逻辑回归（Weighted Logistic Regression），它对所有候选视频进行打分，从高到低排序后将分数较高的一些视频返回给用户。

（2）融合推荐模型

本节使用卷积神经网络（Convolutional Neural Networks，CNN）来学习电影名称的表示。下面会依次介绍文本卷积神经网络以及融合推荐模型。

① 卷积神经网络经常用来处理具有类似网格拓扑结构（Grid-like Topology）的数据。例如，图像可以视为二维网格的像素点，自然语言可以视为一维的词序列。卷积神经网络可以提取多种局部特征，并对其进行组合抽象得到更高级的特征表示。实验表明，卷积神经网络能高效地对图像及文本问题进行建模处理。卷积神经网络主要由卷积（Convolution）和池化（Pooling）操作构成，其应用及组合方式灵活多变，种类繁多。这里以图 6.22 所示的网络进行讲解。

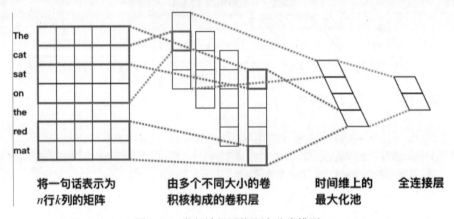

图 6.22 卷积神经网络文本分类模型

假设待处理句子的长度为 n，其中第 i 个词的词向量为 $x_i \in \mathbb{R}^k$，k 为维度大小。

首先，进行词向量的拼接操作：将每 h 个词拼接起来形成一个大小为 h 的词窗口，记为 $x_{i:i+h-1}$，它表示词序列 $x_i, x_{i+1}, \ldots, x_{i+h-1}$ 的拼接，其中，i 表示词窗口中第一个词在整个句子中的位置，取值范围从 1 到 $n-h+1$，$x_{i:i+h-1} \in \mathbb{R}^{hk}$。

其次，进行卷积操作：把卷积核(kernel) $w \in \mathbb{R}^{hk}$ 应用于包含 h 个词的窗口 $x_{i:i+h-1}$，得到特征 $c_i = f(w \cdot x_{i:i+h-1} + b)$，其中 $b \in \mathbb{R}$ 为偏置项(bias)，f 为非线性激活函数，如 Sigmoid。将卷积核应用于句子中所有的词窗口 $x_{1:h}, x_{2:h+1}, \ldots, x_{n-h+1:n}$，产生一个特征图（Feature Map）：

$$c = [c_1, c_2, \ldots, c_{n-h+1}], c \in \mathbb{R}^{n-h+1} \tag{6.17}$$

接下来，对特征图采用时间维度上的最大池化（Max Pooling Over Time）操作得到此卷积核对应的整句话的特征 $\hat{c}$，它是特征图中所有元素的最大值：

$$\hat{c} = \max(c) \tag{6.18}$$

② 在融合推荐模型的电影个性化推荐系统中（见图 6.23），首先使用用户特征和电影特征作为神经网络的输入，其中用户特征融合了 4 个属性信息，分别是用户 ID、性别、职业和

年龄。电影特征融合了 3 个属性信息，分别是电影 ID、电影类型 ID 和电影名称。对于用户特征，将用户 ID 映射为维度大小为 256 的向量表示，输入全连接层，并对其他 3 个属性也做类似的处理。然后将 4 个属性的特征表示分别全连接并相加。对于电影特征，将电影 ID 以类似用户 ID 的方式进行处理，电影类型 ID 以向量的形式直接输入全连接层，电影名称用文本卷积神经网络得到其定长向量表示。然后将 3 个属性的特征表示分别全连接并相加。得到用户和电影的向量表示后，计算二者的余弦相似度作为个性化推荐系统的打分。最后，用该相似度打分和用户真实打分差异的平方作为回归模型的损失函数。

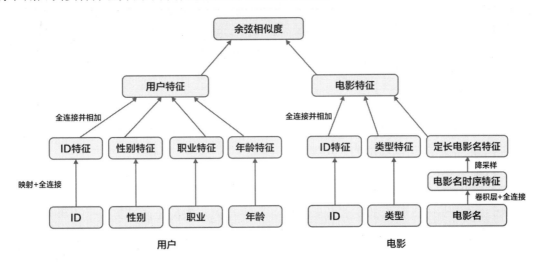

图 6.23　融合推荐模型

2. 数据准备

本项目在 Python2 环境下完成。

以 MovieLens（https://github.com/PaddlePaddle/book/tree/develop/05.recommender_system）百万数据集（ml-1m）为例进行介绍。ml-1m 数据集包含了 6 000 个用户对 4 000 部电影的 1 000 000 条评价（评分范围 1~5 分，均为整数），由 GroupLens Research 实验室搜集整理。

Paddle 在 API 中提供了自动加载数据的模块，数据模块为 paddle.dataset.movielens：

```
import paddle
movie_info = paddle.dataset.movielens.movie_info()
print movie_info.values()[0]
# Run this block to show dataset's documentation
# help(paddle.dataset.movielens)
```

在原始数据中包含电影的特征数据、用户的特征数据和用户对电影的评分。

例如，其中某一个电影特征为：

```
movie_info = paddle.dataset.movielens.movie_info()
print movie_info.values()[0]
```

```
<MovieInfo id(1), title(Toy Story ), categories(['Animation', "Children's",
'Comedy'])>
```

这表示用户 ID 是 1、女性、年龄比 18 岁还小、职业 ID 是 10。

对于每一条训练/测试数据，均为 <用户特征> + <电影特征> + 评分。

例如，我们获得第一条训练数据：

```
train_set_creator = paddle.dataset.movielens.train()
train_sample = next(train_set_creator())
uid = train_sample[0]
mov_id = train_sample[len(user_info[uid].value())]
print "User %s rates Movie %s with Score %s"%(user_info[uid], movie_info[mov_id],
train_sample[-1])
```

```
User <UserInfo id(1), gender(F), age(1), job(10)> rates Movie <MovieInfo id(1193),
title(One Flew Over the Cuckoo's Nest ), categories(['Drama'])> with Score [5.0]
```

用户 1 对电影 1193 的评价为 5 分。

3. 模型配置

下面开始根据输入数据的形式配置模型。首先引入所需的库函数以及定义全局变量。
IS_SPARSE: embedding 中是否使用稀疏更新， PASS_NUM: epoch 数量。

```
from __future__ import print_function
import math
import sys
import numpy as np
import paddle
import paddle.fluid as fluid
import paddle.fluid.layers as layers
import paddle.fluid.nets as nets

IS_SPARSE = True
BATCH_SIZE = 256
PASS_NUM = 20
```

然后为用户特征综合模型定义模型配置：

```
def get_usr_combined_features():
    """network definition for user part"""

    USR_DICT_SIZE = paddle.dataset.movielens.max_user_id() + 1

    uid = layers.data(name='user_id', shape=[1], dtype='int64')
```

```python
usr_emb = layers.embedding(
    input=uid,
    dtype='float32',
    size=[USR_DICT_SIZE, 32],
    param_attr='user_table',
    is_sparse=IS_SPARSE)

usr_fc = layers.fc(input=usr_emb, size=32)

USR_GENDER_DICT_SIZE = 2

usr_gender_id = layers.data(name='gender_id', shape=[1], dtype='int64')

usr_gender_emb = layers.embedding(
    input=usr_gender_id,
    size=[USR_GENDER_DICT_SIZE, 16],
    param_attr='gender_table',
    is_sparse=IS_SPARSE)

usr_gender_fc = layers.fc(input=usr_gender_emb, size=16)

USR_AGE_DICT_SIZE = len(paddle.dataset.movielens.age_table)
usr_age_id = layers.data(name='age_id', shape=[1], dtype="int64")

usr_age_emb = layers.embedding(
    input=usr_age_id,
    size=[USR_AGE_DICT_SIZE, 16],
    is_sparse=IS_SPARSE,
    param_attr='age_table')

usr_age_fc = layers.fc(input=usr_age_emb, size=16)

USR_JOB_DICT_SIZE = paddle.dataset.movielens.max_job_id() + 1
usr_job_id = layers.data(name='job_id', shape=[1], dtype="int64")

usr_job_emb = layers.embedding(
    input=usr_job_id,
    size=[USR_JOB_DICT_SIZE, 16],
    param_attr='job_table',
    is_sparse=IS_SPARSE)

usr_job_fc = layers.fc(input=usr_job_emb, size=16)
```

```
    concat_embed = layers.concat(
        input=[usr_fc, usr_gender_fc, usr_age_fc, usr_job_fc], axis=1)

    usr_combined_features = layers.fc(input=concat_embed, size=200, act="tanh")

    return usr_combined_features
```

如上述代码所示，对于每个用户，输入 4 维特征，其中包括 user_id、gender_id、age_id、job_id。这几维特征均是简单的整数值。为了后续神经网络处理这些特征方便，借鉴 NLP 中的语言模型，将这几维离散的整数值变换成 embedding 取出，分别形成 usr_emb、usr_gender_emb、usr_age_emb、usr_job_emb。

然后，对于所有的用户特征，均输入到一个全连接层（fc）中。将所有特征融合为一个 200 维度的特征。

接着，对每一个电影特征做类似的变换，网络配置为：

```
def get_mov_combined_features():
    """network definition for item(movie) part"""

    MOV_DICT_SIZE = paddle.dataset.movielens.max_movie_id() + 1

    mov_id = layers.data(name='movie_id', shape=[1], dtype='int64')

    mov_emb = layers.embedding(
        input=mov_id,
        dtype='float32',
        size=[MOV_DICT_SIZE, 32],
        param_attr='movie_table',
        is_sparse=IS_SPARSE)

    mov_fc = layers.fc(input=mov_emb, size=32)

    CATEGORY_DICT_SIZE = len(paddle.dataset.movielens.movie_categories())

    category_id = layers.data(
        name='category_id', shape=[1], dtype='int64', lod_level=1)

    mov_categories_emb = layers.embedding(
        input=category_id, size=[CATEGORY_DICT_SIZE, 32], is_sparse=IS_SPARSE)

    mov_categories_hidden = layers.sequence_pool(
        input=mov_categories_emb, pool_type="sum")

    MOV_TITLE_DICT_SIZE = len(paddle.dataset.movielens.get_movie_title_dict())
```

```python
mov_title_id = layers.data(
    name='movie_title', shape=[1], dtype='int64', lod_level=1)

mov_title_emb = layers.embedding(
    input=mov_title_id, size=[MOV_TITLE_DICT_SIZE, 32], is_sparse=IS_SPARSE)

mov_title_conv = nets.sequence_conv_pool(
    input=mov_title_emb,
    num_filters=32,
    filter_size=3,
    act="tanh",
    pool_type="sum")

concat_embed = layers.concat(
    input=[mov_fc, mov_categories_hidden, mov_title_conv], axis=1)

mov_combined_features = layers.fc(input=concat_embed, size=200, act="tanh")

return mov_combined_features
```

电影标题名称（title）是一个序列的整数，代表的是这个词在索引序列中的下标。这个序列会被送入 sequence_conv_pool 层，这个层会在时间维度上使用卷积和池化。所以输出会是固定长度，尽管输入的序列长度各不相同。

最后，定义一个 inference_program 来使用余弦相似度计算用户特征与电影特征的相似性。

```python
def inference_program():
    """the combined network"""

    usr_combined_features = get_usr_combined_features()
    mov_combined_features = get_mov_combined_features()

    inference = layers.cos_sim(X=usr_combined_features, Y=mov_combined_features)
    scale_infer = layers.scale(x=inference, scale=5.0)

    return scale_infer
```

再定义一个 train_program 来使用 inference_program 计算出结果，在标注数据的帮助下来计算误差；定义一个 optimizer_func 来定义优化器。

```python
def train_program():
    """define the cost function"""

    scale_infer = inference_program()
```

```
    label = layers.data(name='score', shape=[1], dtype='float32')
    square_cost = layers.square_error_cost(input=scale_infer, label=label)
    avg_cost = layers.mean(square_cost)

    return [avg_cost, scale_infer]

def optimizer_func():
    return fluid.optimizer.SGD(learning_rate=0.2)
```

4. 训练模型

(1) 定义训练环境

定义训练环境可以指定训练是发生在 CPU 还是 GPU 上。

```
use_cuda = False
place = fluid.CUDAPlace(0) if use_cuda else fluid.CPUPlace()
```

(2) 定义数据提供器

为训练和测试定义数据提供器。提供器读入一个大小为 BATCH_SIZE 的数据。paddle.dataset.movielens.train 每次会在乱序化后提供一个大小为 BATCH_SIZE 的数据，乱序化的大小为缓存大小 buf_size。

```
train_reader = paddle.batch(
    paddle.reader.shuffle(
        paddle.dataset.movielens.train(), buf_size=8192),
    batch_size=BATCH_SIZE)

test_reader = paddle.batch(
    paddle.dataset.movielens.test(), batch_size=BATCH_SIZE)
```

5. 构造训练过程（trainer）

(1) 提供数据

feed_order 用来定义每条产生的数据和 paddle.layer.data 之间的映射关系。比如，movielens.train 产生的第一列数据对应的是 user_id 这个特征。

```
feed_order = [
    'user_id', 'gender_id', 'age_id', 'job_id', 'movie_id', 'category_id',
    'movie_title', 'score'
]
```

(2) 构建训练程序以及测试程序

分别构建训练程序和测试程序，并引入训练优化器。

```
main_program = fluid.default_main_program()
star_program = fluid.default_startup_program()
```

```python
[avg_cost, scale_infer] = train_program()

test_program = main_program.clone(for_test=True)
sgd_optimizer = optimizer_func()
sgd_optimizer.minimize(avg_cost)
exe = fluid.Executor(place)

def train_test(program, reader):
    count = 0
    feed_var_list = [
        program.global_block().var(var_name) for var_name in feed_order
    ]
    feeder_test = fluid.DataFeeder(
    feed_list=feed_var_list, place=place)
    test_exe = fluid.Executor(place)
    accumulated = 0
    for test_data in reader():
        avg_cost_np = test_exe.run(program=program,
                                    feed=feeder_test.feed(test_data),
                                    fetch_list=[avg_cost])
        accumulated += avg_cost_np[0]
        count += 1
    return accumulated / count
```

(3) 构建训练主循环并开始训练

根据上面定义的训练循环数（PASS_NUM）和一些别的参数来进行训练循环，并且每次循环都进行一次测试，当测试结果足够好时退出训练并保存训练好的参数。

```python
# Specify the directory path to save the parameters
params_dirname = "recommender_system.inference.model"

from paddle.utils.plot import Ploter
train_prompt = "Train cost"
test_prompt = "Test cost"

plot_cost = Ploter(train_prompt, test_prompt)

def train_loop():
    feed_list = [
        main_program.global_block().var(var_name) for var_name in feed_order
    ]
    feeder = fluid.DataFeeder(feed_list, place)
    exe.run(star_program)
```

```
    for pass_id in range(PASS_NUM):
        for batch_id, data in enumerate(train_reader()):
            # train a mini-batch
            outs = exe.run(program=main_program,
                           feed=feeder.feed(data),
                           fetch_list=[avg_cost])
            out = np.array(outs[0])

            # get test avg_cost
            test_avg_cost = train_test(test_program, test_reader)

            plot_cost.append(train_prompt, batch_id, outs[0])
            plot_cost.append(test_prompt, batch_id, test_avg_cost)
            plot_cost.plot()

            if batch_id == 20:
                if params_dirname is not None:
                    fluid.io.save_inference_model(params_dirname, [
                        "user_id", "gender_id", "age_id", "job_id",
                        "movie_id", "category_id", "movie_title"
                    ], [scale_infer], exe)
                return
            print('EpochID {0}, BatchID {1}, Test Loss {2:0.2}'.format(
                    pass_id + 1, batch_id + 1, float(test_avg_cost)))

            if math.isnan(float(out[0])):
                sys.exit("got NaN loss, training failed.")
```

开始训练：

```
train_loop()
```

6. 应用模型

（1）生成测试数据

使用 create_lod_tensor(data, lod, place) 的 API 来生成细节层次的张量。data 是一个序列，每个元素是一个索引号的序列。lod 是细节层次的信息，对应于 data。比如，data = [[10, 2, 3], [2, 3]] 意味着它包含两个序列，长度分别是 3 和 2，lod = [[3, 2]]表明其包含一层细节信息，意味着 data 有两个序列，长度分别是 3 和 2。

在这个预测例子中，试着预测用户 ID 为 1 的用户对于电影'Hunchback of Notre Dame'的评分：

```
infer_movie_id = 783
infer_movie_name = paddle.dataset.movielens.movie_info()[infer_movie_id].title
user_id = fluid.create_lod_tensor([[1]], [[1]], place)
```

```
gender_id = fluid.create_lod_tensor([[1]], [[1]], place)
age_id = fluid.create_lod_tensor([[0]], [[1]], place)
job_id = fluid.create_lod_tensor([[10]], [[1]], place)
movie_id = fluid.create_lod_tensor([[783]], [[1]], place) # Hunchback of Notre Dame
category_id = fluid.create_lod_tensor([[10, 8, 9]], [[3]], place) # Animation,
Children's, Musical
movie_title = fluid.create_lod_tensor([[1069, 4140, 2923, 710, 988]], [[5]],
                    place) # 'hunchback','of','notre','dame','the'
```

（2）构建预测过程并测试

与训练过程类似，需要构建一个预测过程。其中，params_dirname 是之前用来存放训练过程中各个参数的地址。

```
place = fluid.CUDAPlace(0) if use_cuda else fluid.CPUPlace()
exe = fluid.Executor(place)

inference_scope = fluid.core.Scope()
```

（3）测试

现在可以进行预测了，要提供的 feed_order 应该和训练过程一致。

```
with fluid.scope_guard(inference_scope):
    [inferencer, feed_target_names,
    fetch_targets] = fluid.io.load_inference_model(params_dirname, exe)

    results = exe.run(inferencer,
                    feed={
                        'user_id': user_id,
                        'gender_id': gender_id,
                        'age_id': age_id,
                        'job_id': job_id,
                        'movie_id': movie_id,
                        'category_id': category_id,
                        'movie_title': movie_title
                    },
                    fetch_list=fetch_targets,
                    return_numpy=False)
    predict_rating = np.array(results[0])
    print("Predict Rating of user id 1 on movie \"" + infer_movie_name +
            "\" is " + str(predict_rating[0][0]))
    print("Actual Rating of user id 1 on movie \"" + infer_movie_name +
            "\" is 4.")
```

本节介绍了传统的个性化推荐系统方法和 YouTube 的深度神经网络个性化推荐系统，并以电影推荐为例，使用 PaddlePaddle 训练了一个个性化推荐神经网络模型。个性化推荐系统几乎涵盖了电商系统、社交网络、广告推荐、搜索引擎等领域的方方面面，而在图像处理、自然语言处理等领域已经发挥重要作用的深度学习技术也将会在个性化推荐系统领域大放异彩。

6.6 PaddlePaddle 平台情感分析

项目源代码地址为 https://github.com/PaddlePaddle/book/tree/develop/06.understand_sentiment。

在自然语言处理中，情感分析一般是指判断一段文本所表达的情绪状态。其中，一段文本可以是一个句子、一个段落或一个文档。情绪状态可以是两类，如（正面，负面）、（高兴，悲伤）；也可以是三类，如（积极，消极，中性）等。情感分析的应用场景十分广泛，如把用户在购物网站（亚马逊、天猫、淘宝等）、旅游网站、电影评论网站上发表的评论分成正面评论和负面评论；或为了分析用户对于某一产品的整体使用感受，抓取产品的用户评论并进行情感分析等。

在自然语言处理中，情感分析属于典型的文本分类问题，即把需要进行情感分析的文本划分为其所属类别。文本分类涉及文本表示和分类方法两个问题。在深度学习的方法出现之前，主流的文本表示方法为词袋模型 BOW（Bag Of Words）、话题模型等；分类方法有 SVM（Support Vector Machine）、LR（Logistic Regression）等。

对于一段文本，BOW 表示会忽略其词顺序、语法和句法，将这段文本仅看作是一个词集合，因此 BOW 方法并不能充分表示文本的语义信息。例如，句子"这部电影糟糕透了"和"一个乏味，空洞，没有内涵的作品"在情感分析中具有很高的语义相似度，但是它们的 BOW 表示的相似度为 0。又如，句子"一个空洞，没有内涵的作品"和"一个不空洞而且有内涵的作品"的 BOW 相似度很高，但实际上它们的意思很不一样。

本节所要介绍的深度学习模型克服了 BOW 表示的上述缺陷，它在考虑词顺序的基础上把文本映射到低维度的语义空间，并且以端对端（End to End）的方式进行文本表示及分类，其性能相对于传统方法有显著的提升。

1. 模型

这里所使用的文本表示模型为卷积神经网络（Convolutional Neural Networks，CNN）和循环神经网络及其扩展。

（1）循环神经网络（Recurrent Neural Networks，RNN）

循环神经网络是一种能对序列数据进行精确建模的有力工具。实际上，循环神经网络的理论计算能力是图灵完备的。自然语言是一种典型的序列数据（词序列）。近年来循环神经网络及其变体（如 Long Short Term Memory，即 LSTM 等）在自然语言处理的多个领域上均表现优异，如语言模型、句法解析、语义角色标注（或一般的序列标注）、语义表示、图文生成、对话、机器翻译等，甚至成为目前效果最好的方法之一。

循环神经网络按时间展开后如图 6.24 所示：在第 t 时刻，网络读入第 t 个输入 x_t（向量表示）及前一时刻隐藏层的状态值 h_{t-1}（向量表示，h_0 一般初始化为 0 向量），计算得出本时刻隐藏层的状态值 h_t，重复这一步骤直至读完所有输入。如果将循环神经网络所表示的函数记为 f，那么其公式可表示为：

$$h_t = f(x_t, h_{t-1}) = \sigma(W_{xh}x_t + W_{hh}h_{t-1} + b_h) \tag{6.19}$$

其中，W_{xh} 是输入到隐藏层的矩阵参数，W_{hh} 是隐藏层到隐藏层的矩阵参数，b_h 为隐藏层的偏置向量（bias）参数，σ 为 sigmoid 函数。

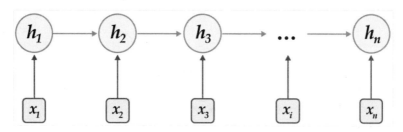

图 6.24 循环神经网络按时间展开的示意图

在处理自然语言时，一般会先将词（one-hot 表示）映射为词向量表示，然后作为循环神经网络每一时刻的输入 x_t。此外，可以根据实际需要的不同在循环神经网络的隐藏层上连接其他层。例如，可以把一个循环神经网络的隐藏层输出连接至下一个循环神经网络的输入构建深层（deep or stacked）循环神经网络，或者提取最后一个时刻的隐藏层状态作为句子表示，进而使用分类模型等。

（2）长短期记忆网络（Long Short Term Memory，LSTM）

对于较长的序列数据，循环神经网络的训练过程中容易出现梯度消失或爆炸现象。为了解决这一问题，Hochreiter S, Schmidhuber J.（1997）提出了 LSTM。

相比于简单的循环神经网络，LSTM 增加了记忆单元 c、输入门 i、遗忘门 f 及输出门 o。这些门及记忆单元组合起来大大提升了循环神经网络处理长序列数据的能力。若将基于 LSTM 的循环神经网络表示的函数记为 F，则其公式为：

$$h_t = F(x_t, h_{t-1}) \tag{6.20}$$

F 由下列公式组合而成：

$$i_t = \sigma(W_{xi}x_t + W_{hi}h_{t-1} + W_{ci}c_{t-1} + b_i)$$
$$f_t = \sigma(W_{xf}x_t + W_{hf}h_{t-1} + W_{cf}c_{t-1} + b_f)$$
$$c_t = f_t \odot c_{t-1} + i_t \odot \tanh(W_{xc}x_t + W_{hc}h_{t-1} + b_c)$$
$$o_t = \sigma(W_{xo}x_t + W_{ho}h_{t-1} + W_{co}c_t + b_o)$$
$$h_t = o_t \odot \tanh(c_t) \tag{6.21}$$

其中，i_t、f_t、c_t、o_t 分别表示输入门、遗忘门、记忆单元及输出门的向量值，带角标的 W

及 b 为模型参数，tanh 为双曲正切函数，$\odot$ 表示逐元素（Elementwise）的乘法操作。输入门控制着新输入进入记忆单元 c 的强度，遗忘门控制着记忆单元维持上一时刻值的强度，输出门控制着输出记忆单元的强度。三种门的计算方式虽然类似，但是具有完全不同的参数，各自以不同的方式控制着记忆单元 c，如图 6.25 所示。

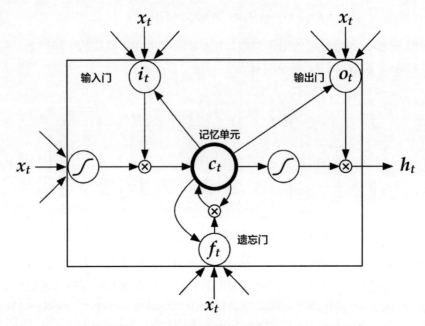

图 6.25　时刻 t 的 LSTM

LSTM 通过给简单的循环神经网络增加记忆及控制门的方式，增强了其处理远距离依赖问题的能力。类似原理的改进还有 Gated Recurrent Unit（GRU），其设计更为简洁一些。这些改进虽然各有不同，但是宏观描述却与简单的循环神经网络一样，即隐状态依据当前输入及前一时刻的隐状态来改变，不断地循环这一过程直至输入处理完毕：

$$h_t = \text{Recurrent}(x_t, h_{t-1}) \tag{6.22}$$

其中，Recurrent 可以表示简单的循环神经网络、GRU 或 LSTM。

（3）栈式双向 LSTM（Stacked Bidirectional LSTM）

对于正常顺序的循环神经网络，h_t 包含了 t 时刻之前的输入信息，也就是上文信息。同样，为了得到下文信息，可以使用反方向（将输入逆序处理）的循环神经网络。结合构建深层循环神经网络的方法（深层神经网络往往能得到更抽象和高级的特征表示），可以通过构建更加强有力的基于 LSTM 的栈式双向循环神经网络来对时序数据进行建模。

如图 6.26 所示（以 3 层为例），奇数层 LSTM 正向，偶数层 LSTM 反向，高一层的 LSTM 使用低一层 LSTM 及之前所有层的信息作为输入，对最高层 LSTM 序列使用时间维度上的最大池化即可得到文本的定长向量表示（这一表示充分融合了文本的上下文信息，并且对文本进行了深层次抽象），最后将文本表示连接至 softmax 构建分类模型。

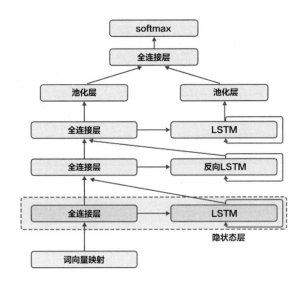

图 6.26　栈式双向 LSTM 用于文本分类

2. 数据集准备

以 IMDB 情感分析数据集为例进行介绍。IMDB 数据集的训练集和测试集分别包含 25 000 个已标注过的电影评论。其中，负面评论的得分小于等于 4，正面评论的得分大于等于 7，满分 10 分。

```
aclImdb
|- test
   |-- neg
   |-- pos
|- train
   |-- neg
   |-- pos
```

PaddlePaddle 在 dataset/imdb.py 中实现了 IMDB 数据集的自动下载和读取，并提供了读取字典、训练数据、测试数据等 API。

3. 配置模型

在该示例中，实现了两种文本分类算法，分别基于文本卷积神经网络以及栈式双向 LSTM。首先，引入要用到的库并定义全局变量：

```
from __future__ import print_function
import paddle
import paddle.fluid as fluid
import numpy as np
import sys
import math
```

```
CLASS_DIM = 2          # 情感分类的类别数
EMB_DIM = 128          # 词向量的维度
HID_DIM = 512          # 隐藏层的维度
STACKED_NUM = 3        # LSTM 双向栈的层数
BATCH_SIZE = 128       # batch 的大小
```

(1) 文本卷积神经网络

构建神经网络 convolution_net。需要注意的是，fluid.nets.sequence_conv_pool 包含卷积和池化层两个操作。

```
#文本卷积神经网络
def convolution_net(data, input_dim, class_dim, emb_dim, hid_dim):
    emb = fluid.layers.embedding(
        input=data, size=[input_dim, emb_dim], is_sparse=True)
    conv_3 = fluid.nets.sequence_conv_pool(
        input=emb,
        num_filters=hid_dim,
        filter_size=3,
        act="tanh",
        pool_type="sqrt")
    conv_4 = fluid.nets.sequence_conv_pool(
        input=emb,
        num_filters=hid_dim,
        filter_size=4,
        act="tanh",
        pool_type="sqrt")
    prediction = fluid.layers.fc(
        input=[conv_3, conv_4], size=class_dim, act="softmax")
    return prediction
```

网络的输入 input_dim 表示的是词典的大小，class_dim 表示类别数。这里使用 sequence_conv_pool API 实现卷积和池化操作。

(2) 栈式双向 LSTM

```
#栈式双向 LSTM
def stacked_lstm_net(data, input_dim, class_dim, emb_dim, hid_dim, stacked_num):

    #计算词向量
    emb = fluid.layers.embedding(
        input=data, size=[input_dim, emb_dim], is_sparse=True)

    #第一层栈
    #全连接层
    fc1 = fluid.layers.fc(input=emb, size=hid_dim)
```

```
#lstm层
lstm1, cell1 = fluid.layers.dynamic_lstm(input=fc1, size=hid_dim)

inputs = [fc1, lstm1]

#其余的所有栈结构
for i in range(2, stacked_num + 1):
    fc = fluid.layers.fc(input=inputs, size=hid_dim)
    lstm, cell = fluid.layers.dynamic_lstm(
        input=fc, size=hid_dim, is_reverse=(i % 2) == 0)
    inputs = [fc, lstm]

#池化层
fc_last = fluid.layers.sequence_pool(input=inputs[0], pool_type='max')
lstm_last = fluid.layers.sequence_pool(input=inputs[1], pool_type='max')

#全连接层，softmax预测
prediction = fluid.layers.fc(
    input=[fc_last, lstm_last], size=class_dim, act='softmax')
return prediction
```

以上的栈式双向LSTM抽象出了高级特征并把其映射到和分类类别数同样大小的向量上。最后一个全连接层的softmax激活函数用来计算分类属于某个类别的概率。重申一下，此处可以调用 convolution_net 或 stacked_lstm_net 的任何一个网络结构进行训练学习（以 convolution_net 为例）。

接下来定义预测程序（inference_program）。预测程序使用 convolution_net 来对 fluid.layer.data 的输入进行预测。

```
def inference_program(word_dict):
    data = fluid.layers.data(
        name="words", shape=[1], dtype="int64", lod_level=1)

    dict_dim = len(word_dict)
    net = convolution_net(data, dict_dim, CLASS_DIM, EMB_DIM, HID_DIM)
    # net = stacked_lstm_net(data, dict_dim, CLASS_DIM, EMB_DIM, HID_DIM,
STACKED_NUM)
    return net
```

这里定义了 training_program（使用从 inference_program 返回的结果来计算误差），以及优化函数 optimizer func。因为是监督学习，所以训练集的标注也在 fluid.layers.data 中定义了。在训练过程中，交叉熵用来在 fluid.layer.cross_entropy 中作为损失函数。

在测试过程中，分类器会计算各个输出的概率。第一个返回的数值规定为cost。

```
def train_program(prediction):
```

```
label = fluid.layers.data(name="label", shape=[1], dtype="int64")
cost = fluid.layers.cross_entropy(input=prediction, label=label)
avg_cost = fluid.layers.mean(cost)
accuracy = fluid.layers.accuracy(input=prediction, label=label)
return [avg_cost, accuracy]    #返回平均cost和准确率acc
```

```
#优化函数
def optimizer_func():
    return fluid.optimizer.Adagrad(learning_rate=0.002)
```

4. 训练模型

（1）定义训练环境

定义训练是在 CPU 上还是 GPU 上：

```
use_cuda = False  #在CPU上进行训练
place = fluid.CUDAPlace(0) if use_cuda else fluid.CPUPlace()
```

（2）定义数据提供器

为训练和测试定义数据提供器。提供器读入一个大小为 BATCH_SIZE 的数据。paddle.dataset.imdb.word_dict 每次会在乱序化后提供一个大小为 BATCH_SIZE 的数据，乱序化的大小为缓存大小 buf_size。读取 IMDB 的数据可能会花费几分钟的时间。

```
print("Loading IMDB word dict....")
word_dict = paddle.dataset.imdb.word_dict()

print ("Reading training data....")
train_reader = paddle.batch(
    paddle.reader.shuffle(
        paddle.dataset.imdb.train(word_dict), buf_size=25000),
    batch_size=BATCH_SIZE)
print("Reading testing data....")
test_reader = paddle.batch(
    paddle.dataset.imdb.test(word_dict), batch_size=BATCH_SIZE)
```

word_dict 是一个字典序列，是词和标注（label）的对应关系，运行下一行可以看到具体内容：

```
word_dict
```

每行如('limited': 1726)的对应关系，该行表示单词 limited 所对应的标注是 1726。

（3）构造训练器

训练器需要一个训练程序和一个训练优化函数。

```
exe = fluid.Executor(place)
prediction = inference_program(word_dict)
```

```
[avg_cost, accuracy] = train_program(prediction)#训练程序
sgd_optimizer = optimizer_func()#训练优化函数
sgd_optimizer.minimize(avg_cost)
```

该函数用来计算训练中模型在 test 数据集上的结果。

```
def train_test(program, reader):
    count = 0
    feed_var_list = [
        program.global_block().var(var_name) for var_name in feed_order
    ]
    feeder_test = fluid.DataFeeder(feed_list=feed_var_list, place=place)
    test_exe = fluid.Executor(place)
    accumulated = len([avg_cost, accuracy]) * [0]
    for test_data in reader():
        avg_cost_np = test_exe.run(
            program=program,
            feed=feeder_test.feed(test_data),
            fetch_list=[avg_cost, accuracy])
        accumulated = [
            x[0] + x[1][0] for x in zip(accumulated, avg_cost_np)
        ]
        count += 1
    return [x / count for x in accumulated]
```

（4）提供数据并构建主训练循环

feed_order 用来定义每条产生的数据和 fluid.layers.data 之间的映射关系。比如，imdb.train 产生的第一列数据对应的是 words 这个特征。

```
# Specify the directory path to save the parameters
params_dirname = "understand_sentiment_conv.inference.model"

feed_order = ['words', 'label']
pass_num = 1   #训练循环的轮数

#程序主循环部分
def train_loop(main_program):
    #启动上文构建的训练器
    exe.run(fluid.default_startup_program())

    feed_var_list_loop = [
        main_program.global_block().var(var_name) for var_name in feed_order
    ]
    feeder = fluid.DataFeeder(
        feed_list=feed_var_list_loop, place=place)
```

```
test_program = fluid.default_main_program().clone(for_test=True)

#训练循环
for epoch_id in range(pass_num):
    for step_id, data in enumerate(train_reader()):
        #运行训练器
        metrics = exe.run(main_program,
                    feed=feeder.feed(data),
                    fetch_list=[avg_cost, accuracy])

        #测试结果
        avg_cost_test, acc_test = train_test(test_program, test_reader)
        print('Step {0}, Test Loss {1:0.2}, Acc {2:0.2}'.format(
            step_id, avg_cost_test, acc_test))

        print("Step {0}, Epoch {1} Metrics {2}".format(
            step_id, epoch_id, list(map(np.array,
                            metrics))))

        if step_id == 30:
            if params_dirname is not None:
                fluid.io.save_inference_model(params_dirname, ["words"],
                                    prediction, exe)#保存模型
            return
```

（5）训练过程处理

在训练主循环里打印了每一步输出，可以观察训练情况。最后，启动训练主循环来开始训练。训练时间较长，如果为了更快地返回结果，可以通过调整损耗值范围或者训练步数，以减少准确率的代价来缩短训练时间。

```
train_loop(fluid.default_main_program())
```

5. 应用模型

（1）构建预测器

和训练过程一样，需要创建一个预测过程，并使用训练得到的模型和参数来进行预测，params_dirname 用来存放训练过程中的各个参数。

```
place = fluid.CUDAPlace(0) if use_cuda else fluid.CPUPlace()
exe = fluid.Executor(place)
inference_scope = fluid.core.Scope()
```

（2）生成测试用输入数据

为了进行预测，任意选取 3 个评论。把评论中的每个词对应到 word_dict 中的 id。如果词

典中没有这个词，就设为 unknown。然后用 create_lod_tensor 来创建细节层次的张量。

```
reviews_str = [
    'read the book forget the movie', 'this is a great movie', 'this is very bad'
]
reviews = [c.split() for c in reviews_str]
UNK = word_dict['<unk>']
lod = []
for c in reviews:
    lod.append([word_dict.get(words, UNK) for words in c])
base_shape = [[len(c) for c in lod]]
tensor_words = fluid.create_lod_tensor(lod, base_shape, place)
```

6. 应用模型并进行预测

现在可以对每一条评论进行正面或者负面的预测。

```
with fluid.scope_guard(inference_scope):

    [inferencer, feed_target_names,
     fetch_targets] = fluid.io.load_inference_model(params_dirname, exe)

    assert feed_target_names[0] == "words"
    results = exe.run(inferencer,
                      feed={feed_target_names[0]: tensor_words},
                      fetch_list=fetch_targets,
                      return_numpy=False)
    np_data = np.array(results[0])
    for i, r in enumerate(np_data):
        print("Predict probability of ", r[0], " to be positive and ", r[1],
              " to be negative for review \'", reviews_str[i], "\'")
```

本节以情感分析为例，介绍了使用深度学习的方法进行端对端的短文本分类，并且使用 PaddlePaddle 完成了全部相关实验。同时，简要介绍了两种文本处理模型：卷积神经网络和循环神经网络。

6.7 本章小结

机器学习是驱动人工智能领域突破性发展的核心技术。AlphaGo 战胜人类围棋冠军、人脸识别、大数据挖掘都和机器学习密切相关。本章在 PaddlePaddle 平台基于 Python 语言从理论到实践介绍了手写体数字识别、图像分类、词向量等机器学习实战案例。

参考文献

[1] Tom Mitchell.Machine Learning[M]曾华军，张银奎，等译.北京：机械工业出版社，2003.

[2] 姚旭，王晓丹，张玉玺，等.特征选择方法综述[J].控制与决策，2012，27（2）：161-166.

[3] Jiawei Han. 数据挖掘概念与技术 [M].北京：机械工业出版社，2012.

[4] XU Rui，Donald Wunsch11．survey of clustering algorithm[J]．IEEE．Trans actions on Neural Networks，2005，16（3）：645-678.

[5] 贺玲，吴玲达，蔡益朝．数据挖掘中的聚类算法综述[J]．计算机应用研究，2007，24（1）：10-13.

[6] 孙吉贵，刘杰，赵连宇．聚类算法研究[J]．软件学报，2008，19（1）：48-61.

[7] 马晓艳，唐雁．层次聚类算法研究[J]．计算机科学，2008，34（7）：34-36.

[8] Bart van Merriënboer, Dzmitry Bahdanau, Vincent Dumoulin, Dmitriy Serdyuk, David Ward e-Farley, Jan Chorowski, and Yoshua Bengio. Blocks and Fuel: Frameworks for deeplearning [J]. arXiv preprint arXiv:1506.00619 [cs.LG], 2015.

[9] Hanke, M., Halchenko, Y. O., Sederberg, P. B., Hanson, S. J., Haxby, J. V. & Pollmann, S. PyMVPA: A Python toolbox for multivariate pattern analysis of fMRI data[J]. Neuroinformatics, 2009，7：37-53.

[10] Hanke, M., Halchenko, Y. O., Haxby, J. V., and Pollmann, S. Statistical learning analysis in neuroscience: aiming for transparency[J]. Frontiers in Neuroscience，2010，4（1）：38-43.

[11] Haxby, J. V., Guntupalli, J. S., Connolly, A. C., Halchenko, Y. O., Conroy, B. R., Gobbini, M. I., Hanke, M. & Ramadge, P. J.. A Common, High-Dimensional Model of the Representational Space in Human Ventral Temporal Cortex[J]. Neuron, 2011(72)：404–416.

[12] Bengio Y, Ducharme R, Vincent P, et al. A neural probabilistic language model[J]. journal of machine learning research, 2003, 3（Feb）：1137-1155.

[13] Mikolov T, Kombrink S, Deoras A, et al. Rnnlm-recurrent neural network language modeling toolkit[C]//Proc. of the 2011 ASRU Workshop. 2011: 196-201.

[14] Mikolov T, Chen K, Corrado G, et al. Efficient estimation of word representations in vector space[J]. arXiv preprint arXiv:1301.3781, 2013.

[15] Maaten L, Hinton G. Visualizing data using t-SNE[J]. Journal of Machine Learning Research, 2008, 9(Nov): 2579-2605.

[16] https://en.wikipedia.org/wiki/Singular_value_decomposition.

[17] https://en.wikipedia.org/wiki/Linear_regression.

[18] Friedman J, Hastie T, Tibshirani R. The elements of statistical learning[M]. Springer, Berlin: Springer series in statistics, 2001.

[19] Murphy K P. Machine learning: a probabilistic perspective[M]. MIT press, 2012.

[20] Bishop C M. Pattern recognition[J]. Machine Learning, 2006，128.

[21] LeCun, Yann, Léon Bottou, Yoshua Bengio, and Patrick Haffner.Gradient-based learning applied to document recognition. Proceedings of the IEEE 86, 11 (1998)：2278-2324.

[22] Wejéus, Samuel. A Neural Network Approach to Arbitrary SymbolRecognition on Modern Smartphones.2014.

[23] Decoste, Dennis, and Bernhard Schölkopf.Training invariant support vector machines. Machine learning 46, 1-3 (2002)：161-190.

[24] Simard, Patrice Y., David Steinkraus, and John C. Platt. Best Practices for Convolutional Neural Networks Applied to Visual Document Analysis. ICDAR,2003，3：958-962.

[25] Salakhutdinov, Ruslan, and Geoffrey E. Hinton. Learning a Nonlinear Embedding by Preserving Class Neighbourhood Structure. AISTATS,2007, 11.

[26] Cireşan, Dan Claudiu, Ueli Meier, Luca Maria Gambardella, and Jürgen Schmidhuber. "Deep, big, simple neural nets for handwritten digit recognition. Neural computation 22, 12 (2010)：3207-3220.

[27] Deng, Li, Michael L. Seltzer, Dong Yu, Alex Acero, Abdel-rahman Mohamed, and Geoffrey E. Hinton. Binary coding of speech spectrograms using a deep auto-encoder. Interspeech, 2010：1692-1695.

[28] Kégl, Balázs, and Róbert Busa-Fekete. Boosting products of base classifiers. Proceedings of the 26th Annual International Conference on Machine Learning, 2009：497-504.

[29] Rosenblatt, Frank. The perceptron: A probabilistic model for information storage and organization in the brain. Psychological review 65, 6 (1958)：386.

[30] Bishop, Christopher M. Pattern recognition. Machine Learning 128 (2006): 1-58.

[31] D. G. Lowe, Distinctive image features from scale-invariant keypoints. IJCV,2004, 60（2）：91-110.

[32] N. Dalal, B. Triggs, Histograms of Oriented Gradients for Human Detection, Proc. IEEE Conf. Computer Vision and Pattern Recognition, 2005.

[33] Ahonen, T., Hadid, A., and Pietikinen, M. Face description with local binary patterns: Application to face recognition.2006, PAMI, 28.

[34] J. Sivic, A. Zisserman, Video Google: A Text Retrieval Approach to Object Matching in Videos, Proc. Ninth Int'l Conf. Computer Vision, 2003：1470-1478.

[35] B. Olshausen, D. Field, Sparse Coding with an Overcomplete Basis Set: A Strategy Employed by V1?. Vision Research, 1997, 37：3311-3325.

[36] Wang, J., Yang, J., Yu, K., Lv, F., Huang, T., and Gong, Y. Locality constrained Linear Coding for image classification. CVPR, 2010.

[37] Perronnin, F., Sánchez, J., & Mensink, T. Improving the fisher kernel for large-scale image classification. ECCVj, 2010（4）.

[38] Lin, Y., Lv, F., Cao, L., Zhu, S., Yang, M., Cour, T., Yu, K., and Huang, T. Large-scale

image clas- sification: Fast feature extraction and SVM training. CVPR, 2011.

[39] Krizhevsky, A., Sutskever, I., and Hinton, G. ImageNet classification with deep convolutional neu- ral networks. NIPS, 2012.

[40] G.E. Hinton, N. Srivastava, A. Krizhevsky, I. Sutskever, and R.R. Salakhutdinov. Improving neural networks by preventing co-adaptation of feature detectors. arXiv preprint arXiv:1207.0580, 2012.

[41] K. Chatfield, K. Simonyan, A. Vedaldi, A. Zisserman. Return of the Devil in the Details: Delving Deep into Convolutional Nets. BMVC, 2014.

[42] Szegedy, C., Liu, W., Jia, Y., Sermanet, P., Reed, S., Anguelov, D., Erhan, D., Vanhoucke, V., Rabinovich, A., Going deeper with convolutions. CVPR, 2015.

[43] Lin, M., Chen, Q., and Yan, S. Network in network. Proc. ICLR, 2014.

[44] S. Ioffe and C. Szegedy. Batch normalization: Accelerating deep network training by reducing internal covariate shift. ICML, 2015.

[45] K. He, X. Zhang, S. Ren, J. Sun. Deep Residual Learning for Image Recognition. CVPR, 2016.

[46] Szegedy, C., Vanhoucke, V., Ioffe, S., Shlens, J., Wojna, Z. Rethinking the incep-tion architecture for computer vision. CVPR, 2016.

[47] Szegedy, C., Ioffe, S., Vanhoucke, V. Inception-v4, inception-resnet and the impact of residual connections on learning. arXiv:1602.07261, 2016.

[48] Everingham, M., Eslami, S. M. A., Van Gool, L., Williams, C. K. I., Winn, J. and Zisserman, A. The Pascal Visual Object Classes Challenge: A Retrospective. International Journal of Computer Vision.2015, 111（1）：98-136.

[49] He, K., Zhang, X., Ren, S., and Sun, J. Delving Deep into Rectifiers: Surpassing Human-Level Performance on ImageNet Classification. ArXiv e-prints, February 2015.

[50] http://deeplearning.net/tutorial/lenet.html.

[51] https://www.cs.toronto.edu/~kriz/cifar.html.

[52] http://cs231n.github.io/classification/.

[53] Learning Multiple Layers of Features from Tiny Images. Alex Krizhevsky. 2009.

[54] Bengio Y, Ducharme R, Vincent P, et al. A neural probabilistic language model[J]. journal of machine learning research, 2003, 3(Feb)：1137-1155.

[55] Mikolov T, Kombrink S, Deoras A, et al. Rnnlm-recurrent neural network language modeling toolkit[C]//Proc. of the 2011 ASRU Workshop. 2011：196-201.

[56] Mikolov T, Chen K, Corrado G, et al. Efficient estimation of word representations in vector space[J]. arXiv preprint arXiv:1301.3781, 2013.

[57] Maaten L, Hinton G. Visualizing data using t-SNE[J]. Journal of Machine Learning Research, 2008, 9(Nov)：2579-2605.

[58] https://en.wikipedia.org/wiki/Singular_value_decomposition.

[59] P. Resnick, N. Iacovou, etc. GroupLens: An Open Architecture for Collaborative Filtering

of Netnews. Proceedings of ACM Conference on Computer Supported Cooperative Work. CSCW, 1994: 175-186.

[60] Sarwar, Badrul, et al. Item-based collaborative filtering recommendation algorithms. Proceedings of the 10th international conference on World Wide Web. ACM, 2001.

[61] Kautz, Henry, Bart Selman, and Mehul Shah. Referral Web: combining social networks and collaborative filtering. Communications of the ACM 40.3, 1997: 63-65.

[62] Peter Brusilovsky. The Adaptive Web. 2007: 325.

[63] Robin Burke. Hybrid Web Recommender Systems, The Adaptive Web. 377-408.

[64] Yuan, Jianbo, et al. Solving Cold-Start Problem in Large-scale Recommendation Engines: A Deep Learning Approach. arXiv preprint arXiv:1611.05480, 2016.

[65] Covington P, Adams J, Sargin E. Deep neural networks for youtube recommendations[C]//Pro-ceedings of the 10th ACM Conference on Recommender Systems. ACM, 2016: 191-198.

[66] Kim Y. Convolutional neural networks for sentence classification[J]. arXiv preprint arXiv:1408.5882, 2014.

[67] Kalchbrenner N, Grefenstette E, Blunsom P. A convolutional neural network for modelling sentences[J]. arXiv preprint arXiv:1404.2188, 2014.

[68] Yann N. Dauphin, et al. Language Modeling with Gated Convolutional Networks[J]. arXiv preprint arXiv:1612.08083, 2016.

[69] Siegelmann H T, Sontag E D. On the computational power of neural nets[C]//Proceedings of the fifth annual workshop on Computational learning theory. ACM, 1992: 440-449.

[70] Hochreiter S, Schmidhuber J. Long short-term memory[J]. Neural computation, 1997, 9（8）: 1735-1780.

[71] Bengio Y, Simard P, Frasconi P. Learning long-term dependencies with gradient descent is difficult[J]. IEEE transactions on neural networks, 1994, 5（2）: 157-166.

[72] Graves A. Generating sequences with recurrent neural networks[J]. arXiv preprint arXiv:1308.0850, 2013.

[73] Cho K, Van Merriënboer B, Gulcehre C, et al. Learning phrase representations using RNN encoder-decoder for statistical machine translation[J]. arXiv preprint arXiv:1406.1078, 2014.

[74] Zhou J, Xu W. End-to-end learning of semantic role labeling using recurrent neural networks[C]//Proceedings of the Annual Meeting of the Association for Computational Linguistics. 2015.

[75] https://en.wikipedia.org/wiki/Linear_regression.

[76] Friedman J, Hastie T, Tibshirani R. The elements of statistical learning[M]. Springer, Berlin: Springer series in statistics, 2001

[77] Murphy K P. Machine learning: a probabilistic perspective[M]. MIT press, 2012.

[78] Bishop C M. Pattern recognition[J]. Machine Learning, 2006: 128.

[79] Sun W, Sui Z, Wang M, et al. Chinese semantic role labeling with shallow

parsing[C]//Proceedings of the 2009 Conference on Empirical Methods in Natural Language Processing: Volume 3-Volume 3. Association for Computational Linguistics, 2009：1475-1483.

[80] Pascanu R, Gulcehre C, Cho K, et al. How to construct deep recurrent neural networks[J]. arXiv preprint arXiv:1312.6026, 2013.

[81] Cho K, Van Merriënboer B, Gulcehre C, et al. Learning phrase representations using RNN encoder-decoder for statistical machine translation[J]. arXiv preprint arXiv:1406.1078, 2014.

[82] Bahdanau D, Cho K, Bengio Y. Neural machine translation by jointly learning to align and translate[J]. arXiv preprint arXiv:1409.0473, 2014.

[83] Lafferty J, McCallum A, Pereira F. Conditional random fields: Probabilistic models for segmenting and labeling sequence data[C]//Proceedings of the eighteenth international conference on machine learning, ICML. 2001, 1：282-289.

[84] 李航. 统计学习方法[M]. 北京：清华大学出版社，2012.

[85] Marcus M P, Marcinkiewicz M A, Santorini B. Building a large annotated corpus of English: The Penn Treebank[J]. Computational linguistics, 1993, 19（2）：313-330.

[86] Palmer M, Gildea D, Kingsbury P. The proposition bank: An annotated corpus of semantic roles[J]. Computational linguistics, 2005, 31（1）：71-106.

[87] Carreras X, Màrquez L. Introduction to the CoNLL-2005 shared task: Semantic role labeling[C]//Proceedings of the Ninth Conference on Computational Natural Language Learning. Association for Computational Linguistics, 2005：152-164.

[88] Zhou J, Xu W. End-to-end learning of semantic role labeling using recurrent neural networks[C]//Proceedings of the Annual Meeting of the Association for Computational Linguistics. 2015.

[89] Koehn P. Statistical machine translation[M]. Cambridge University Press, 2009.

[90] Cho K, Van Merriënboer B, Gulcehre C, et al. Learning phrase representations using RNN encoder-decoder for statistical machine translation[C]//Proceedings of the 2014 Conference on Empirical Methods in Natural Language Processing (EMNLP), 2014：1724-1734.

[91] Chung J, Gulcehre C, Cho K H, et al. Empirical evaluation of gated recurrent neural networks on sequence modeling[J]. arXiv preprint arXiv:1412.3555, 2014.

[92] Bahdanau D, Cho K, Bengio Y. Neural machine translation by jointly learning to align and translate[C]//Proceedings of ICLR 2015, 2015.

[93] Papineni K, Roukos S, Ward T, et al. BLEU: a method for automatic evaluation of machine translation[C]//Proceedings of the 40th annual meeting on association for computational linguistics. Association for Computational Linguistics, 2002：311-318.

[94] alimans, Tim; Goodfellow, Ian; Zaremba, Wojciech; Cheung, Vicki; Radford, Alec; Chen, Xi. Improved Techniques for Training GANs. arXiv:1606.03498 [cs.LG], 2016.

[95] Radford A, Metz L, Chintala S. Unsupervised Representation Learning with Deep Convolutional Generative Adversarial Networks[J]. Computer Science, 2015.